Electronic Switching, Timing, and Pulse Circuits

McGRAW-HILL ELECTRICAL AND ELECTRONIC ENGINEERING SERIES

FREDERICK EMMONS TERMAN, *Consulting Editor*
W. W. HARMAN, J. G. TRUXAL, AND R. A. ROHRER, *Associate Consulting Editors*

ANGELAKOS AND EVERHART Microwave Communications
ANGELO Electronic Circuits
ANGELO Electronics: BJTs, FETs, and Microcircuits
ASELTINE Transform Method in Linear System Analysis
ATWATER Introduction to Microwave Theory
BENNETT Introduction to Signal Transmission
BERANEK Acoustics
BRACEWELL The Fourier Transform and Its Application
BRENNER AND JAVID Analysis of Electric Circuits
BROWN Analysis of Linear Time-Invariant Systems
BRUNS AND SAUNDERS Analysis of Feedback Control Systems
CARLSON Communication Systems: An Introduction to Signals and Noise in Electrical Communication
CHEN The Analysis of Linear Systems
CHEN Linear Network Design and Synthesis
CHIRLIAN Analysis and Design of Electronic Circuits
CHIRLIAN Basic Network Theory
CHIRLIAN Electronic Circuits: Physical Principles, Analysis, and Design
CHIRLIAN AND ZEMANIAN Electronics
CLEMENT AND JOHNSON Electrical Engineering Science
CUNNINGHAM Introduction to Nonlinear Analysis
D'AZZO AND HOUPIS Feedback Control System Analysis and Synthesis
ELGERD Control Systems Theory
ELGERD Electric Energy Systems Theory: An Introduction
EVELEIGH Adaptive Control and Optimization Techniques
FEINSTEIN Foundations of Information Theory
FITZGERALD, HIGGINBOTHAM, AND GRABEL Basic Electrical Engineering
FITZGERALD AND KINGSLEY Electric Machinery
FRANK Electrical Measurement Analysis
FRIEDLAND, WING, AND ASH Principles of Linear Networks
GEHMLICH AND HAMMOND Electromechanical Systems
GHAUSI Principles and Design of Linear Active Circuits
GHOSE Microwave Circuit Theory and Analysis
GLASFORD Fundamentals of Television Engineering
GREINER Semiconductor Devices and Applications
HAMMOND AND GEHMLICH Electrical Engineering
HANCOCK An Introduction to the Principles of Communication Theory
HARMAN Fundamentals of Electronic Motion
HARMAN Principles of the Statistical Theory of Communication
HARMAN AND LYTLE Electrical and Mechanical Networks
HAYASHI Nonlinear Oscillations in Physical Systems
HAYT Engineering Electromagnetics
HAYT AND KEMMERLY Engineering Circuit Analysis
HILL Electronics in Engineering
JAVID AND BROWN Field Analysis and Electromagnetics
JOHNSON Transmission Lines and Networks
KOENIG, TOKAD, AND KESAVAN Analysis of Discrete Physical Systems
KRAUS Antennas
KRAUS Electromagnetics
KUH AND PEDERSON Principles of Circuit Synthesis
KUO Linear Networks and Systems
LEDLEY Digital Computer and Control Engineering

LEPAGE Complex Variables and the Laplace Transform for Engineering
LEPAGE AND SEELY General Network Analysis
LEVI AND PANZER Electromechanical Power Conversion
LEY, LUTZ, AND REHBERG Linear Circuit Analysis
LINVILL AND GIBBONS Transistors and Active Circuits
LITTAUER Pulse Electronics
LYNCH AND TRUXAL Introductory System Analysis
LYNCH AND TRUXAL Principles of Electronic Instrumentation
LYNCH AND TRUXAL Signals and Systems in Electrical Engineering
McCLUSKEY Introduction to the Theory of Switching Circuits
MANNING Electrical Circuits
MEISEL Principles of Electromechanical-Energy Conversion
MILLMAN AND HALKIAS Electronic Devices and Circuits
MILLMAN AND TAUB Pulse, Digital, and Switching Waveforms
MINORSKY Theory of Nonlinear Control Systems
MISHKIN AND BRAUN Adaptive Control Systems
MOORE Traveling-Wave Engineering
MURDOCH Network Theory
NANAVATI An Introduction to Semiconductor Electronics
OBERMAN Disciplines in Combinational and Sequential Circuit Design
PETTIT AND McWHORTER Electronic Switching, Timing, and Pulse Circuits
PETTIT AND McWHORTER Electronic Amplifier Circuits
PFEIFFER Concepts of Probability Theory
REZA An Introduction to Information Theory
REZA AND SEELY Modern Network Analysis
ROGERS Introduction to Electric Fields
RUSTON AND BORDOGNA Electric Networks: Functions, Filters, Analysis
RYDER Engineering Electronics
SCHILLING AND BELOVE Electronic Circuits: Discrete and Integrated
SCHWARTZ Information Transmission, Modulation, and Noise
SCHWARZ AND FRIEDLAND Linear Systems
SEELY Electromechanical Energy Conversion
SEIFERT AND STEEG Control Systems Engineering
SHOOMAN Probabilistic Reliability: An Engineering Approach
SISKIND Direct-Current Machinery
SKILLING Electric Transmission Lines
STEVENSON Elements of Power System Analysis
STEWART Fundamentals of Signal Theory
STRAUSS Wave Generation and Shaping
SU Active Network Synthesis
TERMAN Electronic and Radio Engineering
TERMAN AND PETTIT Electronic Measurements
THALER Analysis and Design of Nonlinear Feedback Control Systems
THALER AND BROWN Analysis and Design of Feedback Control Systems
TOU Digital and Sampled-Data Control Systems
TOU Modern Control Theory
TRUXAL Automatic Feedback Control System Synthesis
TUTTLE Electric Networks: Analysis and Synthesis
VALDES The Physical Theory of Transistors
VAN BLADEL Electromagnetic Fields
WEEKS Antenna Engineering
WEINBERG Network Analysis and Synthesis

SECOND EDITION

Electronic Switching, Timing, and Pulse Circuits

JOSEPH M. PETTIT
Professor of Electrical Engineering
Dean, School of Engineering
Stanford University

MALCOLM M. McWHORTER
Professor of Electrical Engineering
Stanford University

McGRAW-HILL BOOK COMPANY
New York San Francisco St. Louis Düsseldorf London
Mexico Panama Sydney Toronto

Electronic Switching, Timing, and Pulse Circuits
Second Edition

Printed in the United States of America.

Library of Congress catalog card number: 78-114292

34567890 KPKP 7987654

07-049726-5

Preface

The subject matter of this book is more far-reaching than can be suggested in a brief title. This is a book about circuits in which the electronic device, either tube or transistor, is used primarily as a switch, with waveforms that are for the most part nonsinusoidal. This is in contrast to small-signal sine-wave amplification, or even large-signal power amplification where a sine wave is distorted only a few percent at most. In the circuits discussed here, voltages or currents are switched abruptly between predetermined levels, and they may remain at a given level either indefinitely or for short precise periods of time, perhaps for only a few millionths of a second. The switching may initiate other waveforms, such as one in which a voltage rises at a constant rate so that time may be measured or controlled by adjusting the voltage levels which govern the start and finish of the rise. Electronic systems employing predominantly nonsinusoidal waveforms are in widespread use; their applications include television, radar, pulse communication systems, digital computers, and pulse-type instrumentation (even within the electrical power and machinery field).

The electronic devices used in contemporary pulse, logic, and timing circuits are almost exclusively solid-state devices. This second edition has reflected that trend and the vacuum tube has been all but eliminated, with the feeling that older texts such as the first edition can be used for the rare instances requiring vacuum tubes. From the plethora of solid-state devices, those showing the most likelihood of remaining useful in the future have been chosen for study. The monolithic silicon integrated circuit is not singled out to be studied separately, but many of the examples are relevant, and the underlying principles for the circuits using discrete components are applicable to their integrated counterparts. Therefore, this text forms a basic circuit background for anyone interested in integrated electronics.

This book deals with a basic type of analysis for circuits which is essential to all the systems. The analysis is different from that used with sine waves, although there is an interrelationship which can be demonstrated in a formal though cumbersome manner through Fourier series or integrals. The analysis featured here, however, is more direct, and it can be described as the use of "linear segments" or "piecewise linear" approximation to represent the graphical characteristics of a tube or transistor during discrete subperiods of the operating cycle of the circuit. Within each subperiod the voltage waveforms are generally single time-constant exponentials, associated physically with the charging or discharging of a capacitor.

The emphasis is on the use of the graphical characteristics of the electron device, whether transistor or tube. It is assumed that the reader has had an introductory electronics course from which he has already gained an understanding of the physical processes in the devices and how these lead to the particular graphical results. These graphs (the so-called *static characteristics*) are useful in most practical situations where the speed of circuit actions is much slower than the electron dynamics within the devices. A discussion is provided, however, of the inherent limitations on switching speed which the devices introduce.

Another feature of the analysis is that extensive use is made of numerical examples. There are several reasons for doing this. First, successful analysis of complex electronic circuits requires skillful approximation, with attention focused on the dominant elements. This can only be done when numerical values are known. Second, the reader gains some feeling for orders of magnitude of voltages, speeds, etc., which are most often encountered in contemporary circuits. Third, the complete literal algebraic equations for the circuits can be cumbersome indeed, and they would contribute little but mathematical completeness to a presentation which is otherwise aimed at simplification. Finally, the student can become familiar with standard nomenclature and abbreviations. This also helps to give a feel for the range of practical element values which can be used.

The reader is taken from a rather basic level of electrical circuit theory to a substantially higher-level understanding of circuits which can only be described by such words as ingenious and subtle. The prerequisites in mathematics and circuit theory are such that the text can be used in either a fourth- or fifth-year course in the usual electrical engineering curriculum. The material has been taught at Stanford University, where omitting much of Chapters 8 and 9 provides a one-quarter course. The course can be extended to one semester by using all the material and by allowing more time for the discussion of examples and applications. The aim has been to keep this material as compact as possible to help fit it into the already crowded curriculum.

This book can be used at the senior level in an up-to-date electrical engineering curriculum, where it could serve as a companion text to another book featuring the more classical subjects of sine-wave or "radio" engineering.

The variety of subjects falling within this broad and fascinating field presents a difficult problem of organization. The task would be simpler if all circuits came from a single area of application such as television or computers, but this is not the case. Nor is it easy to classify circuits according to function or type, because as yet there is a lack of well-established categories. Moreover, in different applications the same circuit may perform different functions; for example, generating a long pulse of adjustable duration in one case or a short delayed pulse with adjustable delay time in another.

The problem is not helped by the differing terminology for the same circuits in different areas of application: Either the same circuit may have different names according to the area of application, or the same name may be given to different circuits by different people! For example, the *dc restorer circuit* used in television may be called a *clamping circuit* in radar. On the other hand, a *clamping diode* in a computer circuit would be called a *clipper* in television. Clearly a guide is needed for the uninitiated. Moreover, the terminology has not yet yielded to standardization in another respect. The literature is sprinkled with colorful jargon, which in many cases is descriptive and useful, and hence used in this book wherever suitable.

In the end, a compromise organization has been developed with the educational process in view, proceeding from the simple to the complex and from specific examples to general cases.

The authors must acknowledge the contributions of two groups of people who laid a foundation for the understanding of switching, timing, and pulse circuits. These were the World War II staff of the M.I.T. Radar School in Boston and a group in Britain at the Telecommunications Research Establishment, for whom the principal spokesman was F. C.

Williams. These groups were concerned primarily with radar, but the analyses and points of view developed are widely applicable. The work of Reuben Lee provided the earliest clear analysis of the pulse transformer, which appears in Chapter 7. Bibliographical reference to these and other authors will be found at the ends of appropriate chapters.

Finally, the writers are indebted to their students and teaching colleagues at Stanford, particularly Dr. W. R. Kincheloe, and to Dr. Robert M. Scarlett who was influential in the first edition.

JOSEPH M. PETTIT
MALCOLM M. McWHORTER

Contents

Preface vii

1 SWITCHING AND TIMING SYSTEMS **1**

1-1 Transmission-Line Fault Locator 1
1-2 Industrial Counting 3
1-3 Nuclear Research Instrumentation 4
1-4 Other Applications 6

2 SOME CIRCUIT FUNDAMENTALS **7**

2-1 Transients in Linear RC Circuits 7
2-2 Waveform-Shaping Circuits Using Linear RC Elements 13
2-3 Useful Circuit Theorems 18
2-4 Helpful Techniques 22

3 **ELECTRON DEVICES AS SWITCHING ELEMENTS** 30

3-1 Ideal Switches 31
3-2 Electron Devices as Switches 35
3-3 Semiconductor Diodes 37
3-4 Thermionic Diodes 41
3-5 Nonideal Diodes in Circuits 41
3-6 Diode Switching Time 44
3-7 Three-Terminal Devices—The Transistor 45
3-8 Transistors as Separately Actuated Switches 46
3-9 Linear Transistor Operation—Region II 55
3-10 The Emitter Follower 65
3-11 The PNP Transistor 72
3-12 Common-Base Transistor Characteristics 74
3-13 The Field-Effect Transistor (FET) 77
3-14 Other Types of Field-Effect Devices 85

4 **SOME EXAMPLES OF *RC* CIRCUITS INCORPORATING ACTIVE DEVICES** 92

4-1 A Clamping or dc Restorer Circuit 92
4-2 Decoupling Filter 96
4-3 *RC* Coupling Circuit 101

5 **STABLE STATES AND REGENERATIVE SWITCHING. MULTIVIBRATORS** 108

5-1 Introduction to the Multivibrator 108
5-2 The Bistable Multivibrator 109
5-3 The SET-RESET Binary 112
5-4 Switching Speed in the Binary 114
5-5 Counting Circuits Using Binaries 116
5-6 The Emitter-Coupled Binary 123
5-7 The Monostable Multivibrator 128
5-8 An Emitter-Coupled Monostable Multivibrator 134
5-9 The Transistor Astable Multivibrator 140
5-10 A Field-Effect Transistor Multivibrator 144
5-11 Other Forms of Multivibrators 147
5-12 Frequency Division and Synchronization in Multivibrators 149
5-13 Stability of the Multivibrator Period 151

6 **CIRCUITS FOR GENERATING LINEAR VOLTAGE SLOPES** 157

6-1 Simple *RC* Integrator 158
6-2 Linear Sweep Generators Using Current Sources 159

6-3 The Miller Integrator 162
6-4 Improved Miller Integrator 169
6-5 The Bootstrap Sweep Generator 174
6-6 A Practical Bootstrap Circuit 176

7 CIRCUITS CONTAINING INDUCTORS OR TRANSFORMERS 180

7-1 The RL Circuit 180
7-2 The RLC Circuit 184
7-3 The Ringing Oscillator 191
7-4 Other RLC Circuit Configurations 193
7-5 The Pulse Transformer 194
7-6 The Step-Up Transformer 195
7-7 Decay of the Pulse Top 196
7-8 Pulse Trailing Edge 197
7-9 The Step-Down Transformer 199
7-10 Further Comments on the Pulse Transformer 199
7-11 Transformers in Regenerative Circuits 200
7-12 Regenerative Pickoff (Comparator) Circuits 200
7-13 The Monostable Blocking Oscillator 203
7-14 The Astable Blocking Oscillator 206
7-15 The Effect of the Core Material Upon the Pulse Length 211
7-16 Other Blocking Oscillator Configurations 214
7-17 Uses of the Blocking Oscillator 215

8 NEGATIVE-RESISTANCE SWITCHING CIRCUITS 219

8-1 Negative-Resistance and Trigger Devices 219
8-2 General Characteristics of Negative-Resistance Devices 219
8-3 Types of Negative-Resistance Devices 223
A. Two-Terminal NRD 224
B. Three-Terminal Negative-Resistance Devices 228
8-4 Some Circuits Using Two-Terminal NRD 233
8-5 A Tunnel-Diode Astable Oscillator 239
8-6 A Unijunction Oscillator and SCR Power Control 241
8-7 A Simple TRIAC Full-Wave ac Controller 245

9 LINES AND PULSE-FORMING NETWORKS 256

9-1 Transient Behavior of Transmission Lines 257
9-2 Applications 264

9-3 Transmission Lines Used to Produce Pulse Delay 268
9-4 Pulse-Forming Networks and Artificial Lines 270

Appendixes 279

Index 285

Electronic Switching, Timing, and Pulse Circuits

1
Switching and Timing Systems

Before we commence the detailed analysis of switching and timing circuits, it is desirable to look ahead and picture some of the electronic systems in which these circuits will be indispensable as "building blocks." Only a few examples are given, but these should serve to stimulate the interest of the reader and encourage him in his reading of the later chapters, where the various elements of these circuits will emerge from the systematic analysis.

1-1 TRANSMISSION-LINE FAULT LOCATOR

Suppose that one has a telephone or power transmission line as shown in Fig. 1-1, and at some unknown distance from the sending end of the line there has occurred a fault in the form of a short circuit. If the location of this fault, i.e., the distance from the sending station, could be ascertained electrically, it would expedite the dispatching of a repair crew to the scene.

A circuit for accomplishing this is shown in Fig. 1-1*a*. A *switching*

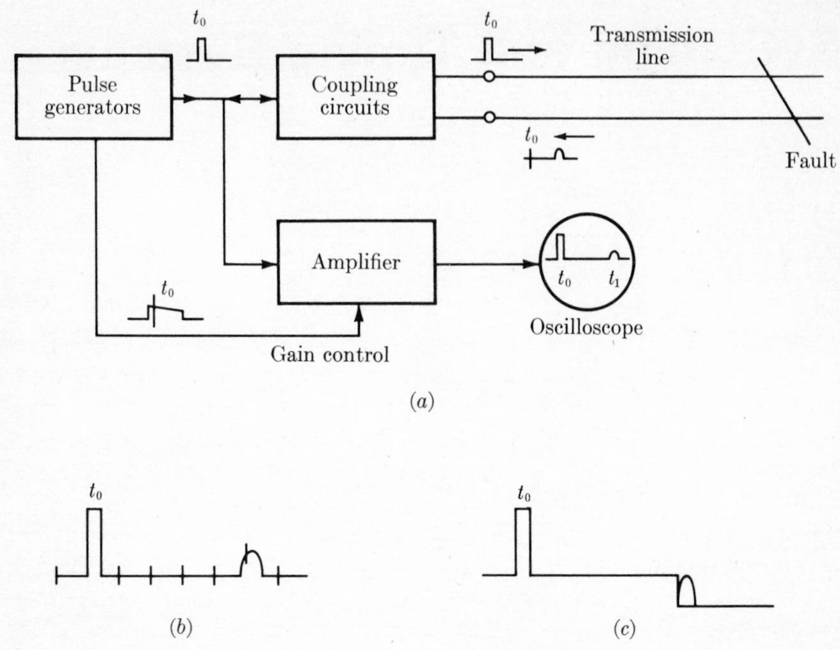

Fig. 1-1 Example of a pulse and timing circuit: transmission-line fault locator. (*a*) Basic circuit. (*b*) Timing pulses. (*c*) Adjustable step.

action produces a voltage pulse of short duration which is sent down the transmission line and is reflected at the fault. Both the transmitted and reflected pulses are passed through the amplifier and displayed on the cathode-ray oscilloscope. Since the operator will know the velocity of transmission on the line, the time interval between the transmitted and received pulses will permit a direct measurement of the distance to the fault.

Because the transmitted pulse will be much larger than the reflected pulse, a gain-control circuit is provided whereby the amplifier gain can be reduced during the transmitted pulse. A separate pulse for this purpose is provided from the pulse generator, and as can be noted from the figure, this gain-control pulse slightly precedes the main transmitted pulse in order that the amplifier gain may be reduced just prior to the time of the transmitted pulse. Full gain is restored in the amplifier before the reflected pulse is received.

Another variation on such a gain-control circuit is to provide a waveform in the control circuit such that the gain of the amplifier steadily increases with time along an exponential curve, to allow for the fact that echoes from greater distances are smaller, and hence require more gain in the amplifier.

There remains the timing problem, namely, measuring the time interval between the two pulses as displayed on the oscilloscope. Figure 1-1*b* shows a conventional means for accomplishing this type of timing measurement. An oscillator is incorporated in the system, which generates pulses at a known rate, say, 100,000/s. Thus the interval between pulses is the reciprocal of 100,000, or 10 μs. In the example illustrated, the reflected pulse occurs approximately 49 μs after the transmitted pulse. If the velocity of transmission on the line is such that the two-way travel time of a pulse is 0.1 mile/μs, the distance to the fault would be 4.9 miles.

In Fig. 1-1*c*, there is an alternative arrangement shown in which a step is introduced into the base line of the display. The position of this step along the horizontal base line can be adjusted by the operator with a control calibrated in either time or distance. In Chap. 7 we discuss circuits by way of which such a step can be produced at an accurately controlled time by means of the adjustment of a dc voltage.

It is interesting to note that the basic elements of a radar system are incorporated in the system shown in Fig. 1-1. In radar a pulse of energy is transmitted at a radio frequency from a directive antenna, aimed like a searchlight, and an echo is produced at a distant target in the same manner as the pulse is reflected in the transmission line in the figure. Similar techniques are used to determine the range of a target. The direction of the target is of course indicated by the direction in which the antenna is pointed.

1-2 INDUSTRIAL COUNTING

Electronic circuits that switch from one state to an opposite state in response to pulses of current or voltage form the basis of devices for counting and digital computing.

The counting circuit provides a visual indication of the number of electrical pulses which have been applied to its input terminals during a specified interval of time. The indication is usually in the form of illuminated numerals, as illustrated in Fig. 1-2, where the count is shown to be 144. The counter is provided with a starting and stopping arrangement and also a reset to restore the indication to zero after the indicated count has been read. The sequence of operations can be controlled manually, or as in the application shown in Fig. 1-2, the cycle of events is controlled automatically. Automatic operation is particularly valuable in a situation such as the one represented in Fig. 1-2, illustrating an industrial operation in which a continuous flow of objects, such as medicinal capsules or steel washers, are emerging from a production machine along a conveyor. It is desired to package these objects in quantities of one

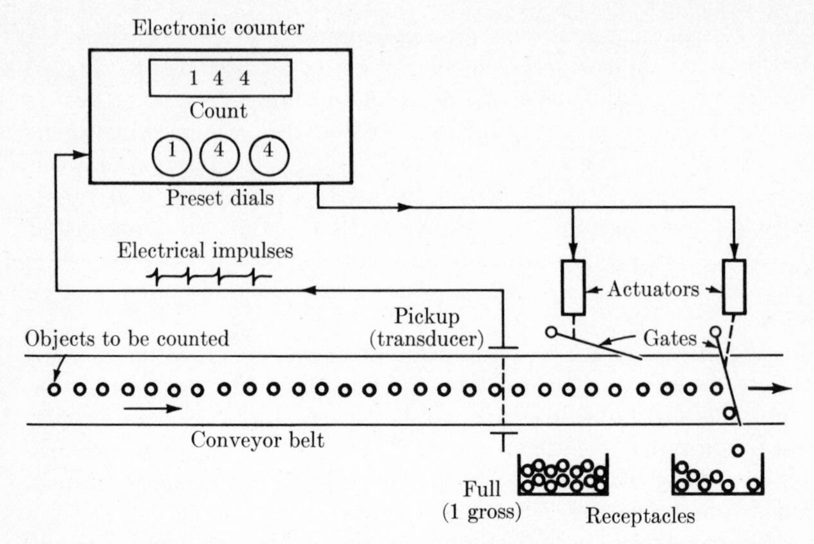

Fig. 1-2 Preset counter; an application of switching-type circuits. Electrical impulses from the passing objects are counted, and every time the count reaches 144 the conveyor is switched to the opposite receptacle.

gross (144). A "pickup" or transducer is arranged to provide an electrical impulse for each object passing by. If the objects are magnetic, the transducer can be an inductive pickup coil of wire; if not, a light beam and photocell combination can be used. These pulses go into the counter-circuit input as shown in the figure.

The counting circuit is of the *preset* variety, in which an output signal occurs when the count reaches a specified value—in this case 144. The operator simply rotates the preset dials to the desired numbers, and the circuit does the rest. Each output signal switches the conveyor gate. Thus after one receptacle or package is filled, the objects flow to the other one, and the full receptacle is replaced by an empty one in time for the next filling.

1-3 NUCLEAR RESEARCH INSTRUMENTATION

Nuclear physics is one area of technical activity in which pulse-type circuits have reached a high degree of perfection. An example of a problem in this area is illustrated in Fig. 1-3.

In a nuclear experiment particles are produced with a distribution of energies that is to be determined. A special transducer, known to the physicist as a *proportional counter*, provides an electrical impulse for each particle, with the amplitude of the pulse being proportional to the energy of the particle. The physicist wants a distribution curve such as is shown in Fig. 1-3*a* which gives the relative number of particles at each energy level.

Such a result can be obtained from the circuits described in this book connected in an arrangement illustrated in Fig. 1-3*b*. Pulses are shown arriving at the input to the circuits. Each of these pulses has a different energy level, and the circuit has been set so that only pulses between energy or amplitude levels A and B are counted. One basic building block of this arrangement is a circuit known as a *discriminator*. Such a circuit responds to pulses of amplitude greater than some arbitrary level, and for each of these delivers an output pulse of fixed amplitude and duration. In the illustration, discriminator A transmits pulses above the amplitude

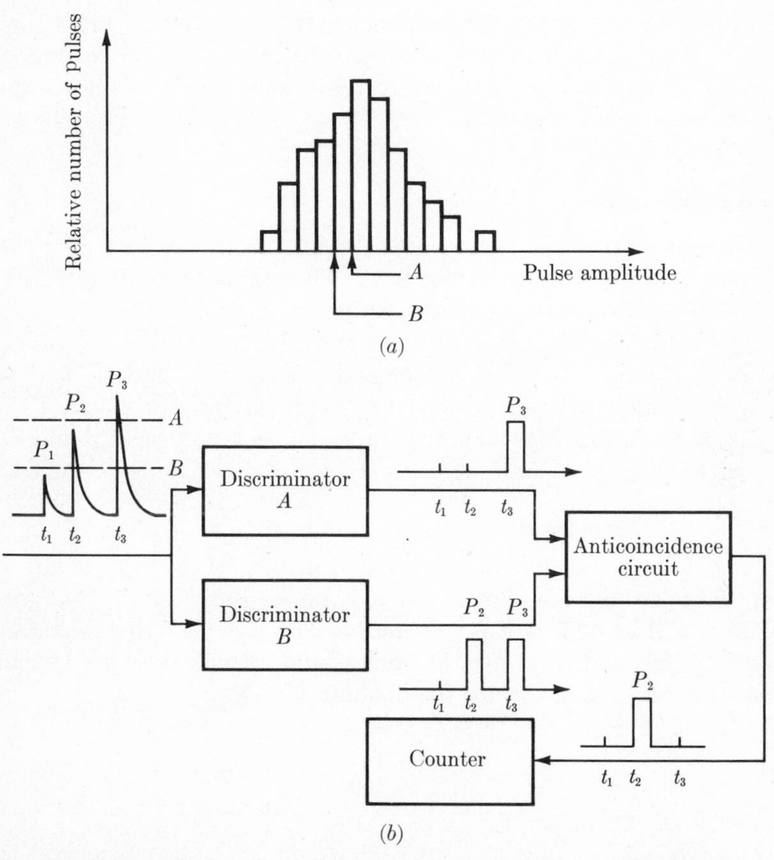

Fig. 1-3 Pulse analysis with the differential discriminator. Only pulses between the voltage levels A and B are counted. By adjusting the levels progressively across the amplitude range of interest, a distribution curve can be measured as in (*a*).

level A, whereas discriminator B transmits those above level B. Thus discriminator A provides an output in response to pulse P_3, whereas discriminator B provides output pulses corresponding both to P_2 and P_3. Pulse P_1 is of too small amplitude to pass through either discriminator.

There remains the task of separating out the pulses which lie between the levels A and B. This is accomplished by one of a family of circuits known as *coincidence circuits.* The particular form needed here is known as an *anticoincidence circuit,* and is one which provides an output pulse if there is an input pulse at *either* A or B, but not if there is a simultaneous one at both inputs. Thus the pulse P_2 produces an output, but P_3 does not. Accordingly, the desired pulse P_2, which is the only one of the three lying between the amplitude levels A and B, is passed on to the counter circuit.

In order to obtain the distribution curve, the levels A and B are kept at a fixed separation or increment, and then for successive settings a pulse count is made for some arbitrary time interval. From these data the desired curve can be plotted.

1-4 OTHER APPLICATIONS

There are many other interesting applications illustrating the need for the circuits we shall examine in this book. The scanning circuits in both television transmitters and receivers are further examples of such applications. In radar systems almost all the circuits except those in the radio-frequency portions of the transmitter and receiver can best be understood by the kind of analysis to be developed in subsequent chapters.

All such applications comprise arrangements of basic building blocks, which, if understood properly, can lead to new arrangements and new systems in the future. The ability to create new arrangements varies greatly among individuals and seems to be strongly dependent upon native aptitude. Such aptitude probably will not be greatly enhanced by a reading of this book, but rather it is hoped that the analysis presented here will provide a superior *understanding* of the functioning of the basic circuits, and that with this understanding a creative person can proceed more effectively toward the invention of new systems.

2
Some Circuit Fundamentals

2-1 TRANSIENTS IN LINEAR RC CIRCUITS

The major portion of this book is devoted to the use of transistors and other devices under highly nonlinear operating conditions. Such operation is essential to the successful functioning of the circuits to be studied. On the other hand, for the purpose of analysis it is useful to consider these circuits in successive discrete intervals of time, during each of which the circuit can be reduced to simple *linear* equivalent circuits. Linear circuits are simpler than nonlinear circuits because of the familiar and well-developed mathematics of linear differential equations.

In this book we shall be more concerned with *transient* conditions, rather than with steady-state sine waves. That is to say, the waveforms of current and voltage in the circuits described in this book are not of the sine-wave form, which results from the steady-state solution of circuit differential equations. Instead, the waveforms are usually combinations of decaying exponentials, as may be illustrated by the currents and voltages resulting from the closing of the switch in Fig. 2-1. Familiarity

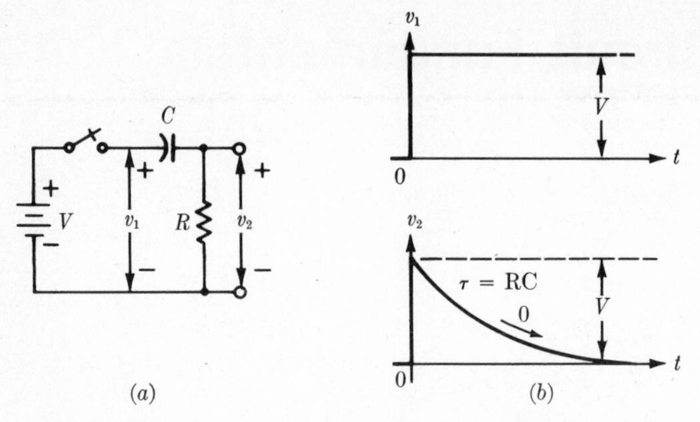

Fig. 2-1 Waveforms in the series *RC* circuit. Initial capacitor voltage is zero. Switch is closed at $t = 0$. The arrow labeled 0 indicates the final value of the waveform.

with the behavior of this simple circuit will be the foundation for the understanding of more complicated circuits later on.

Mathematical analysis of the circuit involves a conventional solution of the differential equation of the circuit, and the result is as follows:

$$v_2 = V\epsilon^{-t/\tau} \tag{2-1}$$

where τ = time constant = RC.

Once it has been established that the form of the solution is a decaying exponential, with time constant $\tau = RC$, the other important aspects of the waveforms shown in Fig. 2-1*b* can be readily deduced from simple physical reasoning. Reasoning of this kind is important and useful in the understanding of more complex circuits.

Consider first the voltage v_1, which can be regarded as the *input* voltage applied to the *RC* circuit. If the battery is an ideal one with no internal resistance, then regardless of the current that may flow, the voltage v_1 must equal the battery voltage V as soon as the switch has closed. Moreover, it will remain at the same voltage level at all subsequent times. Hence the voltage v_1 has the form of a *step*. This type of waveform will be encountered frequently later. Various other circuits in addition to the idealized one shown in Fig. 2-1*a* also produce step voltages.

Voltage v_2 may be regarded as the *output* voltage, and is equal to the voltage drop across the resistor. Analysis shows it must have the decaying exponential form with time constant *RC*. The initial and final values are immediately apparent from simple physical reasoning. If it is assumed that the initial voltage on the capacitor is zero, and if it is understood that any change in the voltage will equal $(1/C)\int i\,dt$, the *time integral* of the current flow (divided by the capacitance), then it also will be understood that with a finite current the voltage cannot change until a finite time has elapsed. Hence, at the instant of closing the switch the capacitor voltage remains zero, and the output voltage must equal the battery voltage.

After a long time has elapsed, however, the current flow will charge the capacitor to the full battery voltage V, and the current will diminish to zero. This is the same as saying that direct current does not flow through a capacitor. Clearly, then, the final value of the voltage across the resistor must be zero when the current becomes zero.

Thus the circuit behavior can be summarized by saying that the voltage v_2 jumps instantly from its original value of zero to an initial value of V, following which it decays exponentially with time constant RC to its final value of zero. The three essential features are:

1. The initial value
2. The final value
3. The time constant

The concept of a *time constant* is as important in transient waveforms as is that of *frequency* in steady-state sine-wave analysis. It provides a measure of the relative speed of the circuit action. For instance, if one is concerned with events happening during a 1-s interval, it is quite significant to know whether the time constant of the circuit is, say, 10 ms or perhaps 10,000 s.

The decaying exponential waveform will be encountered often in the material to follow, and it will be advisable to learn a few of its simple properties. Figure 2-2 shows a generalized form of the exponential wave in which the amplitude is expressed in percentage of the initial value, and the time scale is in multiples of the time constant. For values of time near zero, i.e., for intervals of time that are small compared to the time constant, the waveform is nearly a straight line, with an initial slope proportional to $1/\tau$ and negative in sign. After an interval that is

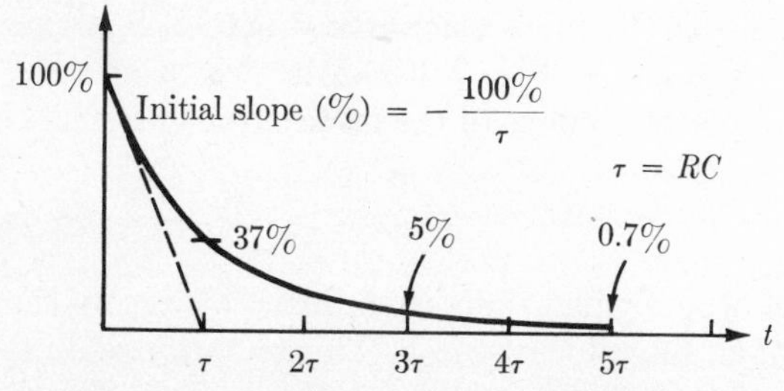

Fig. 2-2 Generalized exponential waveform.

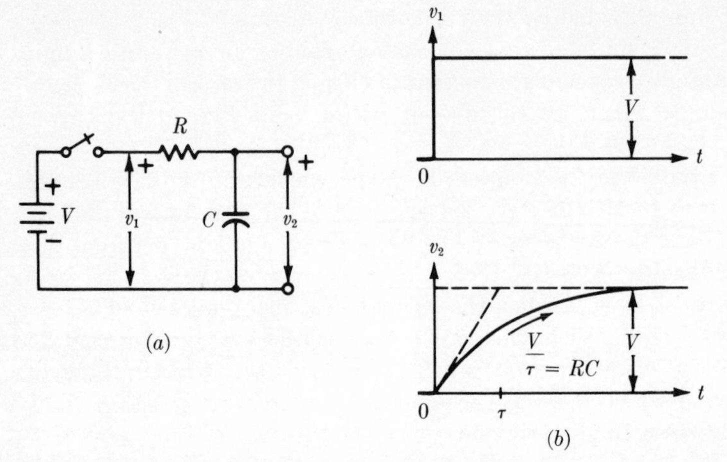

Fig. 2-3 Series *RC* circuit of Fig. 2-1 redrawn to show the output voltage v_2 across the capacitor.

long compared to the time constant, the amplitude approaches zero, and for many practical situations it is convenient to regard the amplitude as essentially zero after a time interval equal to 4τ or 5τ.

Suppose we now consider the voltage across the capacitor rather than across the resistor. In order to establish good habits of analysis, it is desirable to rearrange the circuit as shown in Fig. 2-3. Here the battery and switch again provide the input voltage v_1, and the output voltage v_2 is that across the capacitor. It would be easy to write down the capacitor voltage from our previous knowledge of the resistor voltage, which when added to the capacitor voltage must equal the battery voltage V. It is more instructive, however, to "derive" the capacitor-voltage waveform by direct application of the principles already presented. Since the capacitor voltage is originally zero and since it cannot change instantaneously, the voltage is still zero after the switch is closed. Looking ahead in time, we can see that the final capacitor voltage must be equal to V when the capacitor is fully charged and the circuit current becomes zero. The variation between the initial and final values will be exponential in character and will have the basic time constant $\tau = RC$. When this much is known, the waveform shown in Fig. 2-3*b* can be drawn at once. The exponential is simply an inverted replica of the basic curve of Fig. 2-2. The analytical expression is as follows:

$$v_2 = V(1 - \epsilon^{-t/\tau}) \qquad (2\text{-}2)$$

An important variation occurs when there is an initial charge on the capacitor prior to the time of closing of the switch. This situation is illustrated in Fig. 2-4*a*, where the initial charge causes a voltage V_0 to exist on the capacitor. Analysis of this condition gives the following expression for the output voltage v_2:

$$v_2 = (V_0 + V)\epsilon^{-t/\tau} \tag{2-3}$$

It is evident from the equation and the resulting waveform plotted in Fig. 2-4*b* that the situation is very similar to the circuit of Fig. 2-1, except that the initial value is modified. This initial value is computed as though the charged capacitor acted like a battery of the same initial voltage. On the other hand, the final value is unchanged; the circuit is identical with that of Fig. 2-1, and ultimately the final charge conditions must be the same in both cases. Similarly, the circuit time constant, depending only on R and C, must be the same in both cases.

The analytical result expressed in Eq. (2-3) is again one which can be readily obtained by physical reasoning. When the switch is closed, a finite current will flow at once. However, until this current has flowed for a finite time the capacitor voltage cannot change, and, hence, regardless of the amount of the current, the initial voltage v_2 must equal the sum of the battery voltage V and the capacitor voltage V_0. On the other hand, it should be evident from inspection of the circuit that the final value of v_2 must be equal to zero when the current has decayed to zero. It remains only to justify the fact that the time constant RC of the decaying exponential waveform is still the same as in the earlier case. This is perhaps not intuitively apparent, but experience with physical systems leads one to accept as plausible the fact that the form of the solution of a differ-

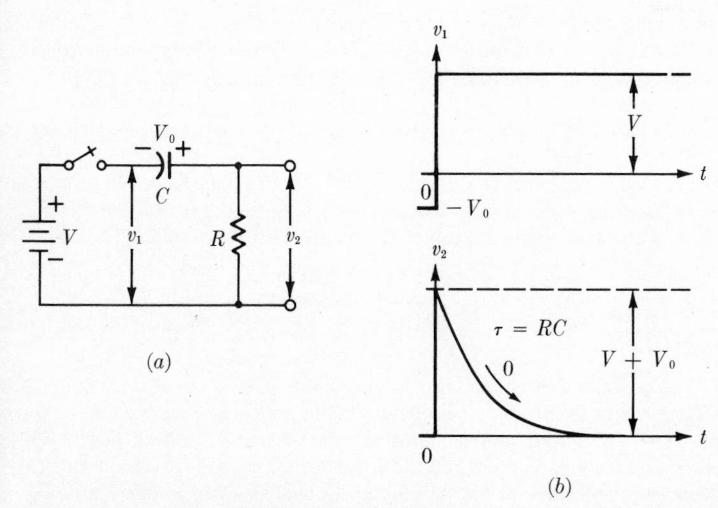

Fig. 2-4 An initial voltage V_0 on the capacitor changes the initial value, but not the final value or the time constant.

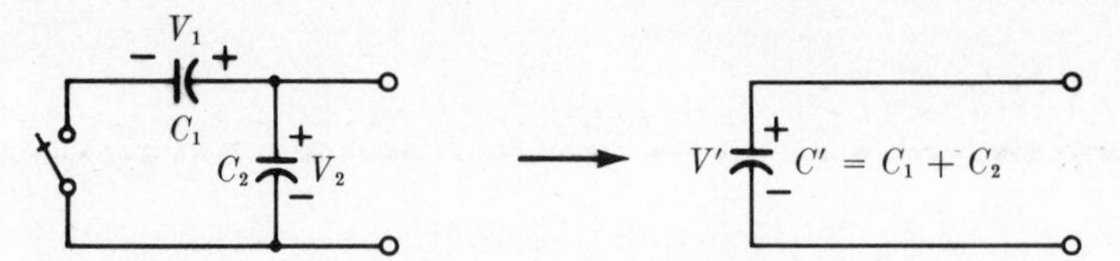

Fig. 2-5 Case of transient in a circuit having only C and no R. Charge readjusts instantaneously upon closing of switch. Infinite current can flow, but only for an infinitesimal time.

ential equation defining the behavior of a physical system depends only upon the system parameters and not upon the initial energy in the system.

From our acquaintance with various circuits we shall learn that many complex circuits are reducible to simple series arrangements comprising a single voltage source connected in series with a single resistance and a single capacitor. Accordingly, it is possible to generalize the results from the cases described in the preceding paragraphs and to express them as three rules that have general utility.

1. The voltage across a capacitor cannot change instantaneously if the charging (or discharging) current is finite or must flow through a resistive circuit.[1]
2. In a series circuit containing only resistors and capacitors where the total series resistance is R and the total series capacitance is C, a sudden change in driving voltage results in exponential transients having the single time constant RC.

[1] Sometimes a switching action produces a sudden reconnection between two capacitors without any apparent series resistance in the circuit. One example involves the interelectrode capacitances in a vacuum tube, which assume new charges when the current is switched on or off in the tube. The equivalent circuit is like that of Fig. 2-5. If the initial voltages are known prior to the closing of the switch, the final voltage can be calculated from a knowledge of the fact that the charge will instantaneously readjust between the two capacitors to provide a new common voltage, while simultaneously preserving the same total charge. The following equations provide the solution for the voltage V' existing after the closing of the switch:

$$\text{Total charge} \quad q = C_1V_1 + C_2V_2$$

$$V' = \frac{\text{total charge } q}{\text{total capacitance}} = \frac{C_1V_1 + C_2V_2}{C_1 + C_2}$$

Since the readjustment of charge can theoretically take place in zero time, it follows that infinite currents can flow, even though the product of current and time (which is proportional to charge) is finite. Infinite current could indeed flow if the connecting wires between the capacitors had zero resistance (and also zero inductance), a condition which is not the actual one of course. Considered practically, however, the action can take place in a time which is negligibly short compared with that required for action in other parts of the circuit.

3. A capacitor with an initial charge is equivalent to an uncharged capacitor of the same capacitance connected in series with a battery equal to the initial capacitor voltage. Thus the initial capacitor voltage acts like a series battery voltage (with zero internal resistance) and may be lumped in with the driving voltages. This initial voltage does not influence the final value of the exponential nor the time constant.

If there is any doubt about calculating the voltages in a circuit, the safest way to begin is to write the equation for the current, which is initially equal to the total driving voltage divided by the total loop resistance. The transient current is then $I(0^+)\epsilon^{-t/\tau}$ where $I(0^+)$ is the initial current. This current multiplied by the resistance R then gives the voltage across each of the resistances in the loop.

2-2 WAVEFORM-SHAPING CIRCUITS USING LINEAR RC ELEMENTS

Before going further with theory and general rules of procedure, it is of interest at this point to show how some practical results can be achieved with even the simple techniques presented thus far. It is frequently necessary to convert one waveform to another, the process generally being called *waveform shaping* or simply *wave shaping*. Three simple cir-

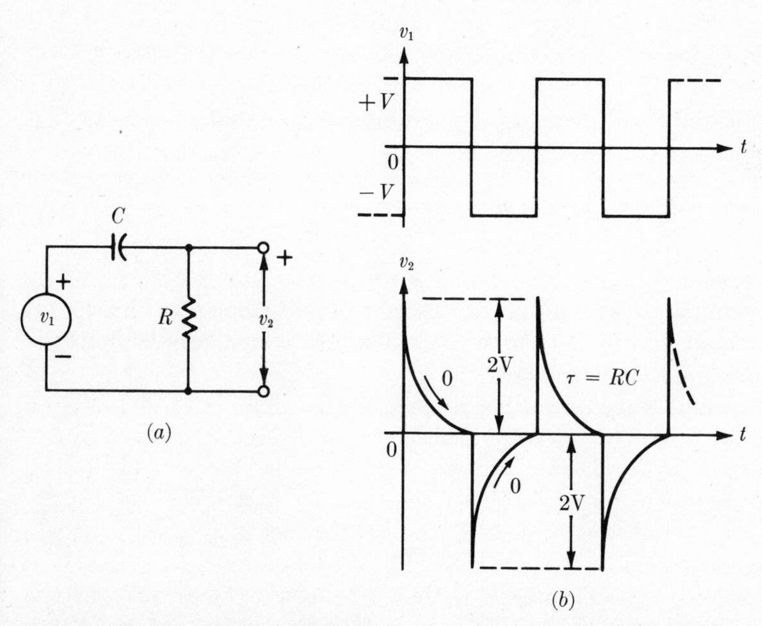

Fig. 2-6 Narrow pulses produced from a square wave by an RC circuit with $\tau = RC$ made small compared to the half period of the square wave.

cuits will serve to demonstrate the use of linear elements for this purpose. They require only R and C, and later will be incorporated into more complicated circuits, which will also use active elements such as transistors. These RC circuits by no means exhaust the possibilities of shaping with such combinations, nor do they give any direct insight into the use of combinations incorporating inductance.

NARROW PULSES FROM RECTANGULAR WAVEFORMS

The circuit of Fig. 2-6 will perform the function of generating narrow pulses, i.e., pulses of short duration. One must say "short" compared to something, and in this case it is the interval between pulses. To accomplish the desired result, the time constant of the RC circuit must be made small compared with this interval. The effect is illustrated in the waveform in Fig. 2-6*b*. The circuit is really the same as that of Fig. 2-1, except that the battery has now been replaced by a square-wave generator of zero internal resistance.

Notice that each transition of the input square wave from one polarity to the other has the same effect as closing the switch in the battery circuit of Fig. 2-1. At each transition, a step of voltage equal to $2V$ is applied to the RC circuit, whereupon an *initial* voltage of $2V$ appears at the output. This voltage quickly decays to zero and remains at this level until the next transition.

Notice that both positive and negative pulses are generated, but means can be provided for eliminating one or the other if this is desired. In addition, it will later prove possible to modify the exponential shape of the pulses into one more nearly rectangular; for that, however, nonlinear elements must be brought in.

LINEAR SLOPE FROM A RECTANGULAR WAVEFORM

To generate a voltage waveform which rises or falls linearly with time during a specified period, it is possible to use the circuit of Fig. 2-3 by simply assigning suitable proportions to the circuit elements. This circuit is shown again in Fig. 2-7, with the waveform which results from the application of a step of voltage.

The output waveform is not a truly straight line, but is actually an exponential curve. The initial portion is nearly straight, however, and is increasingly so if the desired time interval is made shorter and shorter. For many practical purposes, a period equal to, say, $\tau/5$ is sufficiently small. Thus if the desired time is T, the RC product is made at least five times as great, or $5T$.

A note of practical interest is that a penalty is incurred when the straightness of the line is improved by decreasing the time interval. This penalty involves the voltage amplitude of the linear rise as compared to

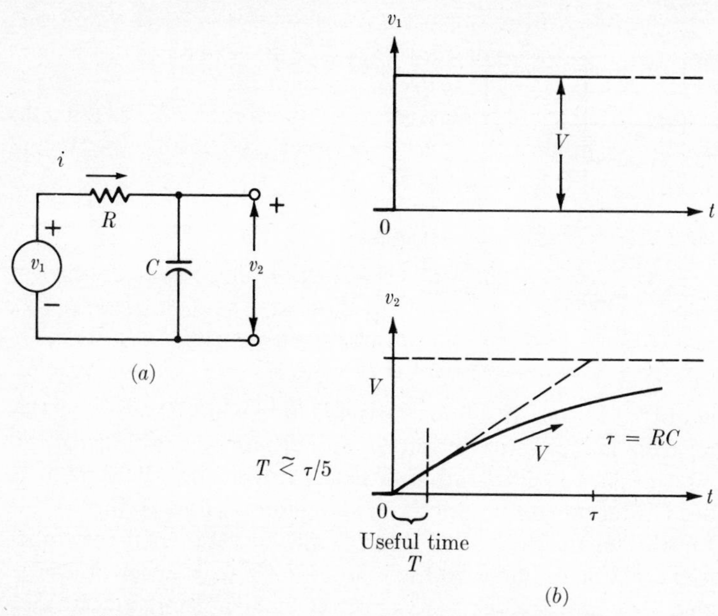

Fig. 2-7 Linear-voltage slope approximated by initial portion of exponential response of RC circuit to a voltage-step input.

that of the supply voltage V. For $T = \tau/5$, the peak value of the linear rise is less than $V/5$.

The circuit of Fig. 2-7 is frequently called an *integrating circuit*, because in some applications this is the function it performs. If the output from the circuit is kept small compared to the input, the circuit current is

$$i(t) = \frac{v_1 - v_2}{R} \approx \frac{v_1}{R} \qquad (\text{if } v_2 \ll v_1) \tag{2-4}$$

The output voltage from the circuit is then approximately

$$v_2(t) = \frac{1}{C}\int_0^t i(t)\,dt \approx \frac{1}{RC}\int_0^t v_1(t)\,dt \tag{2-5}$$

Therefore the output is approximately the time integral of the input voltage as long as the output is small compared to the input. This restriction usually means that the time constant of the circuit must be much longer than the period of the input waveform. Hence, if the step waveform of Fig. 2-7 is repeated in the form of a square wave as in Fig. 2-8, the output waveform would be a "sawtooth" as shown in the figure, which, of course, corresponds to the integral of the input waveform. As mentioned in the preceding paragraph, the sawtooth has a good straight-line

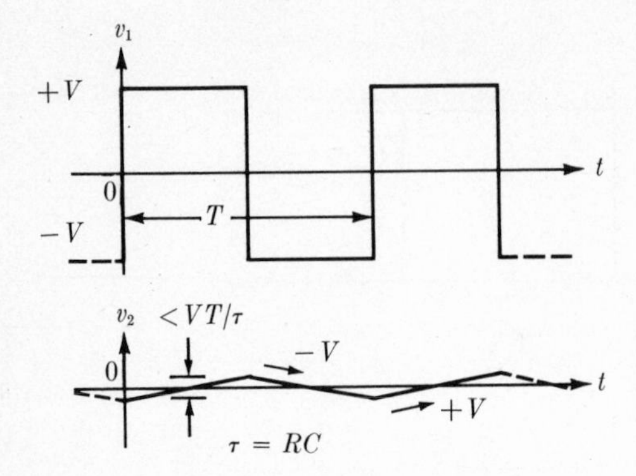

Fig. 2-8 Integrator action of the circuit of Fig. 2-7 in converting a square wave to a sawtooth.

characteristic only if its amplitude is restricted to a small fraction of the input voltage. This is a general limitation on the simple integrating circuit shown, and applies to integration of other waveforms. This *RC* circuit will indeed integrate other kinds of waveforms, and it forms the foundation for the entire field of analog computers. We shall return to the matter of generation of linear voltage slopes in Chap. 6, where amplification is added to eliminate the restriction of small output amplitude.

RECTANGULAR WAVEFORM FROM A LINEAR SLOPE

We return again to the *RC* circuit of Fig. 2-1, but this time we apply a linear voltage *slope*, commencing at $t = 0$, as illustrated in Fig. 2-9. The

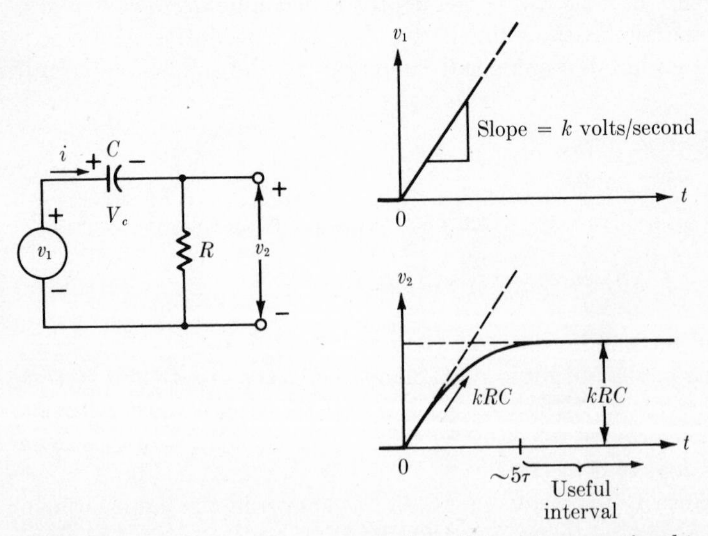

Fig. 2-9 *RC* circuit driven by voltage slope. Output is proportional to slope of input.

corresponding output voltage shown in the figure is given by

$$v_2 = \tau k(1 - \epsilon^{-t/\tau}) \tag{2-6}$$

where $\tau = RC$.

Note that after a sufficient time has elapsed following the start of the slope, perhaps 5τ, a constant output voltage results with a magnitude directly proportional to the slope of the input waveform.

The action of this circuit can be seen more generally if again the output voltage is assumed small compared with the input; then

$$V_c(t) = v_1 - v_2 \approx v_1 \qquad (\text{if } v_2 \ll v_1)$$

$$i(t) = \frac{C\,dV_c}{dt} \approx \frac{C\,dv_1(t)}{dt}$$

and the output voltage is

$$v_2(t) = i(t)R \approx \frac{RC\,dv_1(t)}{dt} \tag{2-7}$$

The circuit therefore acts as a *differentiator* in that the output is the derivative of the input—as long as the output is small compared to the input. This is seen to be true in Fig. 2-9 because for small values of t the output is not small compared to the input, and the output is not yet the approximate derivative of the input either.

The differentiator circuit can convert a sawtooth waveform to a square wave as illustrated in Fig. 2-10. Ideally, each time the slope of the input waveform changes to the new value, the output waveform should

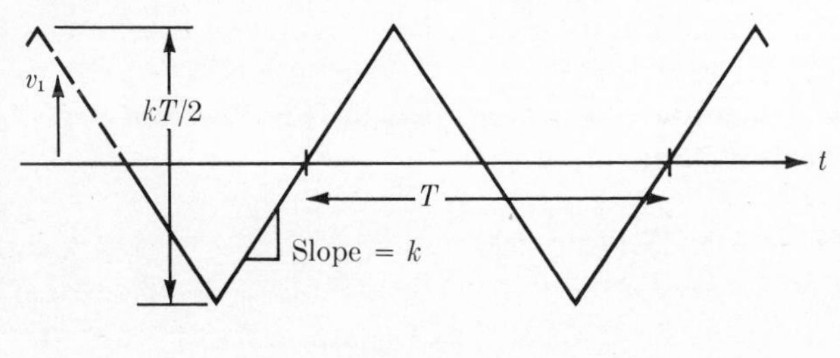

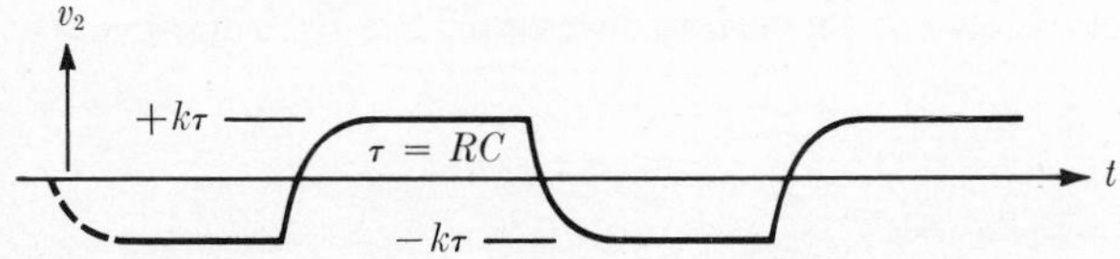

Fig. 2-10 Differentiator action of the circuit of Fig. 2-9 in converting sawtooth wave into a square wave. The circuit is effective only as $\tau \ll T$. Hence, the maximum output amplitude is necessarily small, since $k\tau \ll kT$. (The waveform for v_2 is expanded vertically.)

switch instantaneously to a new value. As was seen in Fig. 2-9, however, a finite transition time of about 5τ is required when the slope of the input suddenly changes. In the example of Fig. 2-10, the transition time can be made arbitrarily small by reducing the time constant of the circuit. Such improvement in the waveform is necessarily accompanied by a reduction in the amplitude of the output voltage compared with that at the input.

2-3 USEFUL CIRCUIT THEOREMS

This book is primarily concerned with the instantaneous values of nonsinusoidal waveforms. Because of the usual preoccupation with sine waves in electrical engineering, it is desirable to remove any uncertainties regarding several familiar circuit theorems that are useful in the work that follows. Also, since these theorems are most frequently used in linear circuits, it is important to point out which of these can be used in nonlinear situations as well.

For instance, Ohm's law for resistive circuits and the differential expressions for inductance and capacitance are always applicable. Set aside, however, the familiar $j\omega L$ and $1/j\omega C$, which are only for the special case of the sinusoidal steady state.

KIRCHHOFF'S RULES (LINEAR OR NONLINEAR CIRCUITS)

These apply, instant by instant, just as they do for direct current and for sine-wave voltages or currents. They are essentially expressions of the geometry (or more precisely, topology) of an electric circuit, where the current can only follow along definite paths. Both forms of Kirchhoff's rules will be useful. Thus, at any instant: (1) The sum of the voltage which rises and drops around a closed mesh is zero. (2) The sum of the currents entering and leaving a node is zero.

SUPERPOSITION THEOREM (LINEAR CIRCUITS ONLY)

This theorem states that if two causes exist simultaneously, the resulting effect is the sum of the individual effects of the two causes. An appropriate example may be seen in Fig. 2-11. The circuit is essentially that of Fig. 2-1,

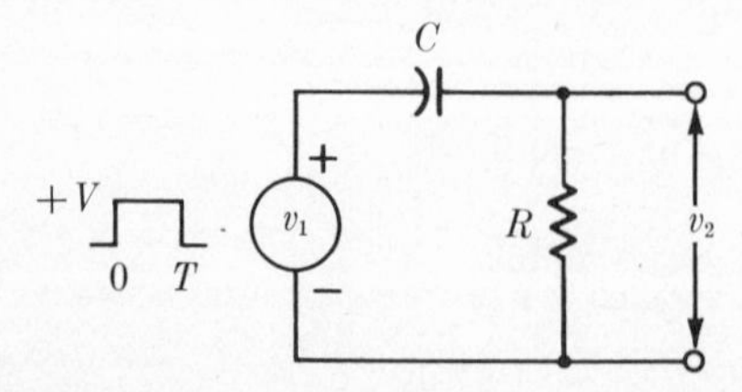

Fig. 2-11 Application of rectangular pulse to *RC* circuit of Fig. 2-1.

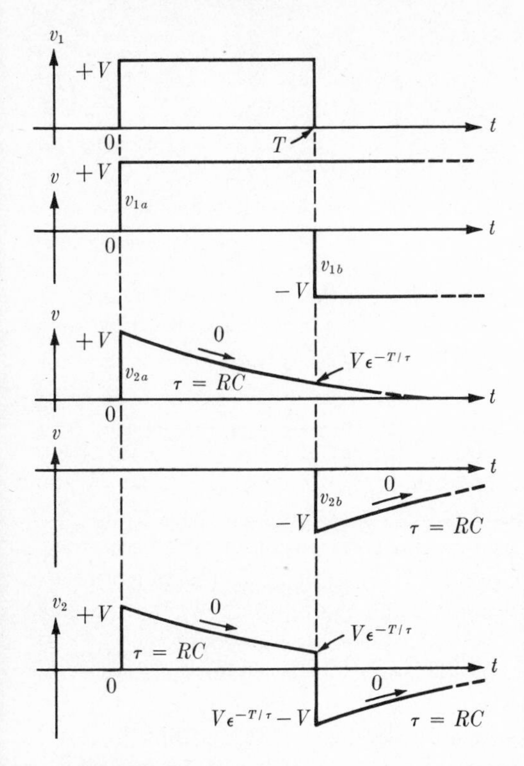

Fig. 2-12 Solution of Fig. 2-11 by superposition. Input waveform v_1 resolved into two step functions v_{1a} and v_{1b} applied separately to the circuit, and the outputs superposed to give v_2.

except that the battery and switch have been replaced by a zero-resistance voltage source providing a single rectangular pulse waveform for v_1. As shown in Fig. 2-12, the voltage pulse can be resolved into two voltage steps of opposite polarity and displaced in time by the pulse duration T. Each of these steps corresponds to the battery-and-switch situation of Fig. 2-1, each therefore producing an exponential output waveform. The complete output waveform is simply the sum of the two individual outputs. The circuit must be a linear one, however, in order to permit this summation.

It is the superposition theorem which permits the analysis of a circuit by resolving complex waveforms into Fourier series. Each term in the series can be regarded as a separate voltage, producing its own output voltage; the summation of these voltages is the total output voltage.

It was stated in the preface that the Fourier-series method is cumbersome compared to the methods that are used in this book. This is true if a detailed calculation must be made. Nevertheless, the *concept* of a Fourier series is useful in visualizing the behavior of a circuit such as that shown in Fig. 2-13. Illustrated there is the so-called "capacitance-compensated voltage divider," the purpose of which is to provide a

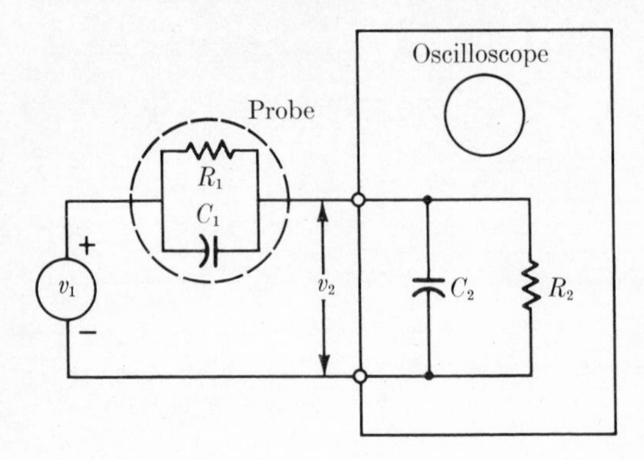

Fig. 2-13 Capacitance-compensated voltage divider. If $R_1C_1 = R_2C_2$, the output voltage v_2 is $v_2 = v_1R_2/(R_1 + R_2)$ for any waveform.

reduction in the amplitude of a voltage waveform, while at the same time preserving the exact shape of the waveform. It is widely used in the input circuits of oscilloscopes, where a simple resistance divider could not be used. The input circuit of the oscilloscope amplifier comprises both the resistance R_2 and the capacitance C_2 due to the input capacitance of the amplifier. The proper voltage divider is obtained as shown in the figure, where the series resistor R_1 is paralleled by a capacitance C_1, and the two RC time constants are made equal. It is a simple exercise to analyze this circuit for the steady-state voltage ratio V_2/V_1, and it will be found that the ratio is independent of frequency. If we think in terms of the Fourier series, we can say that regardless of the input waveform, all Fourier components will undergo the same division ratio, and hence the output waveform will be identical to that of the input.

THEVENIN'S THEOREM (LINEAR CIRCUITS ONLY)

This theorem is used throughout this book as a convenient means for simplifying complex circuits and as a substitute for the writing of simultaneous equations on a mesh or nodal basis. It supplies a direct approach to the end result of providing a circuit containing a single voltage (or current) source, a single capacitance, and a single equivalent series resistance. As already stated, this simple result occurs frequently and provides a convenient measure of the circuit performance in terms of the time constant.

The usual formulation of Thevenin's theorem includes the word *impedance,* which in turn has a *frequency* connotation and must therefore be used with care when we are concerned with instantaneous values of nonsinusoidal waveforms. This is not a problem in many of the practical circuits where the elements to be combined by use of the theorem are all alike, e.g., resistances; in such cases the circuit behavior is independent of frequency. Actually the theorem also can be used with care in nonsinus-

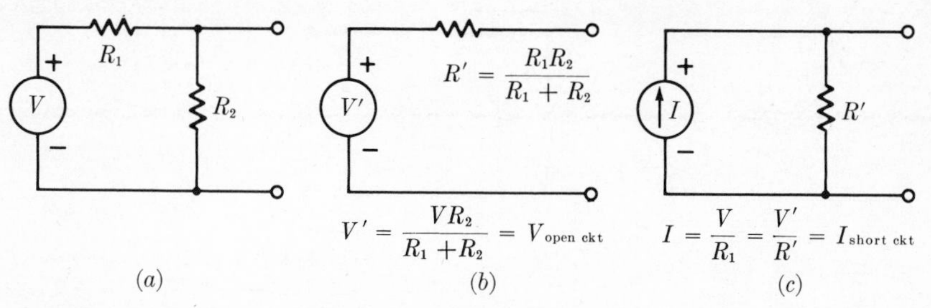

Fig. 2-14 Application of Thevenin's and Norton's theorems to a simple resistive circuit. (*a*) Original circuit. (*b*) Thevenin equivalent circuit. (*c*) Norton equivalent current source.

oidal situations where the circuit elements are not alike, but we shall have no occasion to do this.

Thevenin's theorem can be conveniently stated as follows: A two-terminal active circuit can be replaced by a simple series circuit containing a voltage generator of magnitude equal to the open-circuit voltage of the original circuit, with a series impedance equal to the terminal impedance measured on the original circuit with the independent internal voltage sources short circuited and the current sources open circuited.[1] This is illustrated in Fig. 2-14, where the original circuit is shown in Fig. 2-14*a* and the equivalent in Fig. 2-14*b*.

There is, of course, the alternative form (often distinguished as Norton's theorem), which yields the parallel circuit of Fig. 2-14*c*, comprising a current source having a magnitude equal to the terminal current of the original network with the terminals short circuited, together with the same passive impedance as before, but paralleling the current source.

It would be well for the reader to recall that Thevenin's theorem describes only the equivalent circuit in terms of its *terminal* behavior and cannot be used to evaluate conditions within the circuit. For instance, it will tell nothing about the division of currents *within* the circuit.

[1] An *independent generator* is one whose value does not depend upon the value of the output voltage or current. In any case the impedance can be computed as the ratio of the open-circuit terminal voltage to the short-circuit current, both observed with all generators active.

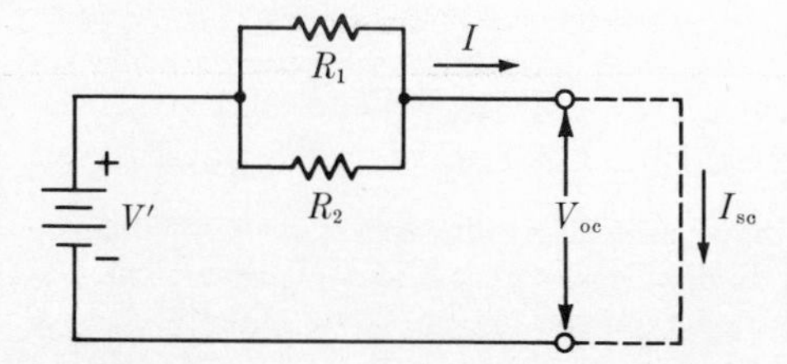

Fig. 2-15 Equivalent representation of Fig. 2-14*b*. This *cannot* be used, however, to determine relative currents in R_1 and R_2. I_{sc} and V_{oc} are the short-circuit current and open-circuit voltage, respectively.

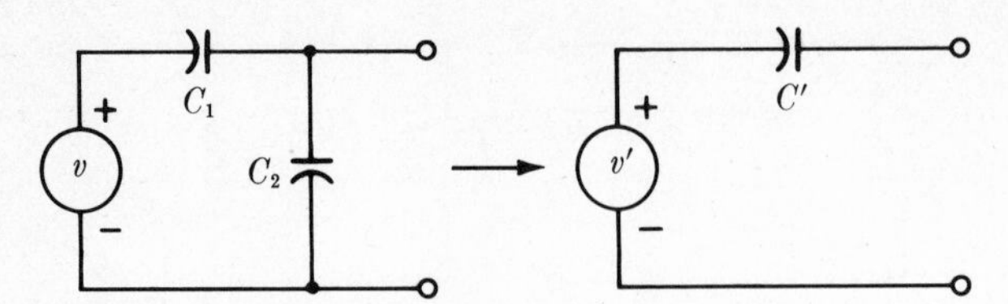

Fig. 2-16 Application of Thevenin's theorem to a source containing two capacitors. $v' = v[C_1/(C_1 + C_2)]$; $C' = C_1 + C_2$.

Thus, while the equivalent circuit of Fig. 2-14*b* can be represented in the form shown in Fig. 2-15, this alternative form should not be used to predict the current in each of the resistors as though the terminal current I divided itself between resistors R_1 and R_2 in proportion to the relative conductance values.

Figure 2-16 gives an example of a situation in which a network of capacitances can be simplified by the use of Thevenin's theorem. A precaution would be appropriate here for the case of initial charges readjusting themselves in accordance with the discussion of Fig. 2-5.

2-4 HELPFUL TECHNIQUES

In the material presented thus far, there have been introduced without announcement certain techniques for working with the problems of switching and waveform-processing circuits. The final task of this chapter is to point out these techniques and to introduce a few new ones, all of which will be valuable when the circuits and waveforms become more complex.

Instantaneous values of current and voltage are the mainstay of these techniques. Try to concentrate upon the rise or fall of a voltage at a given point in the circuit with respect to a reference point, usually called *ground*. Alternatively, the increase or decay of a current can be considered, and an even better example in some circumstances is the ebb and flow of charge itself. There will be little use for sinusoidal terms such as root-mean-square values or for analysis based on frequency.

Schematic diagrams of circuits can be good or bad, in terms of visualizing the essential functioning of the circuit. A good system such as the following is essential.[1]

1. Use a standard set of symbols (as in Figs. 2-1, 2-17, 2-18, and 2-19) for voltage and current sources. Indicate the reference polarity of

[1] Several of these suggestions seem to have been first proposed by F. C. Williams, Introduction to Circuit Techniques for Radio Location, *J. IEEE* (*London*), vol. 93, part IIIA (Proceedings of the Radiolocation Convention), pp. 289–308, 1946.

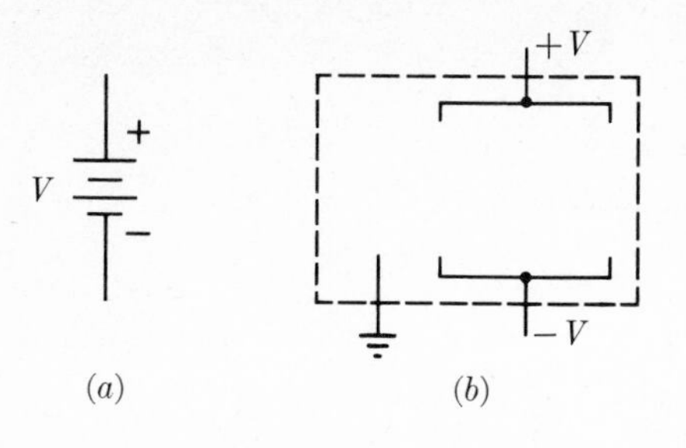

Fig. 2-17 Symbols used for *constant-voltage sources.* Internal-source resistance assumed zero unless shown explicitly as series resistor. (*a*) Symbolized as battery. (*b*) Symbolized as dc bus connections. Potentials are positive or negative with respect to ground.

voltage rises and drops as in Fig. 2-20. Thus when waveforms are drawn for the circuit, the polarities in the waveform diagram can be directly related to those in the circuit diagram.

2. Driving waveforms and dc supply voltages can be assumed to be zero-resistance sources unless a series resistance is shown. The two main types of voltage sources are illustrated in Figs. 2-17 and 2-18. Similarly, the current generator, as in Fig. 2-19, is assumed to have infinite internal resistance unless a parallel resistor is explicitly shown.

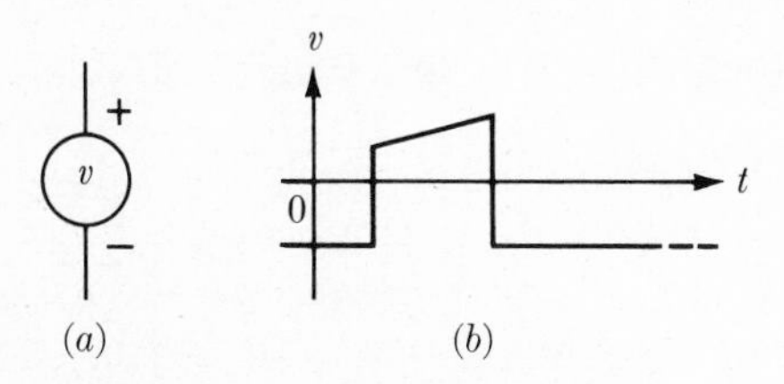

Fig. 2-18 Time-varying *voltage source* with instantaneous amplitude v and zero internal resistance. Polarity marks on the symbol are for *reference;* actual waveform polarity may go positive or negative, but always with respect to the reference polarity. (*a*) Symbol. (*b*) Example of waveform.

3. Most circuits have input and output terminals; label these. Alternatively, indicate the input and output voltages, such as v_1 and v_2 in Fig. 2-1. Frequently there is a common ground connection for an entire circuit, in which case only a single input terminal and a single output terminal may appear.
4. Sketch the waveform in miniature close to the key point in the circuit

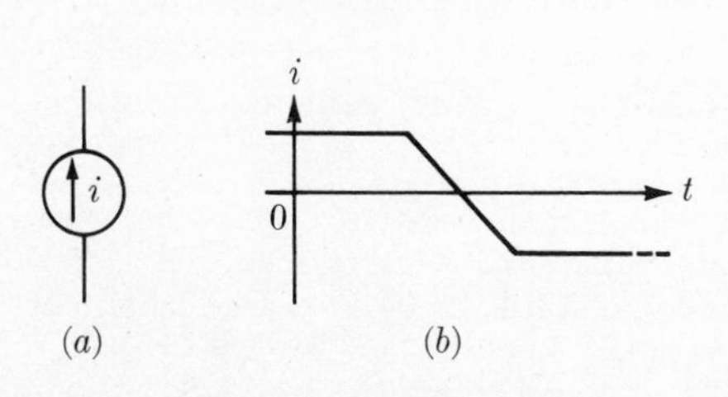

Fig. 2-19 Time-varying *current source,* with instantaneous amplitude i and infinite internal resistance. Arrow on symbol is the *reference* (positive) current direction; actual current may be in either direction, but is to be regarded as positive or negative with respect to the reference. (*a*) Symbol. (*b*) Example of waveform.

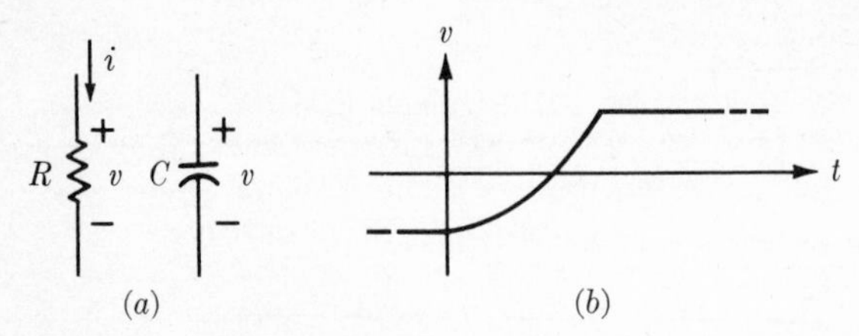

Fig. 2-20 *Voltage drops* across passive elements. Reference polarity marks on the symbols relate circuit voltages to waveform sketches. Actual polarity of a voltage is always with respect to the reference chosen. (*a*) Symbols. (*b*) Example of waveform.

when the waveform is known. Thus, Fig. 2-11 shows the waveform of the driving voltage v_1.

5. Voltage drops occurring in a circuit as a result of current flow need to be labeled for identification, as in Fig. 2-20. Another example is v_2 in Fig. 2-1, which is also the output voltage for that circuit.
6. Direct-current supply buses should be arranged vertically, with the most positive at the top of the diagram. This is illustrated in Fig. 2-17*b* and elsewhere, e.g., in Fig. 4-3.[1]
7. Waveforms should be drawn in adequate detail on a separate graph, with time axes horizontal and with time increasing from left to right. Arrange the various waveforms in a vertical column—input waveform at the top, output waveform at the bottom—with important time references shown down the length of the column by vertical dashed lines. Label the numerical values of key amplitudes, together with time constants and final values of all exponentials, right on the waveforms themselves. An example of this in a simple situation is shown in Fig. 2-1*b*.
8. Use standard nomenclature for the units in the circuit, as shown in Table 1-1 for the common units we shall use. Old and confusing abbreviation systems should not be used; for example, mfd for microfarad could be thought of as millifarad. Use instead μF.

The value of these techniques will become more apparent as the circuits increase in complexity. It is suggested that the reader refer back to this section from time to time. Systematic procedures permit one's full attention to be devoted to the essential functioning of a circuit rather than to ambiguous details and help others who must someday comprehend your creations.

[1] In instances where the most commonly used voltages are negative, the most negative can be placed at the top. However, keeping to an orderly arrangement with most negative on top and most positive on the bottom still helps other people figure out your circuit.

Table 2-1 The international system of units

Quantity	*Unit*	*Symbol*
Length	meter	m
Mass	kilogram	kg
Time	second	s
Temperature	degree Kelvin	°K
Electric current	ampere	A
Luminous intensity	candela	cd
Frequency	hertz	Hz (s^{-1})
Work, energy, quantity of heat	joule	J
Power	watt	W
Electric charge	coulomb	C
Voltage, potential difference, electromotive force	volt	V
Electric resistance	ohm	Ω
Electric capacitance	farad	F
Magnetic flux	weber	Wb
Inductance	henry	H
Magnetic flux density	tesla	T (Wb/m^2)

Submultiple Prefixes

Multiple/ Submultiple	*Prefix*	*Symbol*	*Pronunciation*	*Multiple/ Submultiple*	*Prefix*	*Symbol*	*Pronunciation*
10^{12}	tera	T	tĕr'ȧ	10^{-2}	centi	c	sĕn'tĭ
10^{9}	giga	G	jĭ'gȧ	10^{-3}	milli	m	mĭl'ĭ
10^{6}	mega	M	mĕg'ȧ	10^{-6}	micro	μ	mī'krō
10^{3}	kilo	k	kĭl'ō	10^{-9}	nano	n	năn'ō
10^{2}	hecto	h	hĕk'tō	10^{-12}	pico	p	pē'cō
10	deka	da	dĕk'ȧ	10^{-15}	femto	f	fĕm'tō
10^{-1}	deci	d	dĕs'ĭ	10^{-18}	atto	a	ăt'tō

A CIRCUIT EXAMPLE

Determine the waveform of the capacitor voltage v_C in the circuit of Fig. 2-21 if the switch SW opens at time t_1 after having been closed for an indefinitely long period. The switch then closes again 100 μs later at time

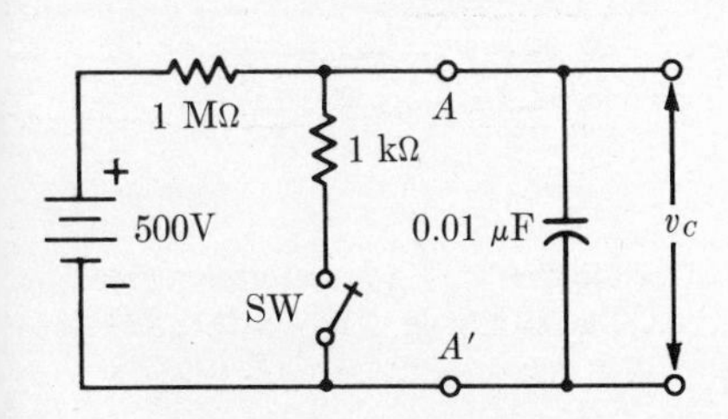

Fig. 2-21 Circuit diagram for the example. The switch SW is closed for $t < t_1$ and $t > t_2$, $t_2 - t_1 = 100\ \mu s$.

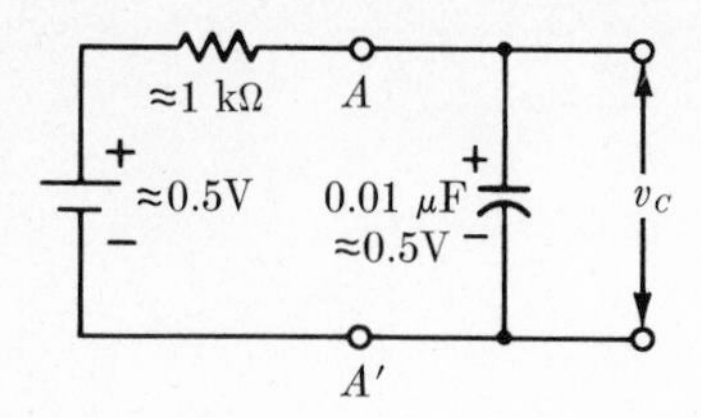

Fig. 2-22 Equivalent circuit for Fig. 2-21 prior to $t = t_1$; switch SW closed. Thevenin's theorem has been used to reduce the circuit to the left of the terminals A-A'. The voltage on the capacitor assumes SW has been closed "a long time."

t_2. Sketch the complete waveform, labeling the key voltage levels, including final levels for exponentials even though the waveform may be interrupted before attaining that final value. Label the time constant associated with each exponential.

Solution The first step is to determine the state of the circuit prior to the first switching action at $t = t_1$. The switch has been closed, so both resistors are in the circuit. Since the network to the left of the capacitor terminals A-A comprises only resistors and a voltage source, it can readily be reduced by use of Thevenin's theorem to a single voltage source and a single resistor in series, as in Fig. 2-22. The result is given with approximate values, the degree of approximation being typical of further numerical work in the book. It may be seen from Fig. 2-22 that the capacitor voltage will be the same as that of the voltage source because of the stipulation that the circuit has been in this condition for a long time.

When the first switching action occurs at $t = t_1$ (switch SW opens), the circuit becomes that of Fig. 2-23. The capacitor will not change voltage during the switching instant but will retain the *initial* voltage of 0.5 V. Thus, Fig. 2-23 depicts the situation at a time which can be designated at $t = t_1^+$, meaning that the switching action associated with the time t_1 has effected the circuit rearrangement but that capacitor voltages have not changed (if only finite currents are available for charging or discharging).

The capacitor will now charge toward a *final* voltage of 500 V. The charging time constant τ_1 is $\tau_1 = (1\ \text{M}\Omega)(0.01\ \mu\text{F}) = 10^{-2}\ \text{s} = 10^4\ \mu\text{s}$. Since the initial value, the final value, and the time constant are known, the waveform following $t = t_1^+$ is completely determined and can be sketched as in Fig. 2-24.

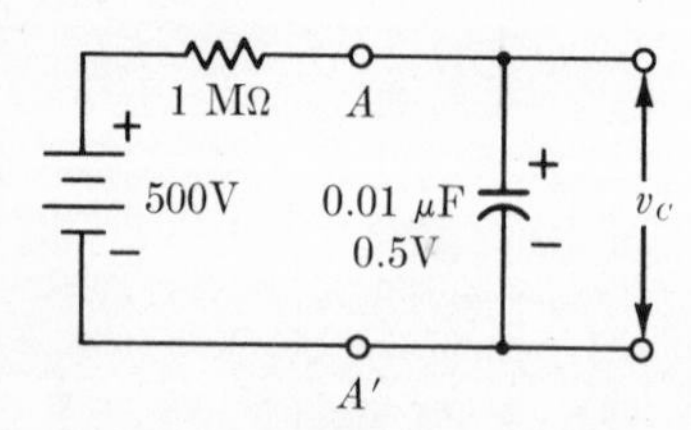

Fig. 2-23 Circuit at $t = t_1^+$, immediately after the switch opens. The circuit is valid until the switch closes again at $t = t_2$.

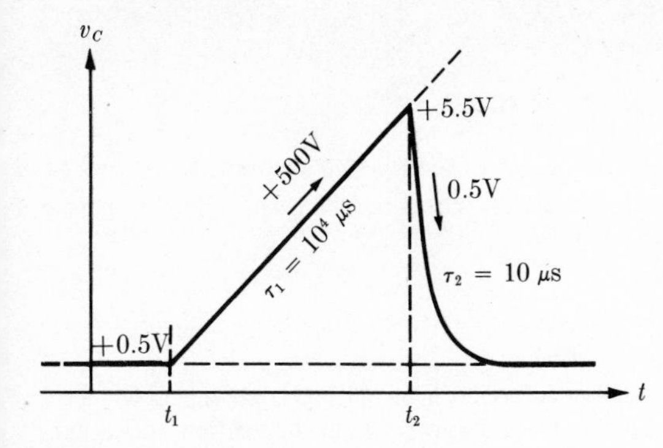

Fig. 2-24 Waveform for the example shown in Fig. 2-21.

As will be noticed immediately from the waveform, the exponential rise of v_C does not proceed very far toward the final value. The time interval of 100 μs until the second switching action occurs at time t_2 is only $\frac{1}{100}$ of the time constant τ_1. Thus the rise is confined to the initial portion of the exponential curve and is virtually a straight line. The slope of this straight line can be determined as in Fig. 2-2, namely

$$\text{Slope} = \frac{\text{final value} - \text{initial value}}{\tau}$$

$$= \frac{500 - 0.5}{10^4} \approx 5 \times 10^{-2}\ \text{V}/\mu\text{s}$$

In the time interval of 100 μs, the voltage will rise by

$$\Delta v_C = (5 \times 10^{-2})(100) = 5\ \text{V}$$

Thus the voltage level reached at the instant just prior to the second switching action (call it $t = t_2^-$) will be

$$v_C(t_2^-) = 0.5 + \Delta v_C = 0.5 + 5 = 5.5\ \text{V}$$

Now the second switching action takes place, and the circuit reverts to its original condition as the switch closes. The capacitor voltage remains at 5.5 V during this instantaneous switching action, and is thus the *initial value* for the exponential discharge which will now take place. The equivalent circuit is shown in Fig. 2-25. The capacitor will discharge

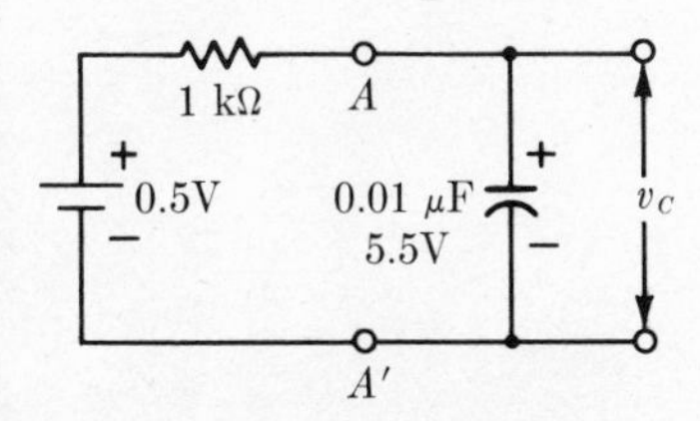

Fig. 2-25 Equivalent circuit at $t = t_2^+$, immediately following the second switching action.

toward 0.5 V with a *time constant* τ_2:

$$\tau_2 = (1\ \text{k}\Omega)(0.01\ \mu\text{F}) = 10^{-5}\ \text{s} = 10\ \mu\text{s}$$

Inasmuch as no further switching actions occur, the capacitor will discharge to its *final value* of 0.5 V, and the waveform sketch can be completed as in Fig. 2-24.

PROBLEMS

2-1. The circuit is initially in a steady-state condition with the switch at A. For $t > 0$ the switch operates periodically—remaining in each position for 1 ms. Compute and sketch v_0 for $0 \leq t \leq 5$ ms.

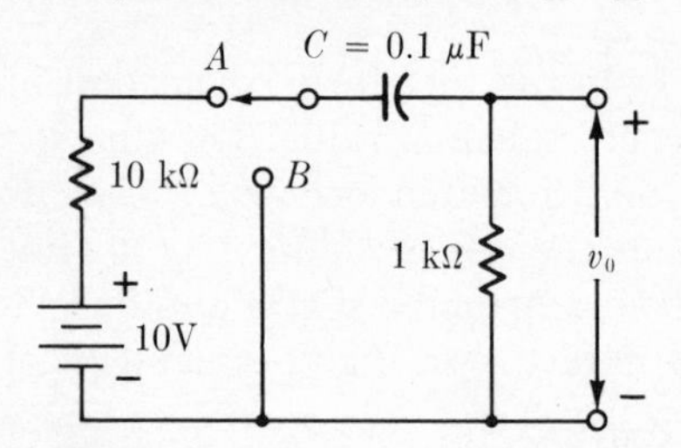

2-2. The circuit operates in steady-state prior to $t = 0$ with $v_1 = 100 \sin \omega t$. The generator v_2 is a step voltage generator ($v_2 = 0$, $t < 0$; $v_2 = 100$ V, $t > 0$). Compute and sketch v_0 for $t \geq 0$. (Do not use differential equations or operational methods for the solution.)

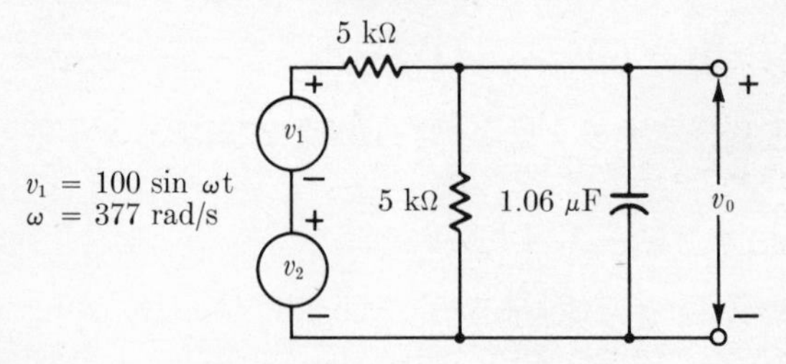

2-3. At $t = 0$ (before the switch closes) $V_{C1} = 50$ V and $V_{C2} = -50$ V. The switch closes at $t = 0$. Sketch and label the voltage $V_{C2}(t)$ for $t > 0$.

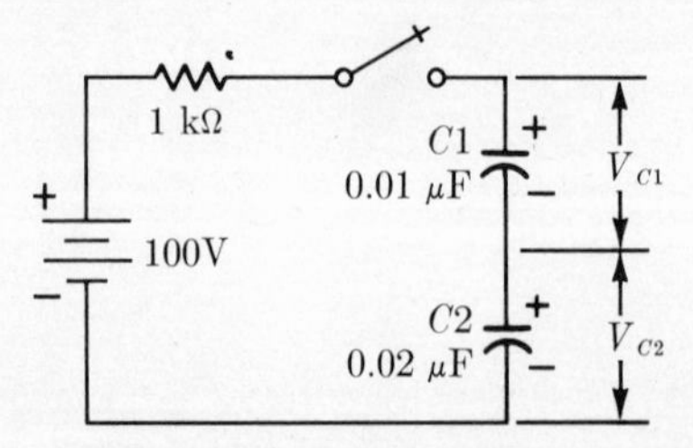

2-4. For $t < 0$ SW1 is open and SW2 is closed. At $t = 0$ SW1 closes and SW2 opens. When $v_2 = 0.5$ V, the switches go back to their original positions. Compute, sketch, and label the waveforms v_1 and v_2.

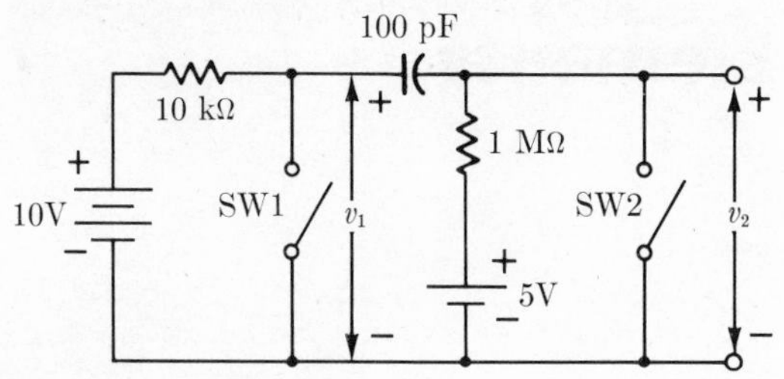

2-5. The switch is closed at all times except for $0 < t < 50$ ms. Calculate, sketch, and label the output waveform v_0.

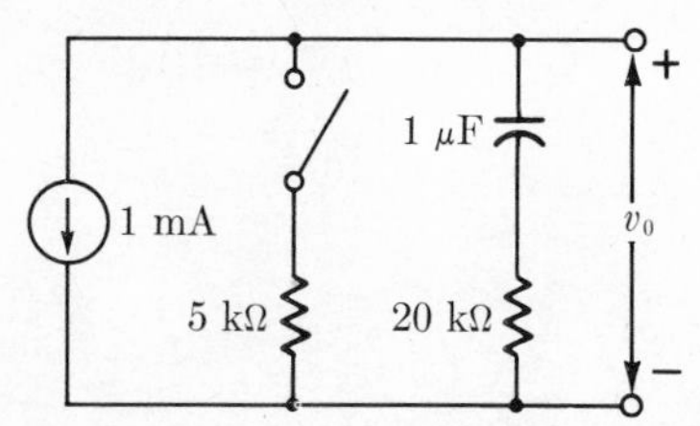

2-6. The circuit shown might represent the collector circuit of an RC coupled transistor. Make a clear sketch to approximate scale of the voltages $v_1(t)$ and $v_2(t)$. Label the waveforms carefully to completely define the voltages.

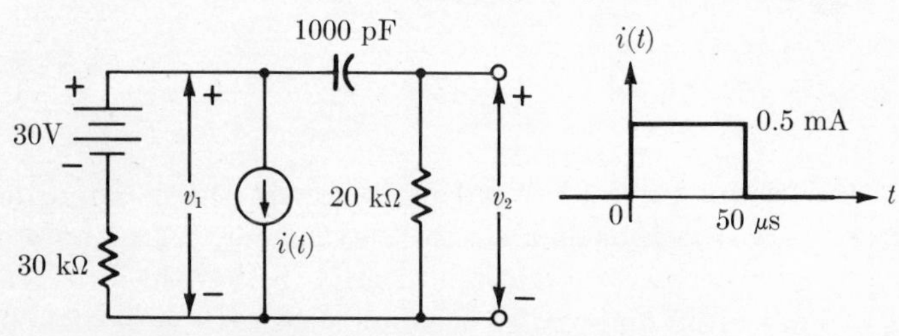

2-7. The initial conditions in the circuit are: Switch in position A, and the charges on $C1$ and $C2$ are zero. For $0 < t < 2$ ms the switch is placed in position B; for $t > 2$ ms the switch is placed in position C. Sketch and label the waveform of $v_0(t)$ for $t > 0$.

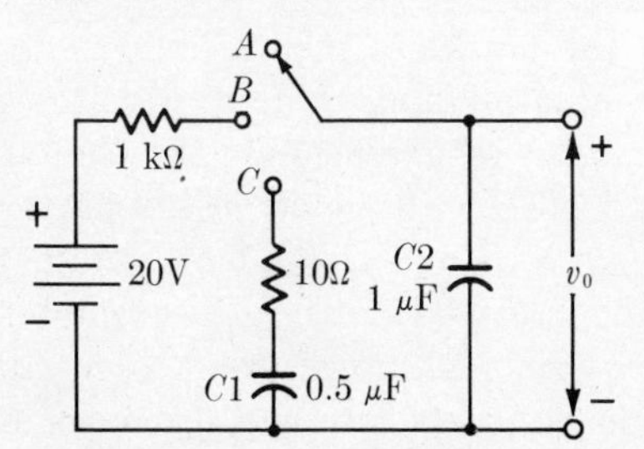

3 Electron Devices as Switching Elements

In the type of circuits to be analyzed in this book the electron device operates either as a switch or as a control device, controlling an output current or voltage in response to an input signal. Either the switching or control can be very rapid and can be accomplished with very little input energy. Since the controlled energy may be large, the concept of power gain or amplification is important. In some cases a small energizing signal may grow or regenerate into a very large output signal from power supplied by a dc or ac power source.

In this chapter we will present the nature of the switching and the amplifying characteristics of semiconductor diodes and transistors in a fashion directly adapted to circuit analysis by extension of the techniques utilized in Chap. 2. Models useful for dealing with the devices in particular regions of operation are developed. One point the student should particularly note is the applicability of the models. In several cases more than one model is developed, and those discussed range from very simple models to complex and exact models. As the student becomes more sophisticated he will more often use the simplest model capable of attaining the precision he needs.

3-1 IDEAL SWITCHES

Two kinds of switches are relevant to the discussion here: the *separately actuated* (Fig. 3-1) and the *self-actuated* (Fig. 3-2). The manually operated knife switch of Fig. 3-1*a* or a telegraph key, for example, is of the separately actuated variety, as is the electric relay of Fig. 3-1*b* when the control winding is in one circuit and the contacts in another. Such a switch is *ideal* if the contact terminals provide a perfect short circuit when closed, as in Fig. 3-1*c*, and a perfect open circuit when not closed, as in Fig. 3-1*d*. Thus, when the contacts are closed, the current through them is determined entirely by the other elements in the circuit in which they are connected, not by the switch itself.

The self-actuated switch is exemplified by the diode, which ideally conducts current in one direction but not in the opposite. An *ideal diode* (Fig. 3-2*a*) is best depicted by its current-voltage characteristic, as in Fig. 3-2*b*. If the polarity of the voltage existing across the rectifier terminals is negative (Region I), the current is zero and the device is an open circuit. If, on the other hand, the polarity tends to be positive (Region II), the current flow can increase to infinity unless constrained by the circuit

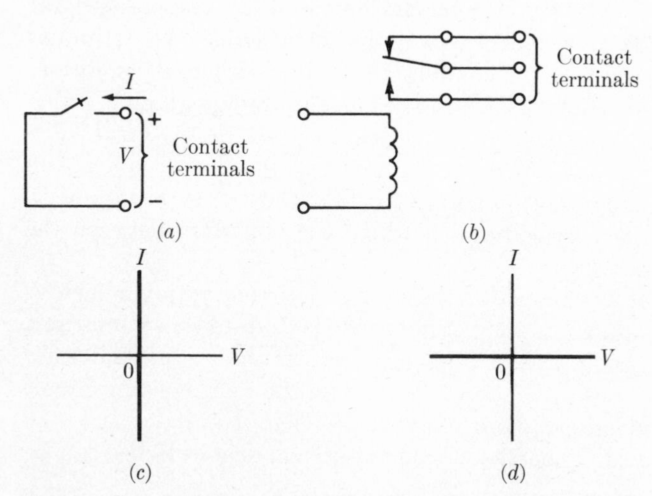

Fig. 3-1 Separately actuated switches. (*a*) Manually operated. (*b*) Electrically operated (relay). (*c*) Current-voltage characteristic: switch *closed*. Note that $V = 0$, independent of I. (*d*) Current-voltage characteristic: switch *open*. Note that $I = 0$. independent of V.

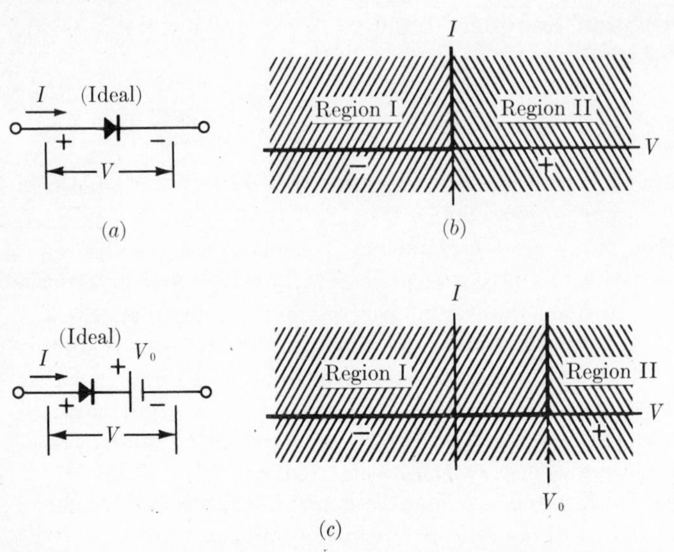

Fig. 3-2 Ideal self-actuated switches. Note that the triangle in the symbol "points" in the forward-current flow direction. (*a*) Symbol for diode (ideal, if so labeled). (*b*) Ideal current-voltage characteristic. (*c*) Ideal diode with offset breakpoint V_0.

in which the diode is connected. Another way of describing the phenomenon is to say that heavy currents can flow in the positive (or "forward") direction and that the voltage drop across the terminals cannot rise above zero, corresponding to a short-circuited condition. The abrupt transition at the origin in the current-voltage graph is called the *breakpoint*. This breakpoint need not be at the origin, but can be displaced as in Fig. 3-2*c*. The main feature is that the device is an open circuit in Region I and a short circuit in Region II, with an abrupt breakpoint determining the voltage level marking the boundary between the two regions.

Functions that can be performed by self-actuated switches such as the ideal diode are numerous and yet not obvious. Indeed they comprise a fair portion of this book, except that *actual* rather than *ideal* diodes are later used. Nevertheless, the principle of operation of a given circuit usually can be understood more easily if the actual switches are *first* considered to be ideal. Then the effects of the nonideal switches can be considered later.

Quite a variety of interesting and useful V-I characteristics may be produced with ideal diodes, current and voltage sources, and resistors. A few of the possible nonlinear V-I curves are shown in Fig. 3-3 together with the circuits which produce them. The circuits are not unique in all cases, i.e., more than one circuit configuration can be developed to give the same V-I characteristics. In all cases the current increases or remains level as V increases. For the current to decrease would require the use of a negative resistance. Note that for each breakpoint in the V-I curve one diode is required.

Another circuit which may be analyzed using the idea of the ideal diode is the series clipper circuit of Fig. 3-4. This circuit transmits any portion of the input waveform more positive than $v_1 = 0$ and "clips" away those portions below. Thus when $v_1 > 0$, the diode is in Region II, as defined in Fig. 3-2, and is a short circuit. The output voltage is necessarily the same as the input, or $v_1 = v_2$. On the other hand, when $v_1 < 0$, the diode is in Region I and is an open circuit. Then the output is disconnected from the input and $v_2 = 0$.

It is not necessary as yet to say anything about the proper magnitude of the resistance R in Fig. 3-4. Choice of magnitude will be considered later in this chapter when actual diodes are used.

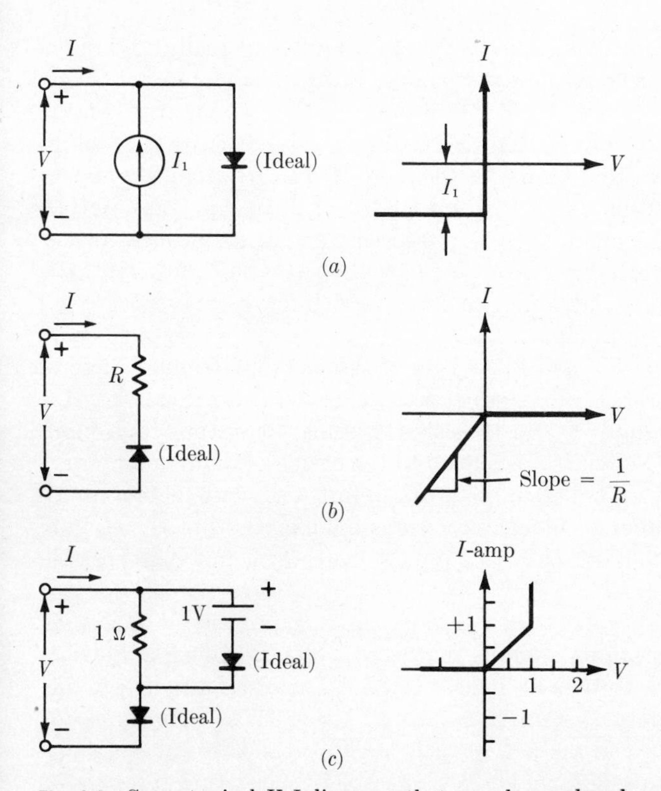

Fig. 3-3 Some typical V-I diagrams that may be produced by ideal diodes, current and voltage sources, and resistors.

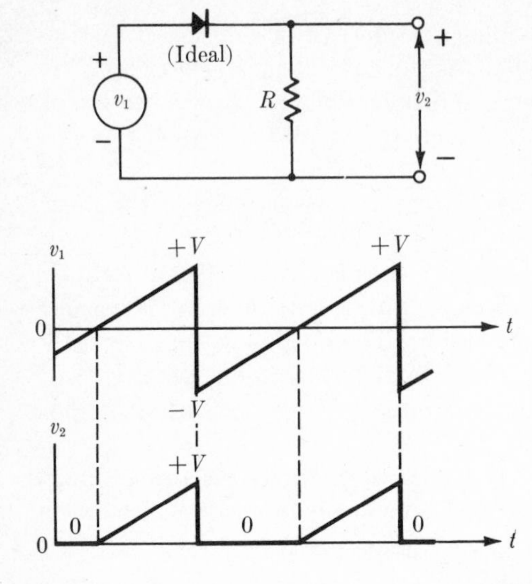

Fig. 3-4 Clipping circuit with ideal self-actuated switch, or diode, used to transmit only those portions of the input waveform above $v_1 = 0$.

Another application, in some ways more subtle but really a straightforward extension of things already said, is in so-called computer "logic circuits," as depicted in Fig. 3-5. Pulses occur at the two inputs, 1 and 2, either simultaneously or not, but identical in voltage (with respect to ground) as shown in the figure. In the case of the OR circuit, which is much like the circuit of Fig. 3-4 except that it has two inputs, an output pulse identical to the input pulse is produced by a signal at one input *or* the other (and at both, of course). With the AND circuit, however, the input voltage sources—assumed to have zero internal resistance—will hold the output at zero unless a pulse of magnitude V (or greater) occurs simultaneously at both 1 *and* 2. As long as either input remains at zero, the diode in that branch will conduct a current supplied through resistor R by the voltage supply V and thus short circuit the output to ground. These OR and AND circuits are central to the design of digital computers and, fortunately, can be made to work quite well with actual semiconductor (or thermionic) diodes in various modifications of the circuits of Fig. 3-5. In that figure only two inputs are shown, but it is possible to have more.

To give an elementary idea of how this type of circuit might be used to do computation, assume that zero volts is called a "logical ZERO" and that V volts is a "logical ONE." These definitions provide the basis for a binary system of counting. Using these definitions for ZERO and ONE we can make the truth table for the OR and AND circuits as shown in Figs. 3-5*a* and 3-5*b*, respectively. This truth table shows nothing more than the waveforms, but is more compact in its expression.

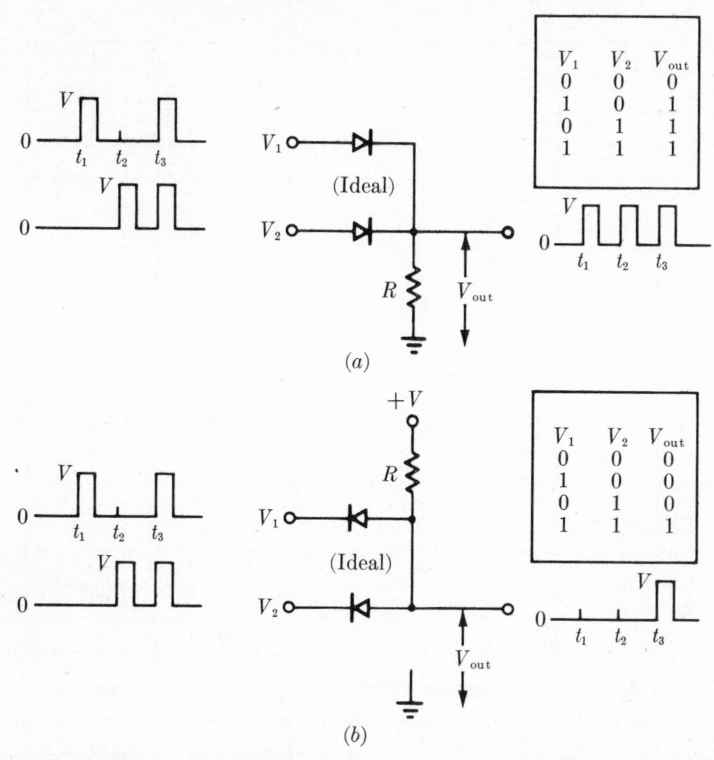

Fig. 3-5 Further applications of the ideal-diode switch in computer-type circuits. An output pulse of amplitude V is transmitted in (*a*) the OR circuit if there is a pulse at input 1 *or* 2 and in (*b*) the AND circuit only if there is a pulse at 1 *and* 2.

Note that interchanging the definitions of ONE and ZERO changes the OR circuit into an AND circuit, and similarly changes the AND circuit into an OR circuit.

3-2 ELECTRON DEVICES AS SWITCHES

Nonideal switches to be considered are the various forms of electronic diodes, including both the semiconductor and thermionic forms. Representative current-voltage characteristics are shown in Fig. 3-6. These actual diodes fall short of the ideal in three separate respects.

1. They do not provide a perfect short circuit in the forward (Region II) direction.
2. Except for the high-vacuum diode, they do not provide a perfect open circuit in the reverse (Region I) direction.

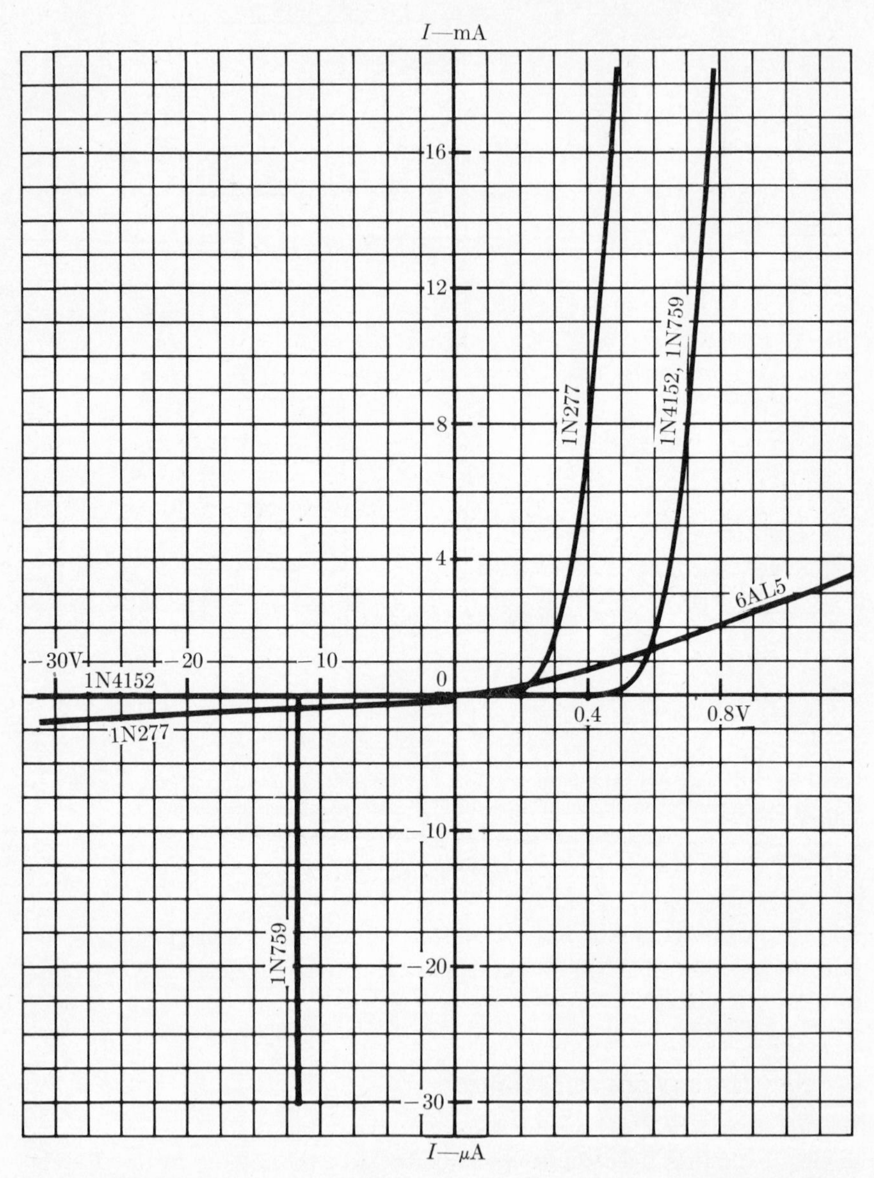

Fig. 3-6 Electronic diodes as switching elements. The 1N277 is a germanium diode; the 1N759 and the 1N4152 are silicon diodes. The 6AL5 curve is for one diode of the two contained in a small vacuum tube. Note the change in scales necessary on different sides of the origin.

3. They possess "inertial" effects, i.e., the voltage or current cannot change instantaneously from one value to another; for instance, certain of the semiconductor devices are relatively slow to switch from "on" to "off" (Region II to Region I).

These defects are not all equally important in a given circuit situation, and the designer can choose from among commercially available devices a unit in which the most critical defects have been minimized. Unfortunately, the nature of the devices is such that improving one defect usually makes the others worse and compromises are necessary.

In order to deal quantitatively with these problems it is necessary first to review the basic nature of semiconductor and vacuum diodes. In modern pulse and timing circuits more than 99 percent of the applications involve semiconductor diodes instead of thermionic diodes unless extremely high voltages or currents are involved; therefore the bulk of this discussion concerns the semiconductor devices. The first edition of this text may be consulted for further details on thermionic devices.

3-3 SEMICONDUCTOR DIODES

GERMANIUM AND SILICON JUNCTION DIODES

The approximate static behavior of either the silicon or germanium *p-n* junction diode is given by

$$I = I_s(\epsilon^{Vq/kT} - 1) \tag{3-1}$$

where one-dimensional current flow in a planar junction is assumed and

I_s = reverse saturation current, amperes
V = applied voltage, volts
q = charge on electron (1.59×10^{-19} C)
k = Boltzmann's constant (1.37×10^{-23} W-s/°K)
T = absolute temperature, degrees Kelvin

At a normal room temperature of about 300°K,

$$I = I_s(\epsilon^{39V} - 1) \tag{3-2}$$

In the forward direction, when V becomes larger than a few tenths of a volt, the forward current becomes simply

$$I \approx I_s\epsilon^{39V} \tag{3-3}$$

Correspondingly, in the reverse direction the current expression becomes

$$I \approx -I_s \tag{3-4}$$

An idealized graphical plot of the current-voltage characteristic in a very small region near the origin is shown in Fig. 3-7*b*. In Fig. 3-7*a* the conventional symbol for the diode is shown together with the corresponding *p-n* structure. A more complete plot for a real silicon unit is shown in Fig. 3-6 by the curves for the 1N4152 and 1N759. In this figure the departure of the diode from an ideal short circuit in the forward direction is quite apparent, although the actual forward voltage drop is small (<1 V). In the reverse direction the 1N4152 shows no visible departure from an ideal diode, but in actuality there is a finite and measureable reverse current on the order of 10^{-8} A (10 nA) at room temperature.

Looking at the forward characteristic more carefully shows that almost no current flows until the forward voltage is increased past about 0.4 V indicating that the current I_s in Eq. (3-1) must be very small. Past the 0.4-V point the current begins to increase very rapidly as an exponential function of V. The steepness of the V-I curve indicates that the *dynamic resistance*, $r_p = dV/dI$, of the diode is very small. Let us differentiate Eq. (3-1) and note that in the useful region $I \gg I_s$:

$$r_p = \frac{dV}{dI} \approx \frac{kT}{qI} \approx \frac{1}{39}I \tag{3-5}$$

Equation (3-5) shows r_p to vary inversely as the diode current and r_p is about 25 ohms at $I = 1$ mA.

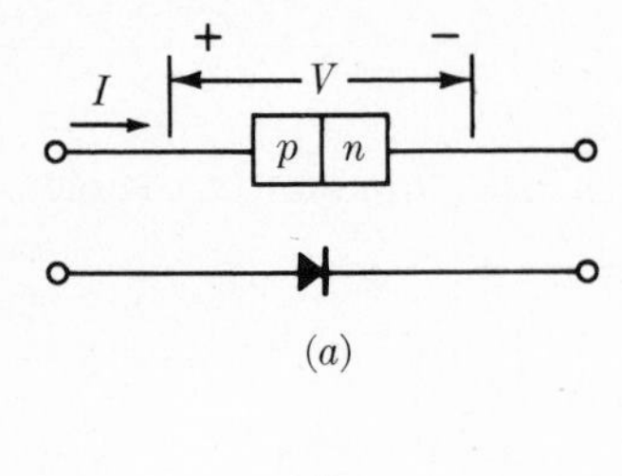

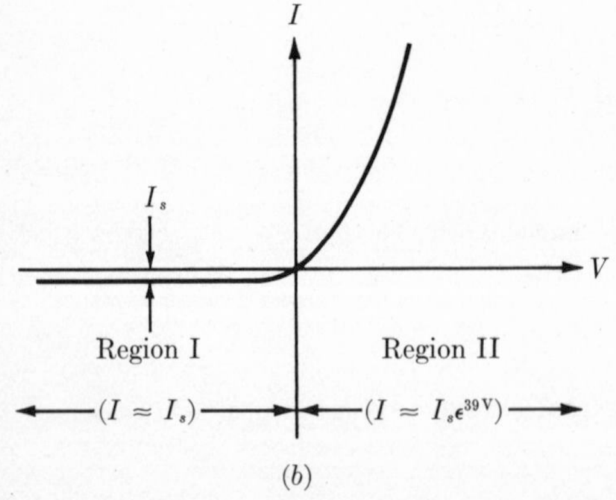

Fig. 3-7 The *p-n* junction diode. (*a*) Polarity convention. The triangle "points" in the direction of forward current flow. (*b*) Current versus voltage in the vicinity of zero voltage.

In the reverse direction the silicon diode has a very small reverse current which increases slowly as V is made more negative until a critical or breakdown voltage is reached. In the case of the 1N759 this voltage is about 11.3 V for the diode shown in Fig. 3-6. The 1N4152 does not break down in the range of reverse voltages shown. The voltage at which this breakdown occurs is controllable in the manufacture of the diode, and a variety of diodes are commercially available with breakdown voltages from about 2 V to 2,000 V. In some applications of a diode, for example in a rectifier circuit, the reverse breakdown is undesirable, and the diode is chosen to have a higher breakdown voltage than any voltage ever appearing across it. In other applications the breakdown is actually used, and the diode is chosen for a specific voltage. Diodes so used are variously called Zener, avalanche, regulator, reference, or just breakdown diodes. Operation of such a diode in the reverse conduction mode will not damage the device if the power dissipation capabilities of the device are not exceeded. The voltage drop at a constant reverse current is very stable over long periods of time in a well-made device, and therefore is often used as a voltage standard in applications such as regulated power supplies. The voltage drop is a function of temperature, however, and critical applications call for diodes which are compensated to give near zero temperature coefficients of voltage. Appendix A gives typical voltage-temperature coefficients as well as other specialized diode data.

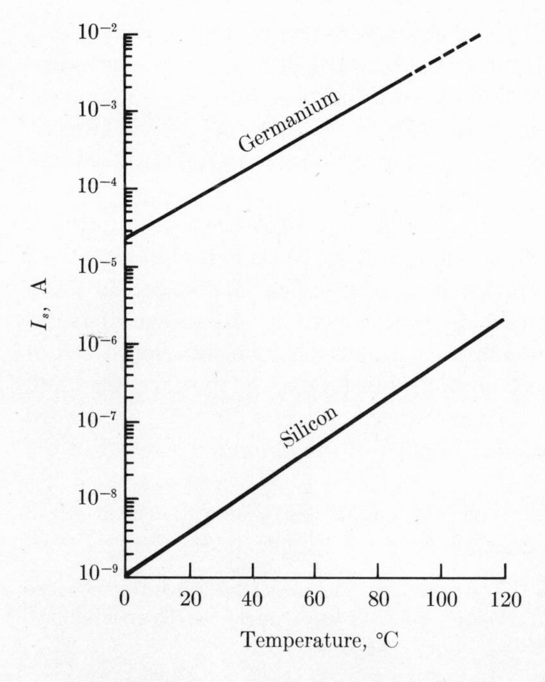

Fig. 3-8 Reverse saturation current I_s for two comparable junction diodes.

The current flowing in Region I where the diode is nominally an open circuit is a function of the operating temperature of the diode as shown in Fig. 3-8. Since I_s is seen to be a straight line on the semilogarithmic graph, I_s must be an exponential function of temperature. This is a natural phenomenon in all semiconductor diodes, but a wide range of reverse currents at a given temperature (usually room temperature is taken for the reference: 25°C) may be obtained by choosing different diode types.

The preceding comments concerning the diode apply to both the silicon and germanium units if certain important differences are noted. First, the current I_s is about four orders of magnitude larger for the germanium unit (Fig. 3-8) thereby making the device a poorer open circuit in Region I. Second, as a consequence of the larger value of I_s, the current begins to rise much sooner as the voltage is raised in the forward direction making the germanium diode a somewhat more ideal switch in Region II. The breakpoint in the forward characteristic occurs at about half the voltage of the breakpoint for the silicon unit. Third, although not shown in the curves of Fig. 3-6, the germanium diode does not break down abruptly in the reverse direction, but instead the I-V curve gets steeper and steeper as more reverse voltage is applied.

OTHER SEMICONDUCTOR DIODE TYPES

Because of the generally superior characteristics of the silicon diode, it is usually preferred unless the lower forward drop of the germanium unit is required. If extremely low forward drop is required, a diode type called a *back diode* may be employed. These conduct appreciable current with even 0.1 V of forward voltage, but are only useful with reverse voltages of 0.5 V or less.

An older type of diode called the *point-contact diode* is somewhat similar in characteristics to the previously described junction diodes, with the exception that the characteristics are somewhat poorer in both Regions I and II. Point contact diodes do have an advantage of faster switching speed than the conventional junction diode, but more recent types of junction diodes have comparable high-speed performance and superior static performance.[1] Therefore the present tendency is toward the use of silicon junction diodes in most of the circuits discussed in the remainder of the book.

Other specialized diode types which are not generally used for a polarity sensitive switch are various forms of trigger diodes such as *four-*

[1] The best high-speed diodes are the back diodes. The next fastest are the so-called "hot-carrier diodes," which have switching times of less than 0.1 ns. The gold-doped, silicon junction diodes have switching times down to 2 ns.

layer diodes and *tunnel diodes*. These will be discussed in conjunction with their circuits.

3-4 THERMIONIC DIODES

The high-vacuum thermionic diode basically comprises a heated cathode to emit electrons into a vacuum and a nearby anode (or plate) to collect the electrons. Electron current can flow easily from cathode to anode (conventional current flows in the opposite direction), but practically zero electron current can flow from anode to cathode, as shown in Fig. 3-6 by the curve labeled 1/2-6AL5. The anode corresponds to the arrow end of the conventional diode symbol. In the forward direction the current is space-charge limited and given by the Child equation

$$I = KV^{3/2} \tag{3-6}$$

where the factor K is called the diode *perveance*, and depends upon the electrode areas and spacings. The simple description given is somewhat modified by the second-order effects due to emission velocity and contact potential. These effects are most noticeable at low voltages and cause the diode to conduct a small current at small negative anode voltages so that the transition between Region I and Region II occurs slightly to the left of the origin.

3-5 NONIDEAL DIODES IN CIRCUITS

As an example of a real diode used in a circuit consider the clipping circuit of Fig. 3-4 with the ideal diode replaced by a 1N4152 silicon diode. The general operation of the circuit will be much as before, but the finite voltage drop across the diode in Region II will cause the output of the circuit v_2 to be less than before. When v_1 is negative, the output will be virtually zero as before, since the silicon diode is nearly ideal in the reverse direction.

To compute the output in the forward direction the familiar *load-line* construction may be used in conjunction with the diode *V-I* curve as shown in Fig. 3-9*b*. The load line is the graphical relationship between v_1 and I as determined by the *circuit*

$$I = \frac{v_1 - v_D}{R} \tag{3-7}$$

This line is easily plotted from the intercept with the current axis, $I = v_1/R$, and the intercept with the voltage axis, $V = v_1$. The slope of the load line is $-1/R$. The other needed data are the diode characteristics, $I = f(v_D)$. The intersection of the two curves is the operating

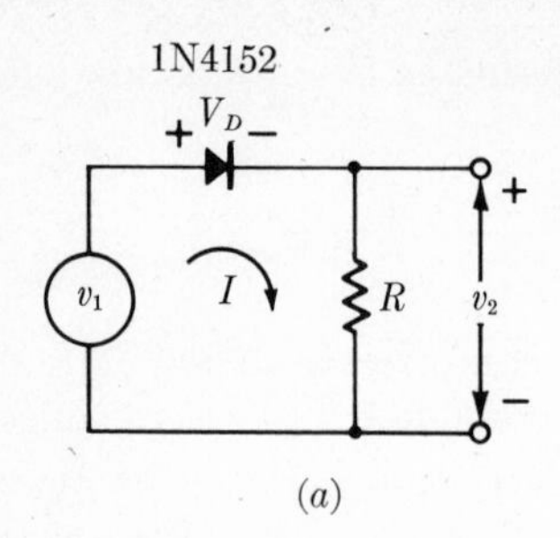

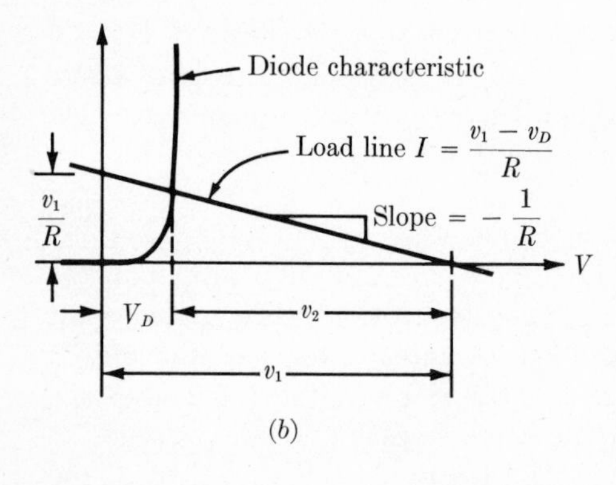

Fig. 3-9 (a) Clipper circuit with silicon diode. (b) Superposition of load line on diode characteristic to find actual operating point.

point for the particular v_1 as shown in Fig. 3-9b. The actual output of the clipper circuit may be calculated graphically by assuming successive values of v_1 and finding v_2. In this case the output will be practically zero until v_1 rises to about 0.5 V; beyond this voltage the output will very nearly follow the input. Choosing a smaller R causes more current to flow for a given v_1 and thereby increases the diode drop somewhat. Hence to make $v_2 \approx v_1$ for $v_1 > 0$, the resistance R should be chosen as large as possible.

Another way to display the performance of this clipper circuit is to plot v_2 as a function of v_1, as shown in Fig. 3-10. The dashed curve is

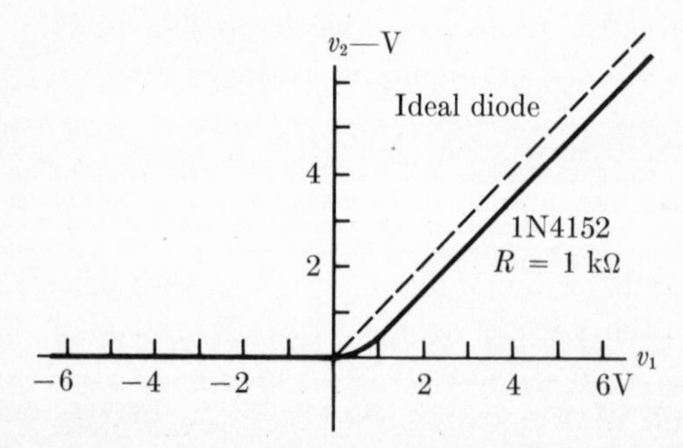

Fig. 3-10 Transfer characteristic of a clipper circuit.

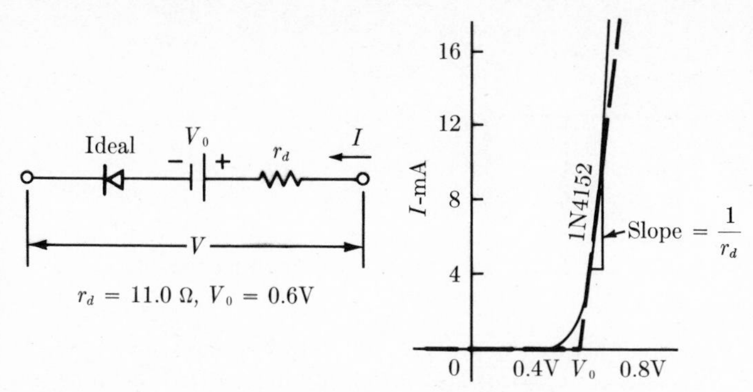

Fig. 3-11 An equivalent circuit for a junction diode.

that for an ideal diode with zero forward drop. The actual curve is displaced to the right by an amount which increases slightly as v_1 increases. Note that changing the diode to a 1N277 would make the circuit a more nearly ideal clipper for $v_1 > 0$, but for $v_1 < 0$ would degrade the performance because of the lower resistance of the germanium diode in Region I, and v_2 would not be as nearly zero as shown for v_1 negative.

Instead of using a graphical analysis for calculating the action of the diode clipper, a circuit which is approximately equivalent to the diode can be used. Figure 3-11 shows an equivalent circuit with an ideal diode, battery, and resistor which together give the dashed-line characteristic shown superimposed upon the actual junction diode curve. For the dashed curve shown the values may be found by picking convenient points on the line chosen to represent the diode. In this case the line has a slope corresponding to an r_d of 11 ohms and an intercept with the voltage axis of 0.6 V. In making such an equivalent circuit, it is imperative to know the approximate V-I region in which the device is to be operated. In the case shown the approximation is very good for $I = 2$ to 15 mA and for $I = 0$, but if the main region of interest was between $I = 0$ and 2 mA, a much better approximation would be $V_0 = 0.45$ V and $r_d = 75$ ohms. In many cases where the exact value of the diode voltage is not too important, the resistance r_d may be taken to be zero.

In Fig. 3-12 the actual diode has been replaced by its equivalent circuit so that the output v_2 may be calculated. For $v_1 < 0.6$ V the output is zero; for $v_1 > 0.6$ V the output is

$$v_2 = \frac{(v_1 - 0.6)R}{R + 11} \tag{3-8}$$

For $R \gg 11$ ohms the output is essentially the input minus 0.6 V.

The use of a *piecewise linear* (i.e., a nonlinear characteristic sub-

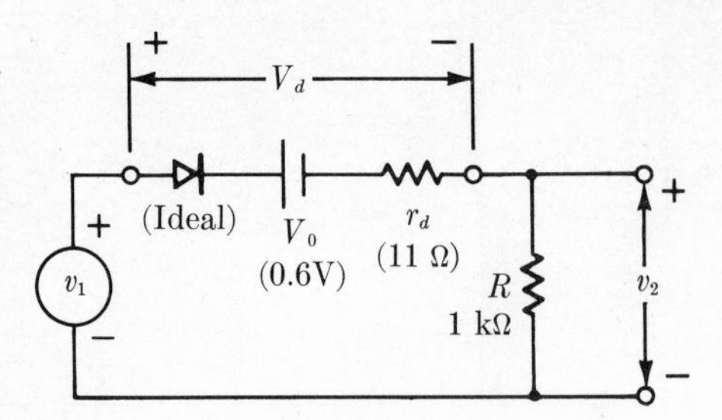

Fig. 3-12 Clipper circuit with equivalent circuit used for the diode.

divided into joined linear segments or "pieces" as in Fig. 3-11) equivalent circuit is especially useful when diodes are combined with reactive elements as discussed in Sec. 4-1.

3-6 DIODE SWITCHING TIME

Although most discussions in this textbook will assume that diode current and voltage can change instantaneously, the fact that diodes and other devices do have finite switching times should be recognized. Switching a semiconductor diode either on or off requires time for internal equilibrium conditions to be obtained. A typical set of waveforms obtained by suddenly switching a current into a junction diode are shown in Fig. 3-13*a*. If the current is at a relatively low level, the junction capacitance dominates, and the voltage across the diode rises slowly to the final static value. At high current levels, transit-time phenomena dominate, and the diode voltage rises quickly to a peak value which is larger than the steady-state value; therefore the diode behaves somewhat as if it had an inductance in series. In the case of the 1N4152 diode with $I_F = 100$ mA, the peak value of forward voltage is about 20 percent greater than the static value which is obtained within 10 ns (10^{-8} s) after application of the current step. The forward switching transient is not of great importance in many circuits, but it may be noticed in some especially critical circuits.

If the diode has been conducting, and the voltage is reversed in an effort to turn it off, the waveforms resulting are shown in Fig. 3-13*b*. The effect of the current carriers being stored in the junction region causes the diode voltage to remain near zero for a short period even though the current through the diode is reversed, as shown in the bottom waveform. The reverse current, in effect, removes the stored charge in the diode, and the voltage across the diode eventually becomes essentially that of the source since the reverse static current is small. Because the amount of charge stored in the diode is proportional to the forward current existing before switching, and because the rate of removal of the stored charge is proportional to the reverse current, the fastest switching

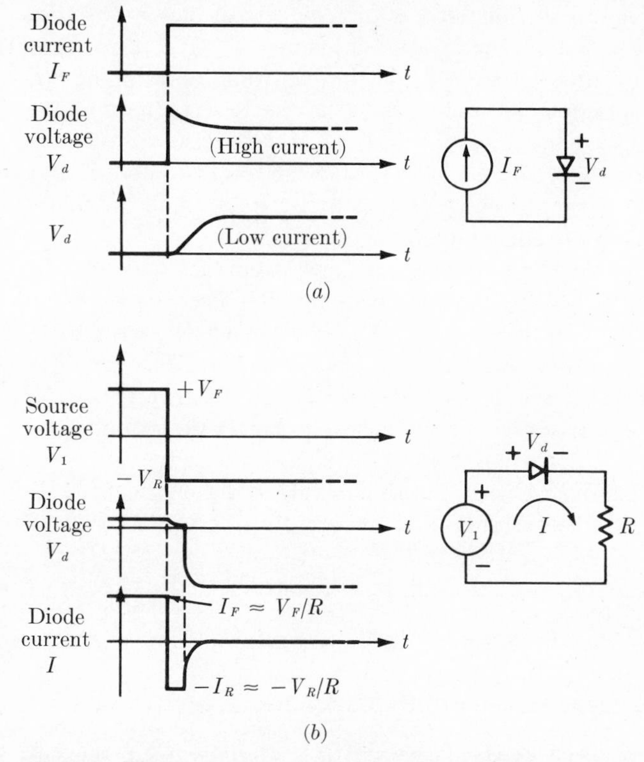

Fig. 3-13 Waveforms occurring during the rapid switching of a junction diode. (*a*) Forward-switching waveforms. (*b*) Reverse-switching waveforms.

takes place when the forward current is relatively small, and when the reverse current I_R is as large as possible. High-speed switching diodes are usually made with semiconductor material doped to reduce the lifetime of minority carriers so that the stored charge is also dissipated by recombination effects. Diodes such as the 1N4152 have reverse recovery times of a few nanoseconds, and the reverse recovery time is no problem in circuits generating pulses even as short as a microsecond. Diodes designed for power rectification may have switching times of many microseconds. and can be a problem even in relatively slow circuits.

3-7 THREE-TERMINAL DEVICES—THE TRANSISTOR

The addition of the third terminal or electrode provides new functions and a degree of flexibility not obtainable with two-terminal devices. Two important aspects of this flexibility are the separation of the input and output circuits leading to the idea of a *controlling* circuit and a *controlled* circuit.

The commonest three-terminal device in operation in switching circuits today is the junction transistor. This device has supplanted the vacuum tube virtually completely because of its small size, ability to operate at very low power, lack of heater, long life, and the practicality of integrating it with other circuit components, such as diodes and resistors, on a single chip of semiconductor material.

Semiconductor field-effect devices are also important and are discussed, but most of the applications we consider use the junction transistor. The junction transistor may be made of either germanium or silicon, and be of either the NPN or PNP structure. The majority of curves and examples are made with NPN silicon transistors for several reasons. Silicon transistors have generally better performance for a given cost, particularly when leakage currents are important, and NPN silicon transistors are used in 99 percent of the applications in integrated circuits. An occasional example is given using PNP transistors to show the change in polarities of the supplies and waveforms. Some unique circuits are possible using both polarities of transistor in a circuit.

3-8 TRANSISTORS AS SEPARATELY ACTUATED SWITCHES

In the case of the transistor, the base-emitter terminals are often employed as the controlling terminals, whereas the collector-emitter terminals are the controlled terminals. If we think in these terms, the appropriate curves to characterize an NPN transistor are shown in Fig. 3-14. The collector characteristics show three relatively distinct regions of operation: Region I where virtually zero collector current flows, and the base current $I_B \leqq 0$. Region II is a region familiar from analysis of linear circuits where the collector current is a more or less linear function of base current. In Region III the transistor conducts both base and collector currents, but the latter is almost unaffected by changes in the base current. The idealized collector characteristic shown in Fig. 3-15 has the three regions labeled and shows the conventional symbol for the transistor, together with the normally accepted nomenclature for voltages and currents. Note that all the currents are defined as flowing *into* the transistor. Region I is also a region of very small (leakage) base current, and is shown in Fig. 3-14 as the region $-7 < V_{BE} < +0.4$ V where $I_B \approx 0$. For voltages more negative than -7 V the base-emitter PN junction breaks down, and the incremental impedance between the terminals becomes very low as in a breakdown diode. The collector current stays near zero even though the base-emitter junction is broken down.

To operate the transistor as an open switch a zero or negative voltage is applied as V_{BE}, thereby putting the transistor into Region I. Nearly zero collector current flows for positive collector voltages not exceeding the breakdown voltage of the transistor and for negative voltages not more negative than V_{BE}. (If the collector is made more negative than the base, the base-collector PN junction becomes forward biased and current flows through it. The transistor is then operating with the normal collector as the emitter, and the normal emitter as the collector.)

In order to operate the transistor as a closed switch a current is supplied to the base of sufficient magnitude to make the transistor *saturate* or operate in Region III. In this region, as in the closed switch of Fig. 3-1*c*, the current flowing through the switch is limited by the external circuits, not by the transistor.

An example will help clarify and introduce some common terms used to describe this mode of operation. The circuit shown in Fig. 3-16*a* is a common form of a two-input NOR gate using *r*esistor-*t*ransistor-*l*ogic (RTL). The circuit is known as a NOR circuit because the application of voltage (logic 1) to either input 1 OR input 2 gives a logic 0 at the output. An OR circuit would give a logic 1 for the same inputs, but the circuit in question gives a NOT 1 output; hence the abbreviation NOT OR—NOR.

The two transistor switches are $Q1$ and $Q2$. Operation of either switch $Q1$ or $Q2$ shorts the output line to ground making $v_o \approx 0$ V. For $t < 0$ both inputs are grounded ($v_1 = v_2 = 0$), and the base currents $I_{B1} = I_{B2} = 0$ also. From the load line drawn on the 2N3642 characteristics for $R_L = 1$ kΩ and $V_{CC} = +5$ V, the operating point is seen to be at A in Fig. 3-16*c*. In this condition both $Q1$ and $Q2$ are OFF, and the collector currents are only a few nanoamperes (at ordinary room temperature). At $t = 0^+$, $v_1 = +2$ V, and base current flows into the base of $Q1$; $Q2$ remains off. The current in the base can be found from a construction very similar to a load line drawn on the I_B-V_{BE} curves for the transistor as shown in Fig. 3-16*d*. There is a small interdependence between V_{BE} for a given I_B and V_{CE}; a good estimate results here by using the curve for $V_{CE} = 0.1$ V. From this construction $I_B = 260$ μA. From Fig. 3-16*c* the operating point in the collector is the intersection of the curve $I_B = 260$ μA and the load line. This point B in Fig. 3-16*c* at $V_{CE} = 0.1$ V and $I_C \approx 5$ mA is in Region III, and the transistor is saturated. Note that very little change in V_{CE} would result from changing the base current to any value greater than about 120 μA because there is very little control of I_C or V_{CE} by I_B in Region III; therefore precision in determining I_B is not required.

From the preceding, $Q1$ is seen to be operating as a nearly ideal switch at point A on the load line when $v_1 = 0$ and at point B when

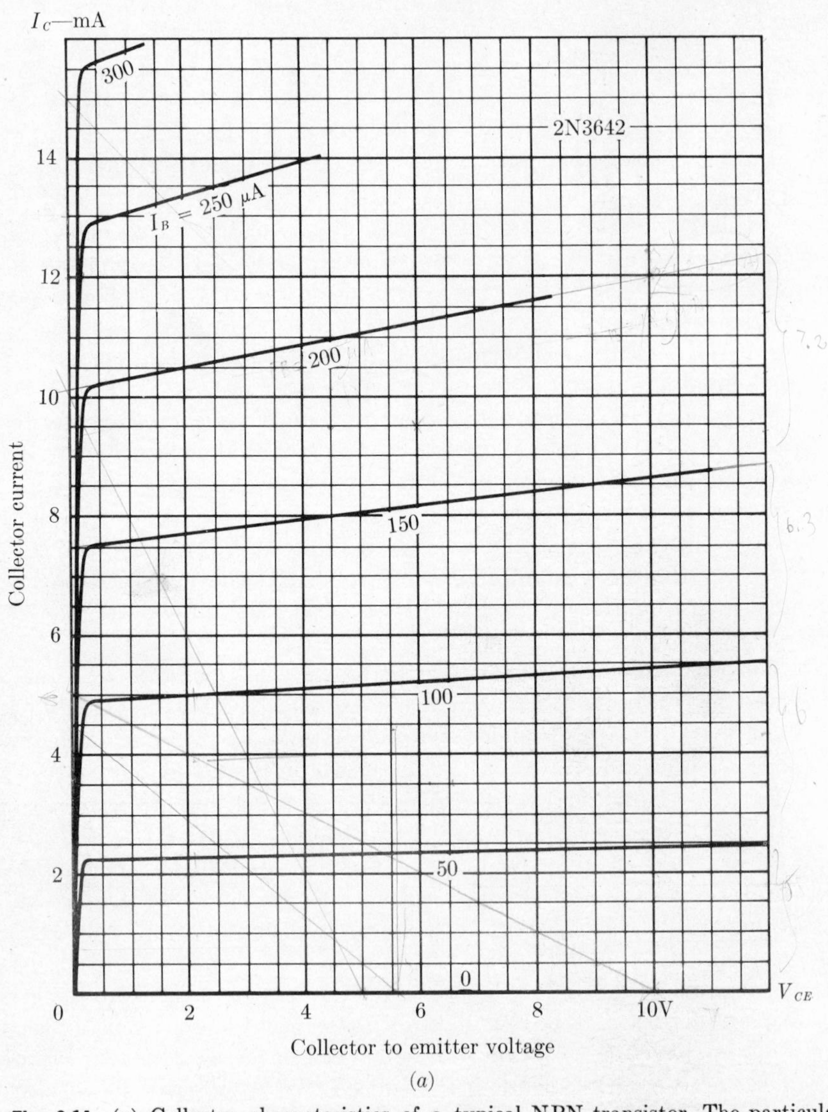

(a)

Fig. 3-14 (a) Collector characteristics of a typical NPN transistor. The particular unit chosen (the type 2N3642) is a silicon epitaxial unit in an epoxy case used in industrial and consumer applications.

$v_1 = 2$ V. Operation in the *linear region*, Region II, is incidental and only occurs for the very brief interval required to switch from one state to another.

The operation of the circuit when the second pulse is applied to the

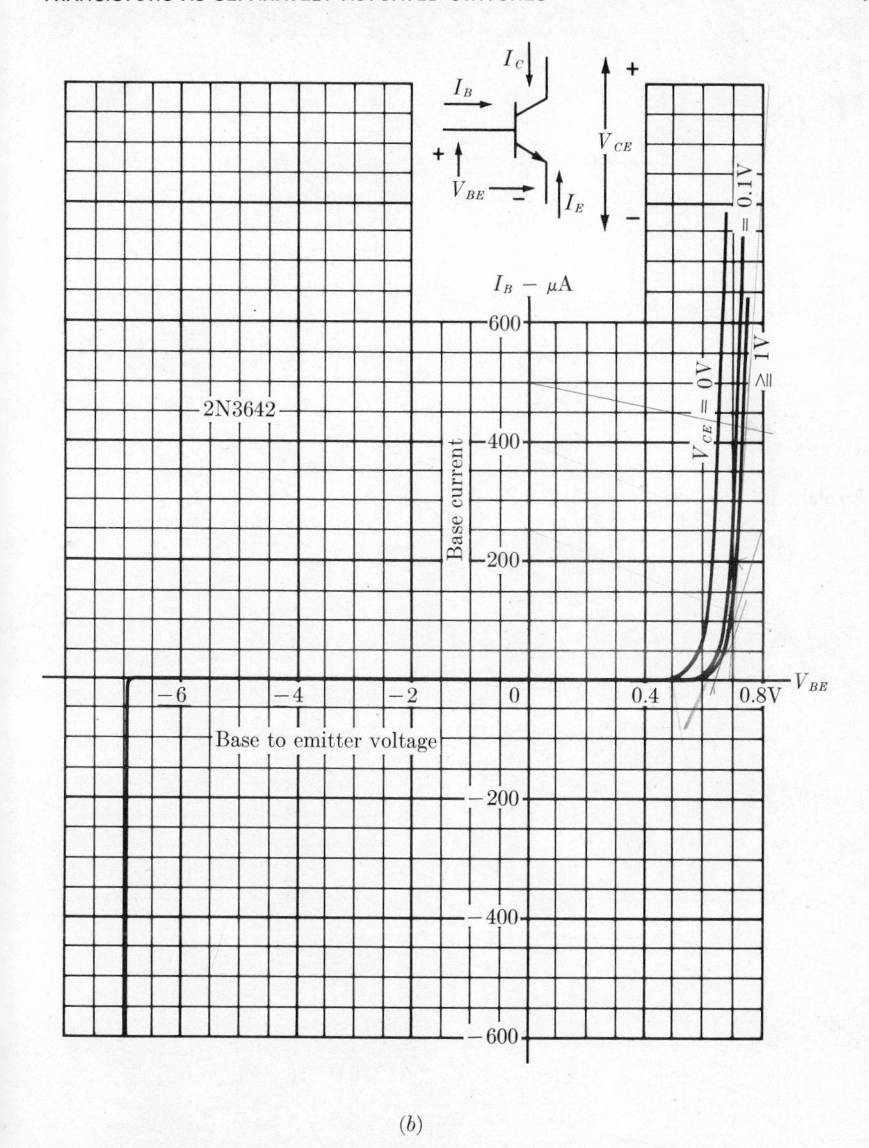

(b)

Fig. 3-14 (b) The base-emitter characteristics of the 2N3642.

base of $Q2$ is exactly the same as described for the operation of $Q1$, except that the currents are now flowing in $Q2$. Again the output v_o during the pulse is about 0.1 V, and in the absence of both v_1 and v_2 is 5 V. In the third pulse interval when a pulse is applied to both inputs simultaneously, the output again goes nearly to zero. The exact value could be found by redrawing the load line for twice the resistance (2 kΩ),

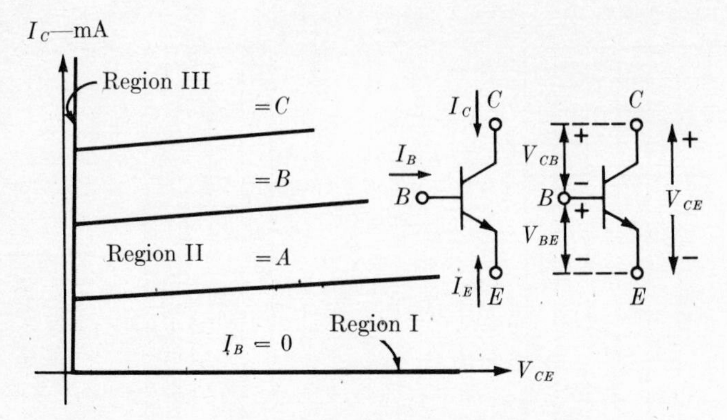

Fig. 3-15 Idealized collector characteristics for an NPN transistor together with the definitions of currents and voltages. These definitions are also valid for PNP transistors.

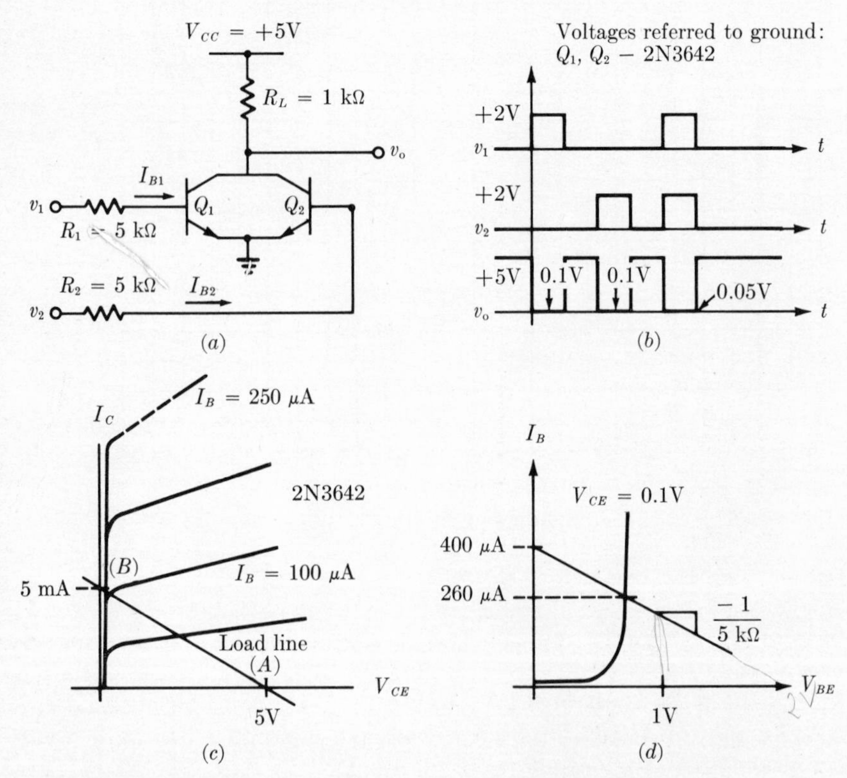

Fig. 3-16 (a) Circuit for a two input NOR gate. (b) Waveforms at input and output. (c) Load-line construction. (d) Construction to find operating base current.

thus effectively doubling the current scale on the transistor characteristics. This would show that V_{CE} for the two transistors in parallel (both conducting) is about 0.05 V. Such a change from $V_{CE} = 0.1$ V would not be noticeable on the scale of Fig. 3-16*d*. The levels of the output from the circuit are sensibly unaffected by changes in the input level as long as the logic 0 state at the input does not rise enough to begin to turn the transistors on, and the logic 1 state at the input is sufficiently high to saturate the transistors.

The minimum driving current I_B required to place the transistor into saturation in a given application is important because this current is usually a load on some previous circuit. Therefore the ratio of collector current to base current at the edge of the saturation region is an important transistor parameter called h_{FE} and defined as

$$h_{FE} \triangleq \left. \frac{I_C}{I_B} \right|_{V_{CE}=K} \tag{3-9}$$

Sometimes h_{FE} is called β_{dc}, the dc current gain. This is usually specified at some low collector voltage K in the 0.5 to 5 V region. In the case of the 2N3642 curve shown, $h_{FE} = 50$ at $I_C = 5$ mA and $V_{CE} = 1$ V. The current gain is a function of I_C and increases as I_C is increased up to some maximum h_{FE}. In the 2N3642 transistor this maximum occurs with a collector current around 100 mA. Above this current h_{FE} rapidly decreases. Typical curves of this and other transistor data appear in Appendix B.

In order for the circuit of Fig. 3-16 to operate properly a minimum h_{FE} of $I_C/I_B = 5/0.26 = 19 = h_{FE(\text{min})}$ is required to insure that operation is in Region III as desired during the pulse input.

For some purposes it is more desirable to use an equivalent circuit for the transistor in Regions I and III than to use the graphical analysis of the preceding. The simplest possible model for the transistor in Region I is illustrated in Fig. 3-17*a* where it is shown to be a perfect open circuit both in the emitter and base circuits. Although this is an imperfect representation, it is very adequate for most applications using silicon transistors, reasonably low impedance circuits, and temperatures up to perhaps 75°C. In applications which do not fit the preceding restrictions the equivalent circuit of Fig. 3-17*b* may be used. The current I_{CEO} is defined as the collector current flowing with zero base current (open-circuited base) and a specified collector voltage. For a typical 2N3642 this current is on the order of a few nanoamperes at 25°C. The current rises exponentially with temperature as shown for the reverse diode current I_s in Fig. 3-8. A useful approximate rule of thumb is that the reverse current will double for each ten-degree rise in junction temperature. Thus, if the current is 15 nA at 25°C, the current at 65°C will be roughly

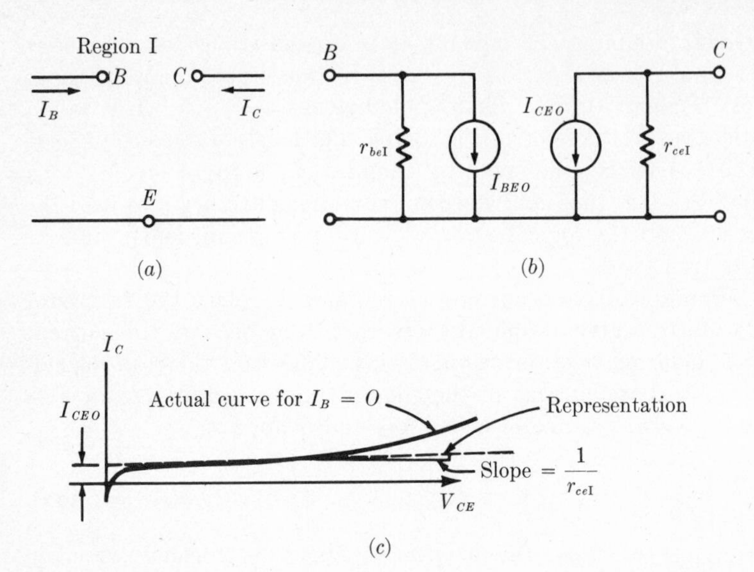

Fig. 3-17 Equivalent circuits for the transistor in Region I (cutoff).

$15 \cdot 2^4 = 240$ nA. The resistor r_{ceI} in the case of the 2N3642 is on the order of 10^9 ohms for $0 < V_{CE} < 10$ V. In many situations where the transistor is *cutoff* or in Region I, a negative V_{BE} will be applied, and the base current will be less than zero. Under these conditions the collector current is even less than for the case $I_B = 0$, and the voltage V_{CE} that may be applied before breakdown occurs is even higher.[1] The model of Fig. 3-17*b* is still all right, but the proper values are different; since the base is not open circuited, I_B is not zero and the equivalent circuit must include this. The base leakage current is shown as I_{BEO}. Both this and r_{beI} are similar in magnitude and temperature behavior to their counterparts in the collector circuit.

In Region III the collector curves seem to merge into one thick, nearly vertical line near the current axis. For many purposes it is adequate to merely use a short circuit to represent the collector circuit and the base circuit as shown in Fig. 3-18*a*. This rather crude equivalent is often good for a first run through the analysis or design of a circuit. The second circuit (Fig. 3-18*b*) is quite adequate for the majority of situations. The voltage V_0 and resistance r_{bIII} correspond to V_0 and r_d for the forward biased diode as shown in Fig. 3-11 and are found in the

[1] The collector current when $I_B \neq 0$ is not properly termed I_{CEO}. For example, if $V_{BE} = 0$, the resulting collector current is I_{CES}, which is smaller than I_{CEO}. Various other situations exist, but the current I_{CEO} in this text refers to the cutoff current flowing, whatever the cutoff conditions. The appropriate value will have to be chosen for the actual conditions.

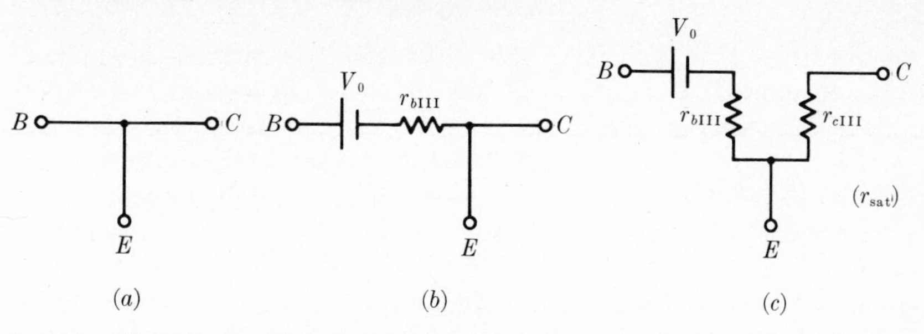

Fig. 3-18 Equivalent circuits for the transistor in Region III (saturation).

same way using the I_B-V_{BE} curves. The last equivalent circuit (Fig. 3-18*c*) is the most complete and includes a resistance r_{cIII}, commonly termed the *saturation resistance,* in the collector circuit. For the 2N3642 operating in the current ranges shown in Fig. 3-14 r_{cIII} is on the order of 15 ohms.

The transistor as a switch will, perhaps surprisingly, operate with current flowing in either direction through the switch. The collector characteristics of Fig. 3-19 are those in Fig. 3-14 magnified about the origin and extended to negative values of V_{CE} and I_C. In the third quadrant the transistor collector is actually operating as the emitter, and the current gain is much smaller, as is evident for the very much smaller I_C

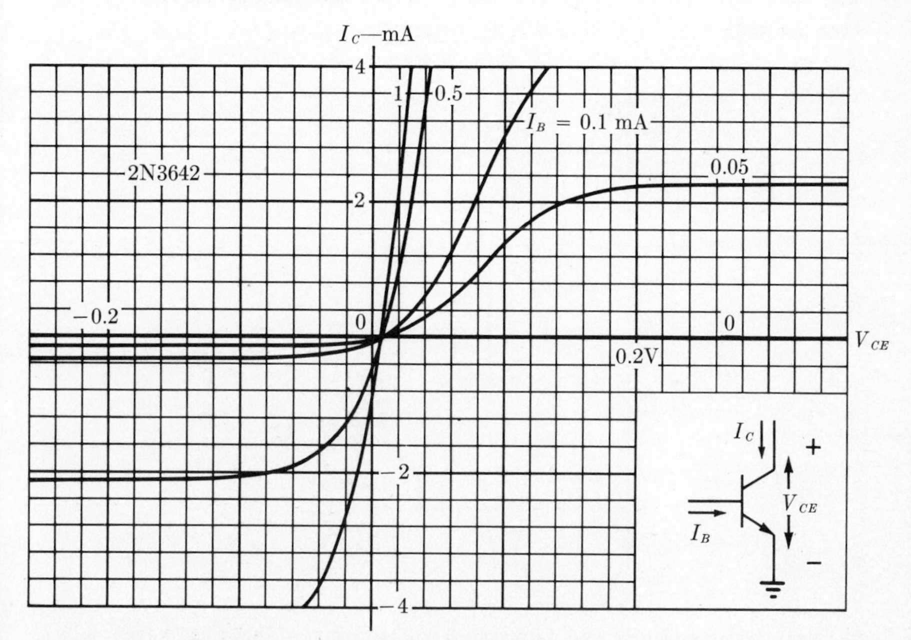

Fig. 3-19 The collector characteristics of the 2N3642 magnified near the origin and extended into the third quadrant.

for a given I_B. However, if a sufficiently large base current is applied, say 1 mA, appreciable current will flow either into or out of the collector terminal with very little voltage drop in the transistor. For the case $I_B = 1$ mA the resistance of the transistor is about 5 ohms. The voltage across the transistor is not exactly zero when $I_C = 0$, but is about 8 mV. The transistor is therefore operating like an ideal switch connected in series with 5 ohms and 8 mV. If even this small voltage cannot be tolerated, as in the case of some modulators, then the transistor emitter and collector terminals may be reversed resulting in the inverted transistor switch which has an *offset voltage* of less than 1 mV instead of 8 mV, as in the normal connection. See Appendix B for curves for this connection.

An example of a circuit utilizing this characteristic of the transistor is the simple modulator circuit of Fig. 3-20. One input is a 1,000-Hz sine wave, and the modulating input is a 500-Hz square wave. During the first

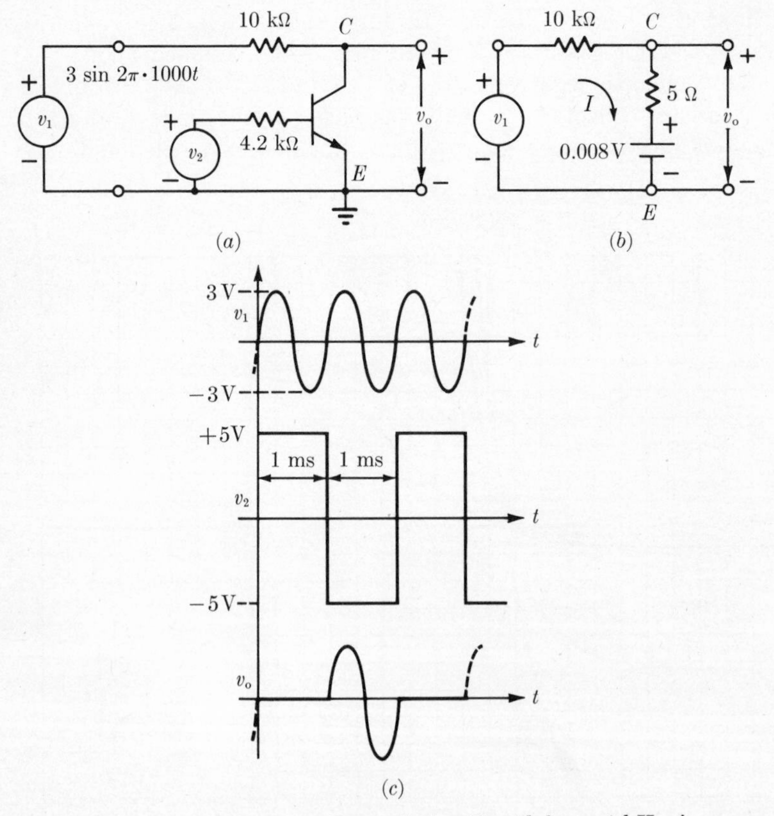

Fig. 3-20 (*a*) A transistor modulator which will modulate a 1-kHz sine wave by a 50-Hz square wave. (*b*) Equivalent circuit for $0 < t < 1$ ms. (*c*) The output waveforms.

half cycle of the square wave the transistor is turned on with a base current of approximately $(5 - 0.8)/4.2\ \text{k}\Omega = 1$ mA. For this base current the equivalent circuit of the transistor is as shown in Fig. 3-20*b*: 5 ohms in series with 8 mV. The current I is then closely equal to $I = v_1/10\ \text{k}\Omega$, and the voltage output is $v_o = 0.008 + 5 \cdot (0.0003 \sin \omega t)$. The output is then the offset voltage, 8 mV, plus a 1.5-mV peak-to-peak sine wave. To the scale of Fig. 3-20*c*, the output for the first millisecond is zero.

At the end of the first millisecond v_2 becomes negative with respect to both the emitter and collector, thus the transistor turns off, making the output $v_o = v_1$. At no time during the period of v_1 does the collector become more negative than the base; therefore the transistor remains off for the entire second millisecond and the output is a single complete sinusoid. At $t = 2$ ms, the transistor again turns on, clamping the output to ground, and the operation of the first millisecond is repeated.

Note that the transistor was required to conduct current out of the collector when $v_1 < 0$ during the first millisecond. A vacuum tube triode can also be used as a low-resistance switch, but it could not have conducted current in both directions as required in this modulator.

3-9 LINEAR TRANSISTOR OPERATION—REGION II

In the previous examples of transistor operation only a very limited operation occurred in the conventional linear operating region, Region II. However, many pulse circuit applications require operation in this region, and we shall need adequate equivalent circuits to describe operation in Region II. The circuits needed are relatively simple but meet one requirement not normally placed upon circuits used for describing the linear region of operation in amplifiers and similar applications, that is, the equivalent must be capable of giving the total instantaneous output voltage, not just the incremental variation from some static operating point. The model in Fig. 3-21 illustrates the additions necessary to the usual small-signal, linear equivalent for a common-emitter transistor. The base-emitter equivalent is the conventional equivalent for a diode comprising the breakpoint voltage V_{II} and the resistance corresponding to the normal h_{ie} (previously called r_d in Fig. 3-11). The circuit is similar to that of Fig. 3-18*c*, but the values are taken from the base-emitter curves with a higher collector voltage typical of Region II operation. The ideal diode symbol usually will be omitted but understood to be present. The circuit is only valid when I_B is positive (for an NPN unit). The collector circuit contains a resistor $1/h_{oe}$ (h_{oe} is a conductance), a constant current generator I_0, and a controlled current generator $h_{fe}I_B$. The V-I curves which are generated by this equivalent circuit are shown in Fig. 3-21*b*. The three parameters produce three needed characteristics

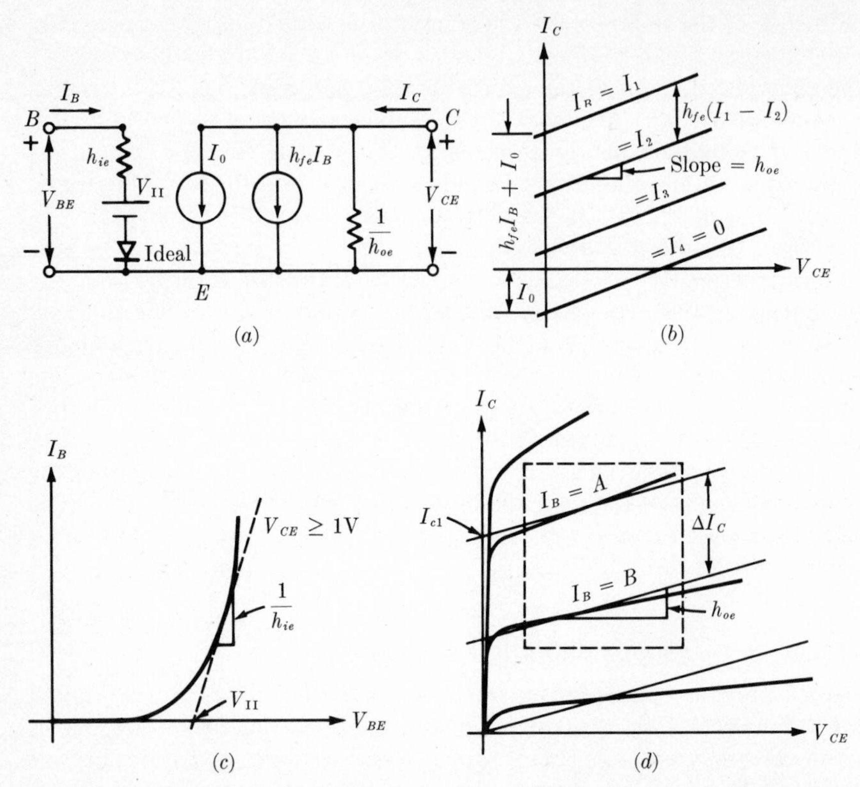

Fig. 3-21 (*a*) Equivalent circuit for a transistor in Region II. (*b*) The idealized collector curves generated by the equivalent circuit of (*a*). (*c*) Representation of the base characteristic. (*d*) Details of drawing idealized characteristics on the real collector curves.

in the curves: The slope of the lines, which are all parallel, is equal to h_{oe} as shown. The spacing of the lines is controlled by h_{fe} (that is, $\Delta I_C = \Delta I_B h_{fe}$), as shown. The vertical position of the lines is controlled by I_0.

To determine these parameters graphically, one must first determine roughly the operating region in the circuit. This is necessary because the transistor, as are all active devices, is inherently a nonlinear device. Therefore the linear equivalent circuit of Fig. 3-21*a* cannot accurately depict the operation of an actual transistor over the complete Region II. Once the operating region is approximately determined, or only guessed at for a first approximation, it can be outlined on the collector characteristics as shown by the dashed rectangle in Fig. 3-21*d*. In this region a set of equally spaced, parallel lines that most closely approximate the real collector characteristics in this region are drawn by eye. For most cases only two lines really need be drawn for an adequate representation. These lines are the light lines for $I_B = A$ and B in Fig. 3-21*d*. From the sketched lines, which are what the equivalent circuit will really represent, the values are taken to determine the transistor parameters. The slope of the light lines is h_{oe}, and the vertical spacing ΔI_C determines h_{fe} such that

$$h_{fe} = \frac{\Delta I_C}{\Delta I_B} = \frac{\Delta I_C}{A - B} \tag{3-10}$$

The last parameter I_0 usually proves the most confusing but can be easily found by noting that the collector current in the equivalent circuit (Fig. 3-21*a*) is

$$I_C = I_0 + h_{fe}I_B + V_{CE}h_{oe} \tag{3-11}$$

By setting $V_{CE} = 0$ the last term is removed, and the resulting current is the intersection of the curve for a given I_B and the current axis

$$I_{C1} = I_0 + h_{fe}(A)\,\big|_{V_{CE}=0} \tag{3-12}$$

Since I_{C1}, h_{fe}, and A are known, I_0 can be found. This current has no particular physical significance and is used only to make the total current come out correctly. In particular, I_0 is *not* one of the cutoff currents, and in fact may be a negative or positive current for an equivalent circuit of the same transistor operating at different points on its characteristics.

One may worry too much about getting a truly accurate equivalent circuit, but two points should be kept in mind. The first is that most electronic circuits would not operate at all without the active device, e.g., a transistor, but the better the circuit is, the less it will depend upon the *exact* characteristics of the device. The second point is perhaps the cause of the first, and is that the characteristics of most active devices vary from device to device, from time to time in the same device, and with the ambient conditions. These variations are typically much greater than is the case with passive devices like resistors and capacitors. Therefore a desirable circuit derives most of its detailed characteristics from its passive elements and as little as possible from the active elements, making the demands upon the representation of the active elements minimal. Such a desirable situation does not always obtain, and varying demands will be made for accuracy. As always, the equivalent circuit should be kept as simple as possible consistent with the requirements.

The greatest simplification possible to Fig. 3-21*a* would be to consider the base-to-emitter circuit a short circuit, as in Fig. 3-18*a*, and the output the single generator $h_{fe}I_B$. Such a representation is adequate if I_B

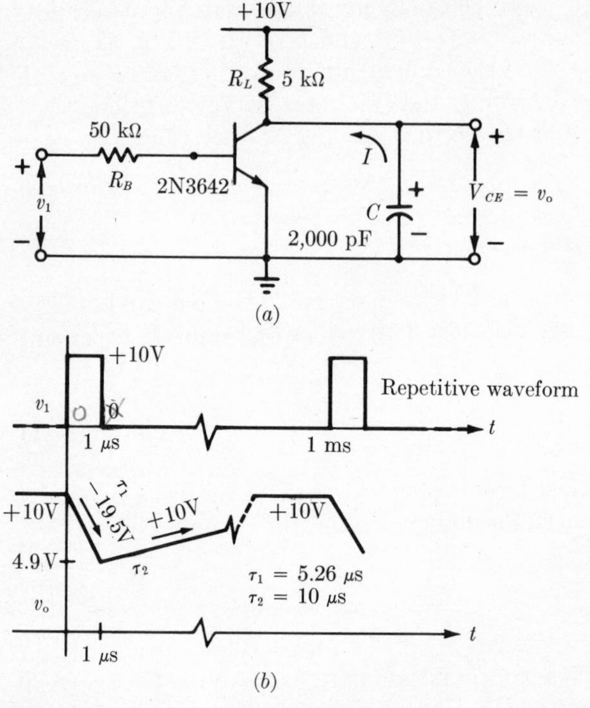

Fig. 3-22 A simple transistor sweep generator.

is determined by the external circuit, and if the external resistance in the collector circuit is small compared to $1/h_{oe}$. An example will serve to clarify some of the preceding ideas and show what simplifications accomplish.

The circuit of Fig. 3-22*a* is a transistor driven by a periodic pulse voltage, and turned off between pulses. The output is the collector voltage obtained by periodically discharging the capacitor through the transistor and charging it through the 5 kΩ resistor. Since the time constant $R_LC = 10\ \mu s$, there is ample time between pulses (999 μs) to completely recharge the capacitor to +10 V. The transistor has no effect during these intervals because it is in Region I (cutoff) between pulses.

At $t = 0^+$ base current flows and an approximate value can be found by regarding the base-emitter circuit as a short circuit:

$$I_B \approx (+10)/(50\ \text{k}\Omega) = 0.2\ \text{mA}$$

The voltage V_{CE} at this time is the same as the capacitor voltage at $t = 0^-$, since the capacitor cannot instantaneously change charge. Therefore the transistor is operating at $V_{CE} = 10$ V, $I_B \approx 200\ \mu A$ at $t = 0^+$. The curves of Fig. 3-14 show that $I_C = 12.3$ mA. All of this current is flowing from the capacitor at $t = 0^+$ because there is zero voltage across R_L at that time. The initial rate of change of v_o is then

$$\frac{dv_o}{dt} = \frac{I(0^+)}{C} = \frac{-12.3 \times 10^{-3}}{2 \times 10^{-9}} \approx -6 \times 10^6 \text{ V/sec} \qquad (t = 0^+)$$
$$\approx -6 \text{ V}/\mu\text{s}$$

At the end of the 1 μs that the pulse is applied to the base, the voltage V_{CE} will have dropped from +10 to about +4 V assuming that the initial rate of change of V_{CE} persisted the whole microsecond. For 1 $\mu s < t < 1$ ms the transistor is turned off and the capacitor recharges from the approximate +4 V determined at the end of the pulse to +10 V with $\tau_2 = R_L C = 10\ \mu$s.

The manner in which the operating point moves on the transistor characteristics during the cycle of operation is of considerable interest, and is important in establishing the regions to be represented by equivalent circuits. A somewhat simplified set of transistor collector characteristics is shown in Fig. 3-23 where only the curves for $I_B = 200\ \mu$A and $I_B = 0$ are shown. At $t = 0^-$, $I_C = 0 = I_B$, and $V_{CE} = +10$ V as shown. At $t = 0^+$ the operating point jumps instantaneously (nearly) upward to the line for $I_B = 200\ \mu$A. Since the capacitor voltage cannot change instantaneously, the line is straight up or V_{CE} = const. As the capacitor begins to discharge, the operating point travels down the line for $I_B = 200\ \mu$A until $t = 1\ \mu$s, when the pulse input is removed. The operating point then again jumps downward along a line of constant voltage until the line for $I_B = 0$ is again reached. After the jump the operating point moves relatively slowly along the voltage axis and given sufficient time to recover reaches the initial starting point at $V_{CE} = +10$ V. From this picture it is obvious that the two heavily lined segments are the two for which equivalent circuits must be developed.

This much of an answer has been obtained without recourse to equivalent circuits, and might suffice if the details of the voltage v_o

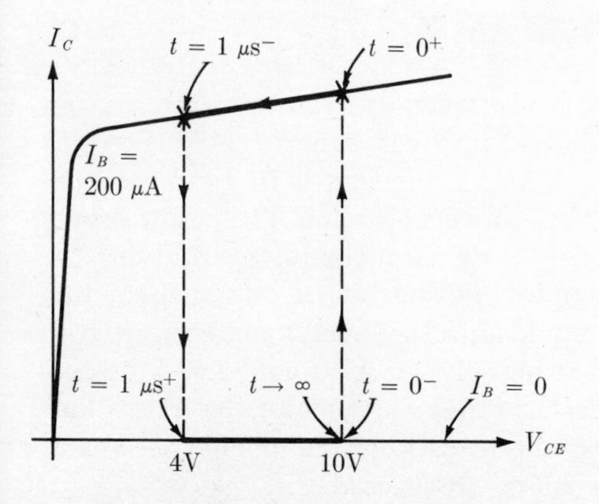

Fig. 3-23 The locus of the operating point of the circuit in Fig. 3-22.

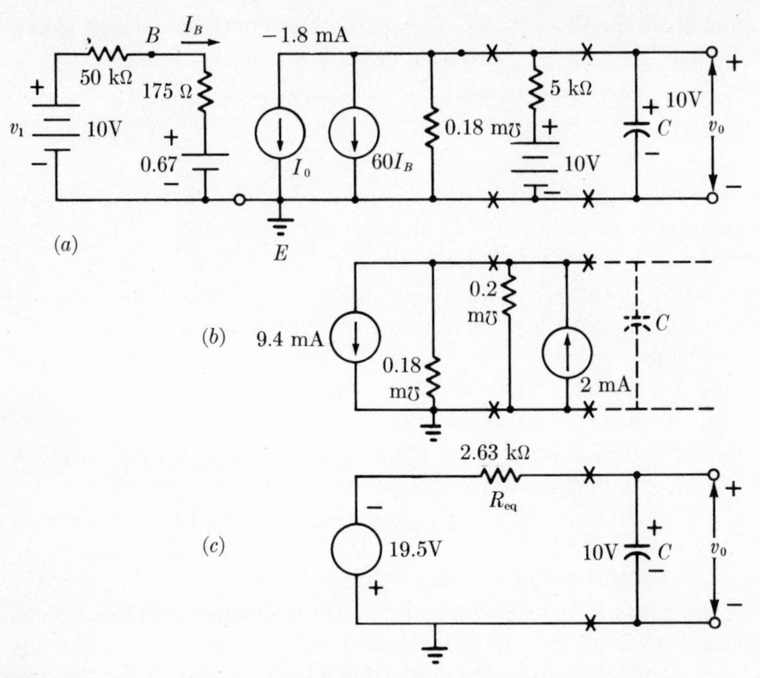

Fig. 3-24 An equivalent circuit for Fig. 3-22 valid for $0 < t < 1$ μs.

were not important. However, for more insight into what is going on in the circuit and to obtain a more accurate answer, replace the transistor by its equivalent circuit as in Fig. 3-24*a*. The values given for the transistor parameters assume the operating region found in the preceding, and the curves of Fig. 3-14. There are two independent circuits to be solved: the first is the base circuit for I_B, the second is to find the output $v_o(t)$. The base current is constant during the pulse, and is

$$I_B = \frac{10 - 0.67}{50 \text{ k}\Omega + 0.175 \text{ k}\Omega} = 0.186 \text{ mA} \qquad (3\text{-}13)$$

To solve the collector circuit, there are several ways to proceed, but an easy way is shown in the transitions of Fig. 3-24*a*, *b*, and *c*. First the value of $h_{fe}I_B = 60(0.186) = 11.2$ mA is found and added to I_0 to give the total generator of 9.4 mA inside the equivalent circuit. The power supply and load resistor R_L are replaced by their parallel equivalent giving the circuit of Fig. 3-24*b*. To get a simple series circuit, the parallel conductances and current generators are added, respectively, and converted to the series equivalent of Fig. 3-24*c* where the −19.5 V and 2.63 kΩ represent the source driving the capacitor during the time of the pulse. This reduced circuit is valid as long as the transistor operates in the area of Region II where the parameters chosen are accurate.

From Fig. 3-24c the details of the waveform $v_o(t)$ for $0 < t < 1$ μs can be obtained by inspection. At $t = 0^+$, $v_o = +10$ V as before and is decreasing toward a final value of -19.5 V with a time constant

$$\tau_1 = R_{eq}C = 2.63\ \text{k}\Omega(2 \times 10^{-9}) = 5.26\ \mu\text{s}.$$

The actual collector voltage does not reach this final value because the operating interval is only a small fraction of the time constant. Also the equivalent circuit would not be valid over the entire voltage range. To find the value of v_o at the end of the pulse, use the actual equation for v_o

$$v_o(t) = -19.5 + 29.5\epsilon^{-t/\tau_1}$$
$$= -19.5 + \frac{29.5}{1.21} = +4.9\ \text{V} \qquad (\text{at } t = 1\ \mu\text{s}) \qquad (3\text{-}14)$$

The details of the waveform v_o are shown in Fig. 3-22b. The waveform for 1 μs $< t <$ 1 ms is the same as before since the transistor is turned off during this time. The time constant $\tau_2 = R_LC = 10$ μs, and the final value is $+10$ V. The final value is nearly reached in $5\tau_2$ ($+9.964$ V at $t = 51$ μs) so that the circuit could repeat the waveform calculated at $t = 0$ by applying another pulse any time after $t \approx 50$ μs.

The fact that the final value of the waveform during the 1-μs interval is less than zero may be disturbing, but should not be since it merely expresses the fact that the transistor is drawing a relatively large current from the capacitor, which could not be sustained on a steady-state basis. Note that extending the pulse to 3 μs would cause the output (according to the circuit of Fig. 3-24c) to drop to -2.8 V. This is not possible since the transistor curves do not extend into the second quadrant, but instead follow the current axis to the origin at low voltages. To resolve this situation the Region II equivalent circuit is used to solve for the time when the collector voltage goes from Region II to Region III. This boundary occurs at a collector voltage $V_{CE} = v_o$ of about 0.14 V. Substituting $v_o = 0.14$ into Eq. (3-14) and solving for t gives $t = 2.13$ μs. For a pulse input lasting longer than this time, the transistor will saturate (operate in Region III), and $v_o \approx 0$ V as shown in Fig. 3-25. To calculate

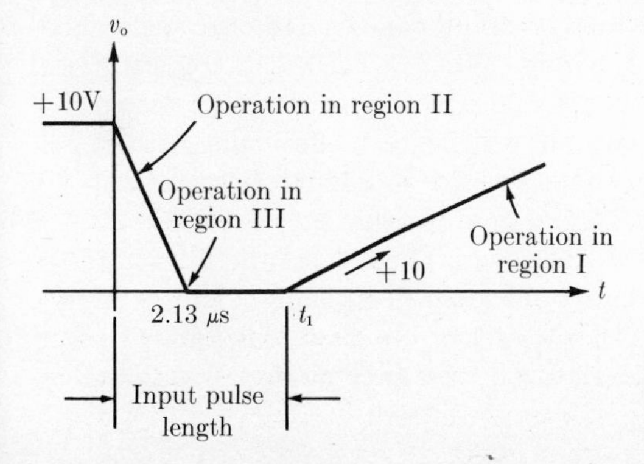

Fig. 3-25 The output from the circuit of Fig. 3-22 when the input pulse is lengthened.

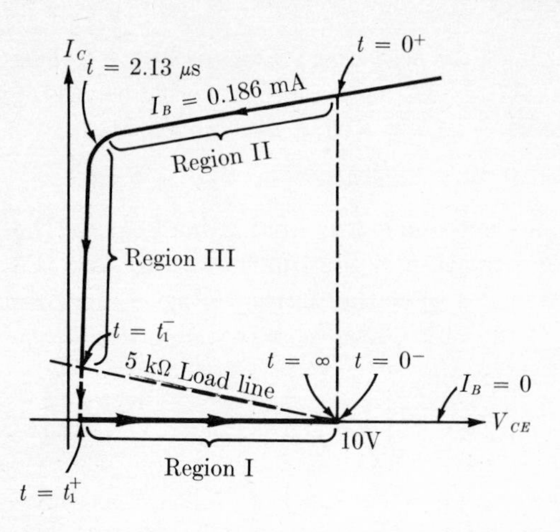

Fig. 3-26 The locus of operation for the circuit of Fig. 3-22 with the longer pulse length of Fig. 3-25.

the behavior in this region an equivalent circuit from Fig. 3-18 should be used.

The operation of the circuit for a long pulse traverses all three regions as shown in Fig. 3-26, which is similar to Fig. 3-23, but the longer pulse length causes the operation to continue down the curve $I_B \approx 200\ \mu A$ past the "knee" of the curve into Region III. If the pulse is long enough, the capacitor current will substantially go to zero, and the operating point will be on the static load line as shown at $t = t_1^-$.

TRANSISTOR CLIPPER CIRCUIT

A second example which also shows the operation of the transistor in the three regions is the clipper, or overdriven amplifier, in Fig. 3-27. The objective of this circuit is to develop a reasonably good square wave from a sinusoidal or other periodic input voltage. In this case the input is a 7 V rms, 60-Hz sine wave, which might serve as a source of a moderately precise timing waveform (about 0.1 percent frequency accuracy for short periods of time in the United States and almost perfect over long periods of time). The output from this kind of circuit can be adequately calculated by computing a few critical points. When $v_1 \leq 0$, the transistor will be off, and $v_o = +10$ V. As v_1 goes positive a point will be reached at $v_1 = v_a \approx 0.6$ V where base current will begin to flow and with it collector current. At some higher input voltage V_b enough base current will flow to saturate the transistor; at that point further increases in input will not materially reduce the collector voltage. These points are illustrated in Fig. 3-27*b*, where the distance between V_a and V_b is exaggerated for clarity.

As described previously, V_a is the base voltage at which base current begins to flow, and from Fig. 3-14 this is seen to be about 0.6 V. To calculate the voltage V_b draw a load line on the collector curves and find at what value of I_B saturation commences as shown in Fig. 3-27c. The required base current is 100 μA, and the required input voltage v_1 is

$$\begin{aligned} v_1 &= I_B R_B + V_{BE} \\ &= 0.1(10\text{k}\Omega) + 0.67 = 1.67 \text{ V} \end{aligned} \tag{3-15}$$

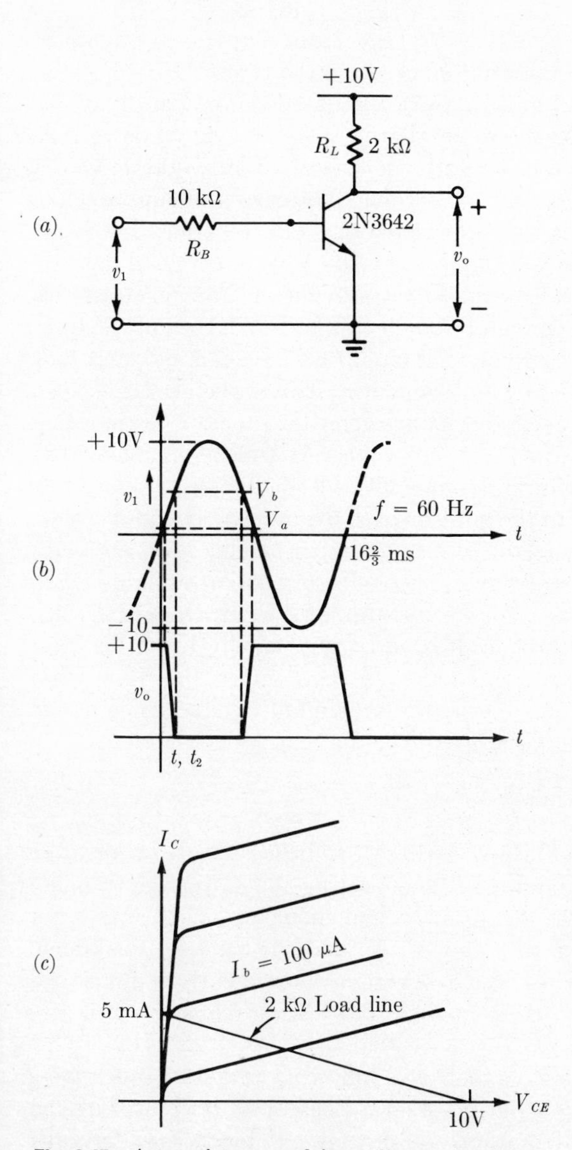

Fig. 3-27 A transistor overdriven clipper circuit.

If the approximate rise and fall times of the square wave are needed, the times at which V_a and V_b occur may be calculated as

$$t_1 = \frac{\sin^{-1}(V_a/10)}{2\pi f} = \frac{\sin^{-1}(0.6/10)}{377} = 0.16 \text{ ms}$$
$$t_2 = \frac{\sin^{-1}(V_b/10)}{2\pi f} = \frac{\sin^{-1}(1.67/10)}{377} = 0.445 \text{ ms}$$

The rise and fall times are $t_2 - t_1 = 0.29$ ms. Note that the positive portion of the square wave is somewhat longer than the zero portion, because the two transition points V_a and V_b are not situated symmetrically about the axis of symmetry of the sine wave, that is, 0 V. If we add a dc voltage of $(V_b + V_a)/2 = 1.14$ V in series with the sine wave, it would make the square wave have equal on and off periods. To make the square wave more nearly square, that is, to have faster rise and fall times the input voltage can be increased; for example, $v_1 = 100$ V peak would give a rise time less than one tenth of that obtained previously. For very fast rise and fall times another clipper can be connected to the output of the first.

One phenomenon occurring in this circuit we have ignored, and that is in the details of the base-emitter waveform. This waveform is clipped also on the positive input cycle by the base-emitter junction so that V_{BE} does not exceed 0.7 V during this half cycle. At the beginning of the negative cycle the transistor is completely off and $V_{BE} = v_1$ until v_1 becomes more negative than the emitter-base breakdown at about -7 V. The base-emitter junction again acts as a clipper at this level and considerable reverse base current flows. The collector current remains off so the previous calculation of output waveform is all right. Operation with the base-emitter junction broken down does not usually hurt the transistor if the current is kept low enough to stay within the dissipation limits of the transistor. In this case the maximum current I_B is about $[-10 - (-7)]/10\ \text{k}\Omega = -0.3$ mA and the dissipation is

$$(0.3 \times 10^{-3})7 = 2.1 \text{ mW}$$

This amount of power would not damage the transistor. However, the whole problem may be eliminated by connecting the cathode of a diode to the base and the anode to ground. The diode prevents V_{BE} from becoming more negative than about 0.7 V, and the reverse breakdown cannot occur. The waveform at the base is then a very nearly symmetrical square wave of about 1.4 V peak-to-peak, but having relatively long rise and fall times.

If the circuit of Fig. 3-27 is to drive a following circuit, it is desirable to know the nature of the output resistance. Because the circuit is driven over all three regions of operation, the output resistance is widely different at different times during a cycle of operation. During the portion of the cycle that the transistor is cut off the output resistance is $R_L = 10$ kΩ. During the time the transistor is in Region II, the resistance is $R_L \parallel 1/h_{oe}$ or about 10 kΩ in parallel with 20 kΩ, or 6.7 kΩ; while the transistor is saturated the output resistance is $R_L \parallel r_{cIII} \approx r_{cIII} \approx 25$ ohms. The fact that the output resistance can vary 400:1 during the cycle can have serious consequences on the output waveform if the load is not a high impedance compared to the maximum output resistance. This problem will be illustrated with regard to the next circuit to be considered.

3-10 THE EMITTER FOLLOWER

A very useful amplifier for pulse circuits is one with a high input resistance, a low output resistance, and a voltage gain of nearly one. The emitter follower fits this description and is indeed a useful circuit if its virtues and limitations are understood. Figure 3-28 shows the circuit drawn with two power supplies, although one can be used providing the input is suitably biased. Since a transistor base current is small, $V_{BN} \approx v_s$, and since V_{BE} is nearly a constant in Region II, $v_o = V_{BN} - V_{BE} \approx v_s - V_{BE}$, and the output signal is approximately the input signal. To find the actual initial operating point of the circuit a procedure involving successive approximations is useful. To begin with, assume $V_{BN} = v_s = 0$; then V_{EN} (the N subscript means referred to ground as V_{BN} is shown in Fig. 3-28) is about -0.7 V (from Fig. 3-14). The emitter current

$$I_E = \frac{[-10 - (-0.7)]}{5 \text{ k}\Omega} = -1.86 \text{ mA}$$

Since the base current is small, $I_C \approx -I_E = 1.86$. The actual base current can now be estimated by referring to Fig. 3-14 for the conditions $I_C = 1.86$, $V_{CE} = 10.7$ V giving $I_B \approx 35$ μA. Now a new calculation may

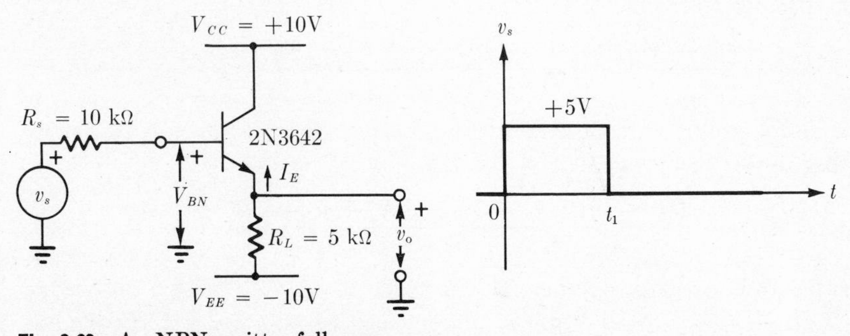

Fig. 3-28 An NPN emitter follower.

be made with a good deal of precision. This calculation proceeds as follows:

$$V_{BN} = v_s - I_B R_s = 0 - 0.035(10\ \text{k}\Omega) = -0.35\ \text{V} \tag{3-16}$$

$$V_{BE}\ (\text{from Fig. 3-14}) = 0.67\ \text{V} \qquad (\text{for } I_B = 35\ \mu\text{A}) \tag{3-17}$$

$$V_{EN} = V_{BN} - V_{BE} = -0.35 - 0.67 = -1.02\ \text{V} \tag{3-18}$$

$$I_E = \frac{-10 - V_{EN}}{5\ \text{k}\Omega} = -1.8\ \text{mA} \tag{3-19}$$

$$I_C = -I_E - I_B = +1.8 - 0.035 = 1.77\ \text{mA} \tag{3-20}$$

The process could in principle be repeated for further accuracy, but little accuracy is gained because the next approximation for base current causes negligible change in the voltages and currents already found. Although the procedure may seem cumbersome, the first approximation can be made quickly mentally and is really quite sufficient if only the region of operation in Region II is desired.

The analysis of the circuit proceeds by replacing the transistor with an equivalent circuit valid about the operating point just found as in Fig. 3-29*a*. The object here is to reduce this circuit to a simple Thevenin equivalent, but, because there is an active generator in the circuit, caution must be exercised in developing the equivalent. Merely short circuiting the voltage sources and open circuiting the current sources to find the equivalent impedance will not work here because the value of one of the generators depends upon what is attached to the output terminals. To determine the equivalent impedance the open-circuit output voltage and short-circuit output current will be determined. Then the output resistance is $Z_{\text{eq}} = V_{\text{oc}}/I_{\text{sc}}$. For analysis the circuit will be split at the point $x - x$, and the effect of the load R_L can be determined later. From Fig. 3-29*b*

$$I = I_B + h_{fe} I_B + I_0 + h_{oe}(V_{CC} - v) = 0 \qquad (\text{for an open circuit}) \tag{3-21}$$

$$I_B = \frac{v_s - v - V_{II}}{R_s'} \tag{3-22}$$

Substituting I_B from Eq. (3-22) into Eq. (3-21) and solving for $v = V_{\text{oc}}$:

$$V_{\text{oc}} = \frac{(v_s - V_{II})(h_{fe} + 1) + I_0 R_s' + V_{CC} h_{oe} R_s'}{h_{fe} + h_{oe} R_s' + 1} \tag{3-23}$$

The short-circuit current can be found from Eqs. (3-21) and (3-22) by letting $v = 0$ such that

$$I = I_{\text{sc}} = \frac{(v_s - V_{II})(h_{fe} + 1) + I_0 R_s' + V_{CC} h_{oe} R_s'}{R_s'} \tag{3-24}$$

The desired output resistance, R_{eq}, is then

$$R_{\text{eq}} = \frac{V_{oc}}{I_{\text{sc}}} = \frac{R_s'}{(h_{fe} + 1) + h_{oe} R_s'} \tag{3-25}$$

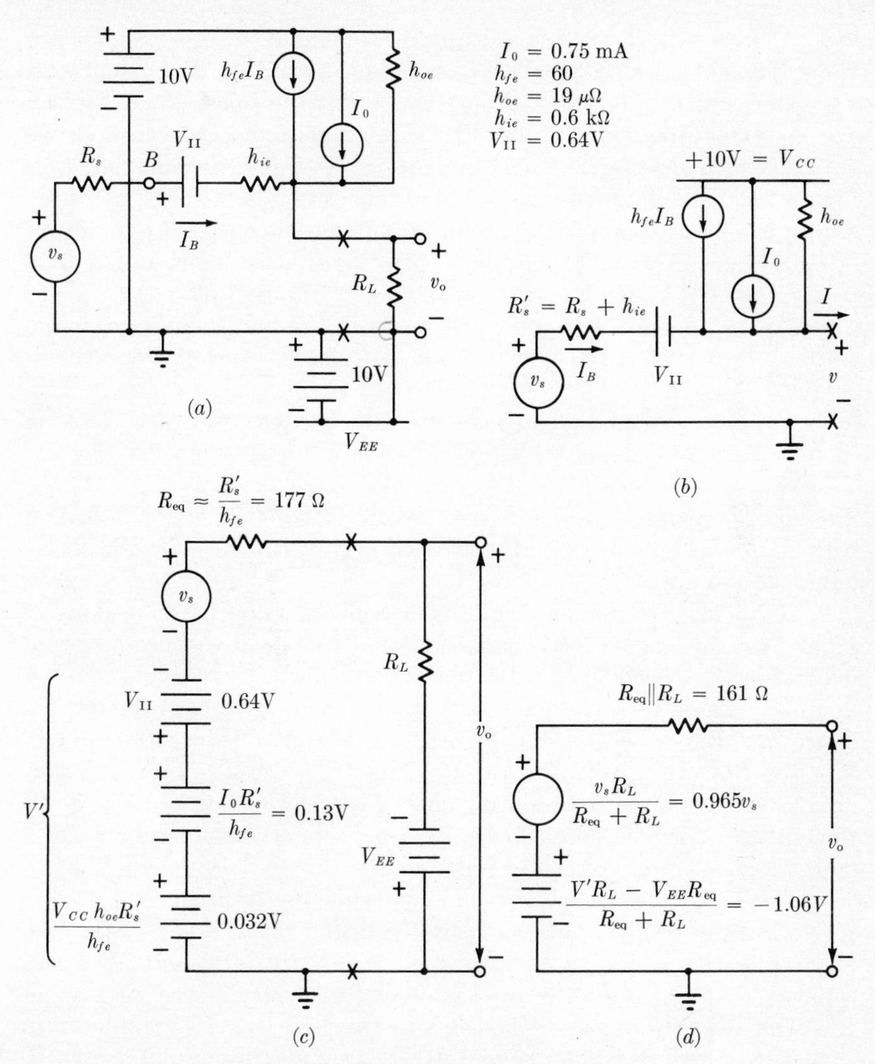

Fig. 3-29 Equivalent circuits for the emitter follower.

In the usual case $h_{oe}R'_s \ll 1 \ll h_{fe}$ so that very adequate approximations for V_{oc} and R_{eq} are

$$V_{oc} \approx v_s + \frac{I_0 R'_s}{h_{fe}} + \frac{V_{CC} h_{oe} R'_s}{h_{fe}} - V_{II} \tag{3-26}$$

$$R_{eq} \approx \frac{R'_s}{h_{fe}} \tag{3-27}$$

These equations are shown schematically in Fig. 3-29c, where the open-circuit voltage is shown to be a variable generator approximately equal to v_s in series with three fixed voltages of which only V_{II} has a significant value. The output resistance is the source R_s divided by the current gain h_{fe} and is about 170 ohms. From this circuit the open-circuit voltage gain is seen to be about unity. The voltage gain and resistance at the output terminals can be found by making a Thevenin equivalent to Fig. 3-29*c*. The total circuit still has an output resistance of about R_s/h_{fe}, a voltage gain of 0.965, and an offset between input and output dc voltage of -1.06 V. The output for the input pulse shown in Fig. 3-28 is $v_o = -1.06$ V for $v_s = 0$, and $v_o = 0.965(5) - 1.06 = 3.77$ V for $v_s = 5$ V. Note the close agreement between the first value of v_o found graphically (-1.02 V) and the one calculated from the equivalent circuit.

A similar calculation may be used to find the resistance looking into the base terminal of the amplifier when the output is loaded by R_L. The approximate result is that the input resistance is $R_{in} \approx h_{fe}R_L$. Therefore the emitter follower behaves like an impedance transformer with $R_{in} \approx h_{fe}R_L$, and output resistance $R_{out} \approx R_s/h_{fe}$, but with nearly unity voltage gain.

If an input signal having a wide voltage range is applied as in Fig. 3-30*a*, the emitter follower will initially be cut off and begin to conduct when V_{BE} becomes greater than about 0.6 V. This occurs with an input of $-10 + 0.6 = -9.4$ V, as shown. The output then follows the input until $V_{EN} \approx V_{CC}$ or V_{CE} becomes close to zero. If we set v_o equal to V_{CC}, the input voltage causing this to occur is $v_o = 10 = -1.06 + 0.965v_s$ or $v_s = 11.5$ V. From the resulting output waveform shown in Fig. 3-30*b* the emitter follower is seen to be a clipper circuit, but one requiring a large range of input voltage to function.

The output resistance of the circuit while the transistor is cut off is $R_L = 5$ kΩ, while the transistor is in the linear region $R_{out} \approx 160$ ohms, and while the transistor is saturated $R_{out} \approx r_{cIII} \approx 25$ ohms. These changes in output resistance are shown in Fig. 3-30*c*.

One use of an emitter follower is to charge a capacitor rapidly from a relatively high resistance source. If, for example, the 10 kΩ source R_s were used directly to charge a 0.01 μF capacitor, the charging time constant would be 100 μs. However, if the capacitor were connected to the output of an emitter follower, the charging-time constant would be only $R_{out}C = 160 \cdot 10^{-8} = 1.6$ μs. This improvement in charging time is often useful, but notice must be taken that in discharging the capacitor the situation will be quite different, and not as favorable. Consider the circuit of Fig. 3-31 where the capacitor is shown connected to the output of the emitter follower. A positive pulse for v_1 rapidly charges C because of the large emitter current flowing in the *permitted* direction in the transistor, that is, in the direction of the arrow. At the end of the pulse the capacitor

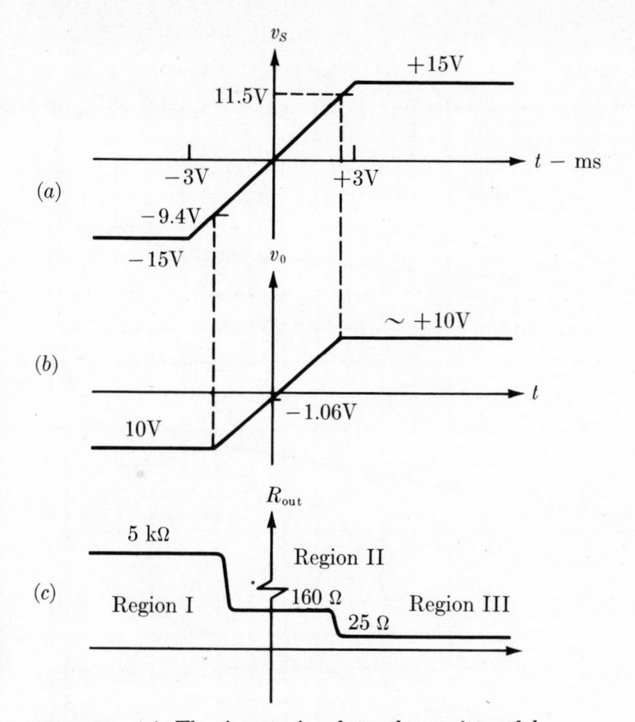

Fig. 3-30 (*a*) The input signal to the emitter follower. (*b*) The output signal and the transfer characteristic. (*c*) The output resistance corresponding to different regions of operation.

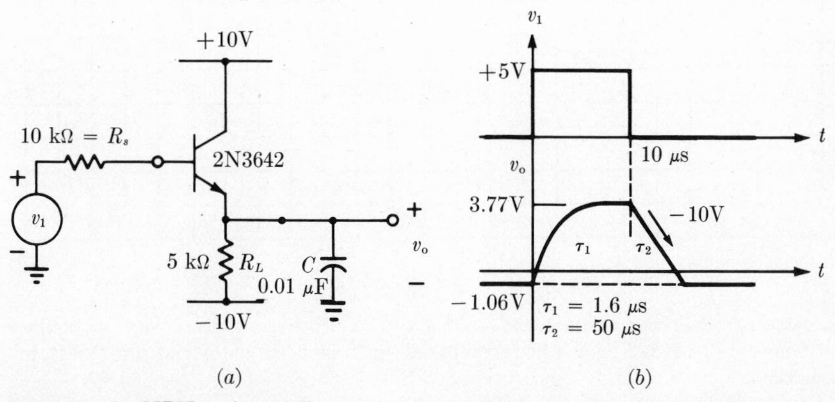

Fig. 3-31 An NPN emitter-follower circuit. (*a*) With capacitor load. (*b*) The input and output waveforms.

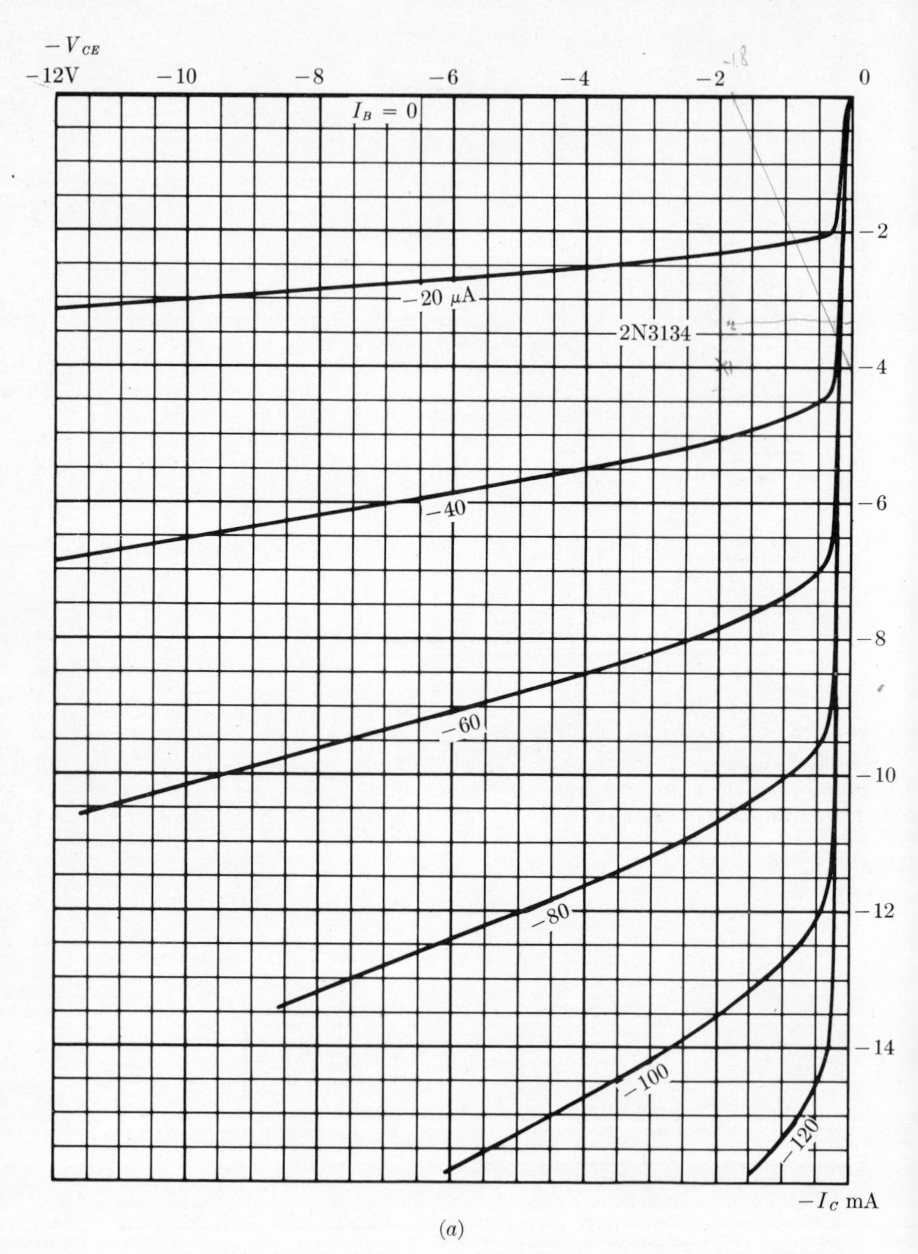

(*a*)

Fig. 3-32 (*a*) Collector characteristics of a typical PNP transistor. The particular unit chosen (2N3134) is a silicon epitaxial unit used in industrial and military applications.

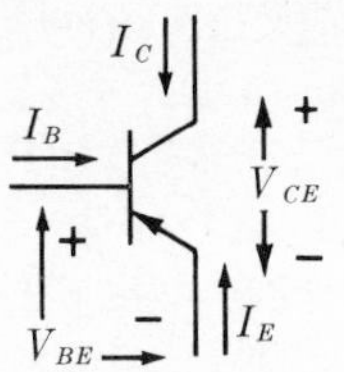

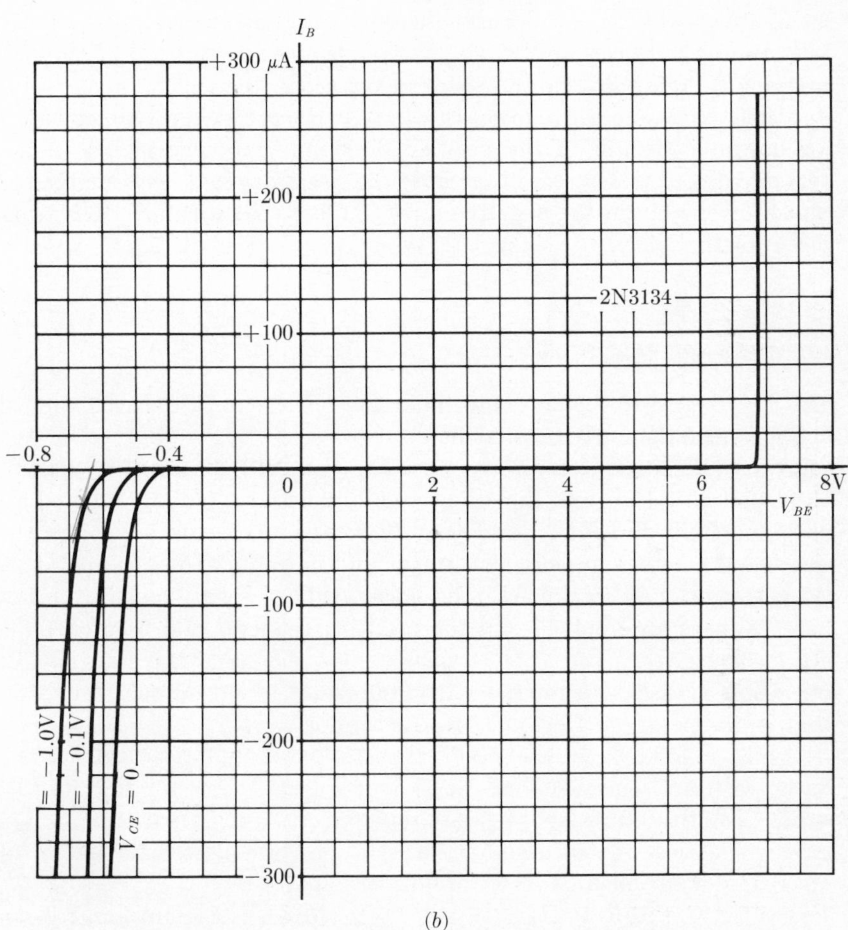

(b)

Fig. 3-32 (*b*) Base-emitter characteristics of a typical PNP transistor (2N3134).

will be very nearly charged to the voltage $v_o = 3.77$ V obtained with no C, since many time constants ($10/1.6 = 6.2\tau_1$) are available for charging. At $t = 10^+$ μs the input returns to zero volts and the capacitor begins to discharge. It cannot discharge through the transistor, however, because current cannot flow into the emitter unless the emitter-base junction breaks down, which it will not with $V_{BE} = -3.77$ V. The capacitor must therefore discharge through $R_L = 5$ kΩ to the -10 V supply. The discharging-time constant $\tau_2 = 5$ k$\Omega(10^{-8}) = 30$ μs with final value of -10 V, as shown in the waveform for v_o. The transistor remains cut off until the output voltage falls to within a few tenths of a volt of -1.06 V. At this voltage level the transistor returns to Region II, and the final value of the waveform changes to the actual final value, -1.06 V, with time constant $\tau_1 = 1.6$ μs. As can be seen very little of the discharging is done with the transistor in the active region.

The NPN emitter follower is seen to perform excellently in charging the output capacitor in the positive direction, but very poorly in the negative direction. One way to reverse the situation, that is, to charge the capacitor quickly in the negative direction, is to substitute a PNP transistor for the NPN.

3-11 THE PNP TRANSISTOR

The PNP transistor behaves much like the NPN, except that the applied voltages and the currents that flow are reversed. Figure 3-32 gives the collector characteristics and symbol for the 2N3134 PNP transistor. Note that all the current directions and voltages are similarly defined for both the PNP and NPN, and the differences are accounted for in the signs of the actual numerical voltages and currents. Thus to operate a PNP transistor in Region II—the linear region—negative collector and base voltages are applied, and the resulting collector and base currents are also negative. The PNP transistor has the same three regions of operation as discussed for the NPN; note that to cut off the transistor, that is, to operate in Region *I*, V_{BE} is zero or *positive*.

Figure 3-33 shows the PNP transistor connected as an emitter follower with a circuit identical to that of Fig. 3-28, except that the dc supply polarities V_{EE} and V_{CC} are reversed. By using an approximation process identical to that used for the NPN emitter follower, the quiescent operating point of this new circuit is found to be $V_{BE} = -0.6$ V, $V_{EN} = v_o = +0.75$ V, $I_E = +1.85$ mA, $I_C = -1.84$ mA, and $I_B \approx -15$ μA. The magnitude of the quiescent output voltage v_o is very nearly identical for the two circuits, even though comparison of the transistor characteristics, for example, h_{fe}, I_0, would show the two transistors to be quite different.

The circuit is driven by the same pulse driving the NPN circuit so that the outputs can be compared. The initial positive jump in v_1 at $t = 0^+$ turns the transistor off, and C charges through the resistor R_L giving a time constant $\tau_1 = 50$ μs. When the output gets within a few

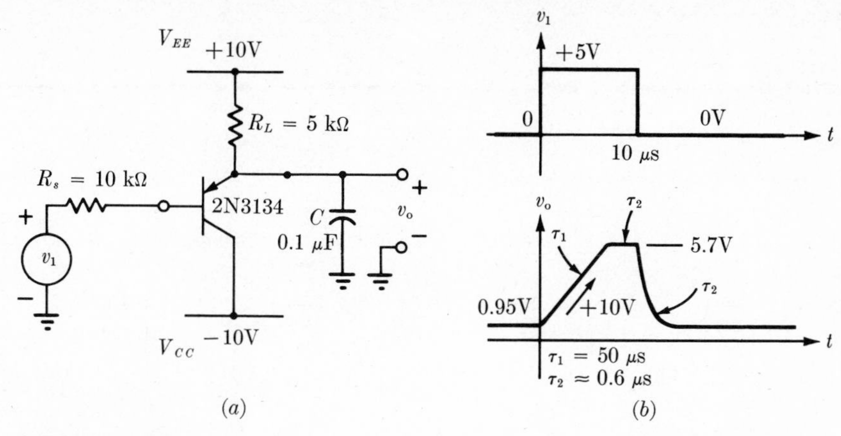

Fig. 3-33 A PNP emitter-follower circuit. (*a*) With capacitor load. (*b*) The output waveforms.

tenths of a volt of the actual final value 5.7 V the transistor begins to conduct. The time constant then changes from 50 μs to $R_{out}C$ = 0.6 μs. (The lower value of R_{out} is caused by the higher h_{fe} of the 2N3134 transistor compared to the 2N3642.) At the end of the pulse the negative-going transition causes the transistor to conduct heavily momentarily and to discharge the capacitor quickly. (The details of the calculation of this circuit are left as a problem in the problem section.)

The PNP emitter follower is seen to produce an output pulse which has a speedy fall time, but poor rise time—the opposite of the situation with the NPN emitter follower. Many times only the rise or the fall is important and the type of emitter follower can be chosen to fit, but occasionally both the rise and fall must be kept fast. For such a situation both types of transistors may be brought into play in what is known as a complementary emitter follower (Fig. 3-34). In this circuit the charging current is supplied through the NPN transistor, and the discharging current by the PNP. When one transistor is conducting, the other is cut off so that the stage is much like a class B amplifier. One difficulty is that for input voltages in the range $+0.5 > v_1 > -0.5$ V neither transistor will conduct. If this dead zone or hysteresis is undesirable, the transistors must be biased in such a manner as to make them conduct a small amount of current with no input voltage applied. Such a bias circuit is shown in Fig. 3-34*b*. The forward biased diodes provide the necessary V_{BE}, whereas the 10-ohm resistors reduce the change in collector current in the transistors with variations in the diode drop from one diode to the next and with temperature changes.

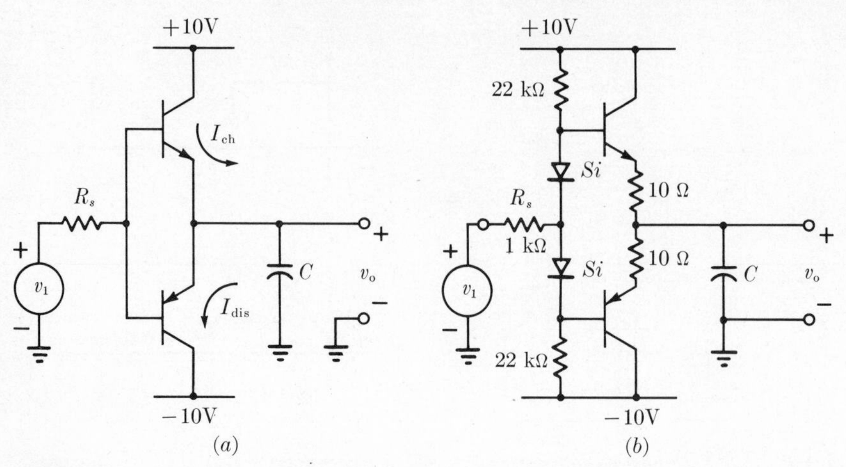

Fig. 3-34 A complementary emitter follower (*a*) and a similar circuit (*b*) biased to eliminate the dead zone near the origin.

3-12 COMMON-BASE TRANSISTOR CHARACTERISTICS

Another set of static characteristics, which is sometimes used to describe a transistor, is the common base set. These provide some additional insight into transistor operation, and may occasionally help to analyze a circuit. Figure 3-35 shows a representative collector and emitter characteristics for a silicon NPN transistor. In the linear region the collector characteristics are seen to be parallel, equally spaced lines which are also closely parallel to the voltage axis. Therefore a very adequate representation for the collector circuit is a current generator with a magnitude proportional to the emitter current, as shown in Fig. 3-36 by the generator $h_{fb}I_E$. The conductance in parallel with this current generator is so small that it cannot be accurately found from the graphical characteristics. The value of h_{fb} is very nearly minus one and may be taken as such for many applications. For more precise calculations $h_{fb} = -h_{fe}/(h_{fe} + 1)$. The emitter-base circuit is again a diode characteristic and is represented by the combination r_{eII} and V_{II}. Note that for an NPN transistor I_E is negative so that $I_C = h_{fb}I_E \approx -1$(negative value of I_E) = the normal positive I_C. One fact is apparent from the curves that might not be obvious without them: namely, the linear region—Region II—continues to very low collector-to-base voltages, in fact, to slightly negative collector voltages.

To illustrate a use for the common-base connection consider the circuit of Fig. 3-37. Here the transistor is used to make a nearly ideal current source, which is diverted into the capacitor at $t = 0$ by opening the switch. While the switch is closed $V_{CB} = 10$ V and $I_E \approx -10/5\ \text{k}\Omega = -2$ mA.

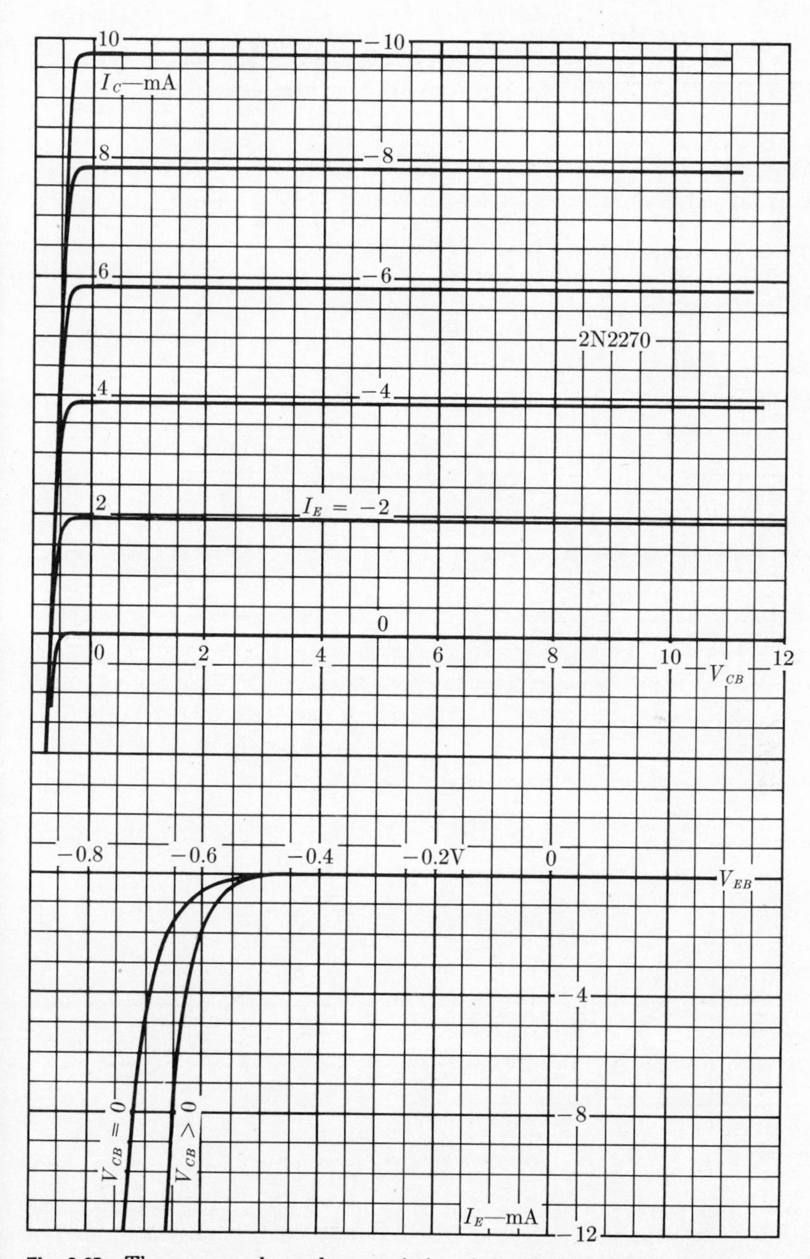

Fig. 3-35 The common-base characteristics of the 2N2270 NPN transistor.

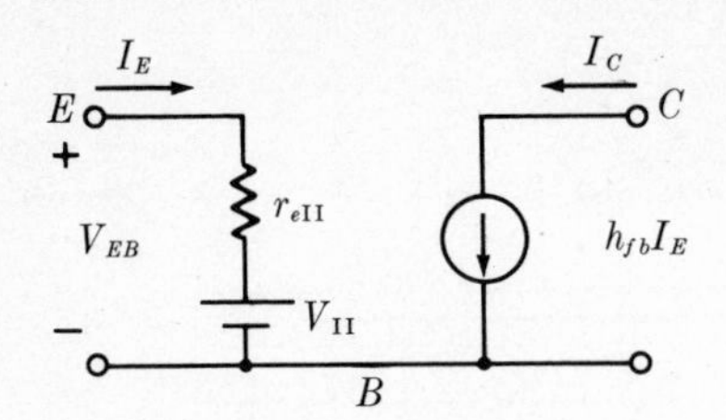

Fig. 3-36 A simple common-base equivalent circuit valid for Region II.

At this current level the curves of Fig. 3-35 show $V_{EB} = 0.6$ V. A more nearly exact value of emitter current is then

$$I_E = (-10 + 0.6)/5\ \text{k}\Omega = -1.88\ \text{mA}$$

The collector curves show $h_{fb} = -0.97$ so that

$$I_C = -0.97\ (-1.88) = +1.82\ \text{mA}$$

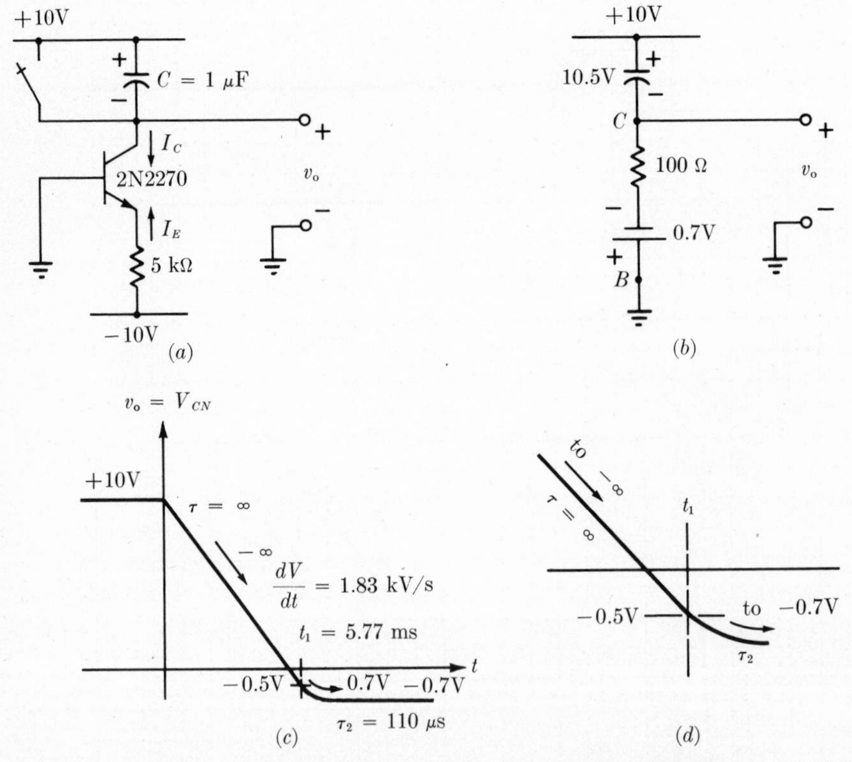

Fig. 3-37 A common base transistor used to make a negative-going ramp. (*a*) The circuit. (*b*) An equivalent circuit valid after t_1. (*c*) The output waveform. (*d*) The region around t_1 expanded.

For $t < 0$ this current is flowing in the closed switch; at $t = 0$ the switch opens and the capacitor begins to discharge at a rate

$$dV/dt = -I_C/C = -(1.82 \times 10^3) = -1.82 \text{ kV/s}$$

This rapid fall in voltage persists until the transistor leaves Region II at a $V_{CB} \approx -0.5$ V. The time required for this to occur is t_1

$$t_1 = \frac{\Delta V}{dV/dt} = \frac{+10 - (-0.5)}{1.82 \times 10^3} = 5.77 \text{ ms} \tag{3-28}$$

After this time the transistor is operating in saturation, and a reasonable equivalent circuit valid at t_1^+ is shown in Fig. 3-37*b*. The circuit shows that the final value of v_o is -0.7 V, as also can be found directly from the collector characteristics (the point $I_C = 0$, $I_E = -1.88$ mA, and $V_{CB} = -0.7$ V). This final portion of the waveform has a short time constant $\tau_2 = 110$ μs. The resulting complete waveform is shown in Fig. 3-37*c*. One important point often missed is that there is no discontinuity in the *slope* of the waveform at t_1 even though there is an abrupt change in time constant with the two equivalent circuits used. This is true because the current drawn by the transistor at $V_{CB} = -0.5$ V, the transition point, is the same for both equivalent circuits chosen. Since the rate of change of capacitor voltage is proportional to the current flowing in it, there can be no change in rate.

3-13 THE FIELD-EFFECT TRANSISTOR (FET)

The field-effect transistor is like the junction transistor in that it can be used to perform similar switching and amplifying functions, but is different in that the input resistance of the FET is very high, in fact as high as the resistance of a reverse-biased diode. The base-emitter resistance h_{ie} of an ordinary transistor is in the range of 10^0 to 10^4 ohms; the FET has an input resistance of 10^8 to 10^{16} ohms depending upon the type. The first described type is a thin semiconducting channel with ohmic contacts at both ends and a region in the middle of the channel with a junction to the opposite kind of semiconducting material as is shown in Fig. 3-38. Two types are possible as shown: one with an n-type *channel* and p-type *gate* region, and a second with a p-type channel and an n-type gate. In this textbook, these two types of field-effect devices are called n-channel and p-channel FET's, respectively.

The characteristic curves for a typical n-channel FET are shown in Fig. 3-39 together with the definitions of the various currents and voltages. For gate voltages $V_{GS} \leqq 0$ V the gate draws nearly zero current, and the current drawn by the drain shows two distinct regions of operation: at low drain voltage the device behaves like a resistor controlled

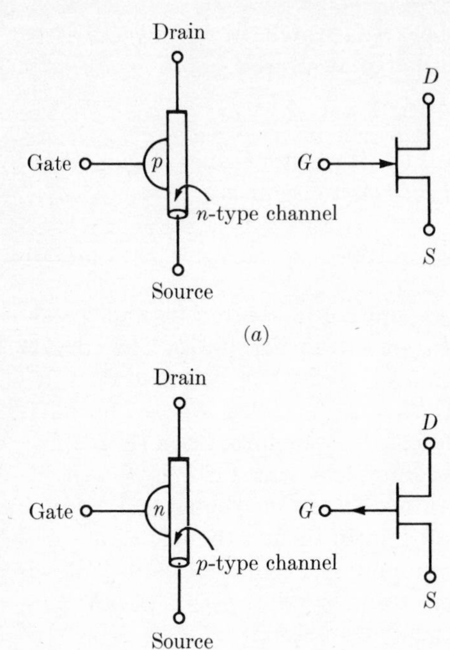

Fig. 3-38 A cross-sectional view of junction FET devices and their symbols. (*a*) *n*-Channel FET. (*b*) *p*-Channel FET.

by the voltage on the gate; at high drain voltage (above about 5 V for the MPS 105) the device behaves like a current generator controlled by the gate voltage. For a sufficiently negative gate voltage the FET becomes nonconducting or cut off. (This is also termed the pinch-off region.) The FET is seen to have two regions similar to the transistor—a cutoff or Region I, and a linear or Region II—but there is no region quite corresponding to the saturation region of the transistor. Application of a positive gate voltage can further reduce the resistance between the drain and source, but at the expense of losing the high input impedance of the device. This is shown more explicitly in Fig. 3-40, where the transfer- and gate-current characteristics are plotted. The drain current at a given V_{DS} varies approximately parabolically with the gate voltage for negative V_{GS} such that

$$I_D = \frac{I_O(V_{GS} - V_{CO})^2}{V_{CO}^2} \tag{3-29}$$

where I_O is the drain current flowing with $V_{GS} = 0$, and V_{CO} is the gate voltage required to cut off the drain current. At low values of V_{DS}, where the device is not behaving like a constant current device, the drain current does not obey the preceding relation, but instead depends almost linearly upon the gate voltage as is shown by the curve for $V_{DS} = 1$ V.

Many FET's are symmetrical devices, that is, essentially the same characteristics will be obtained if the source and drain terminals are interchanged. The devices are usually tested by the manufacturer with a specific end as the source, and may or may not meet the device specifications when the source and drain are interchanged. The fact that the device can be used to conduct either from drain to source, as in Fig. 3-39, or in the reverse direction is shown by the curves of Fig. 3-41, which show operation in both the first and third quadrants. The device is seen to

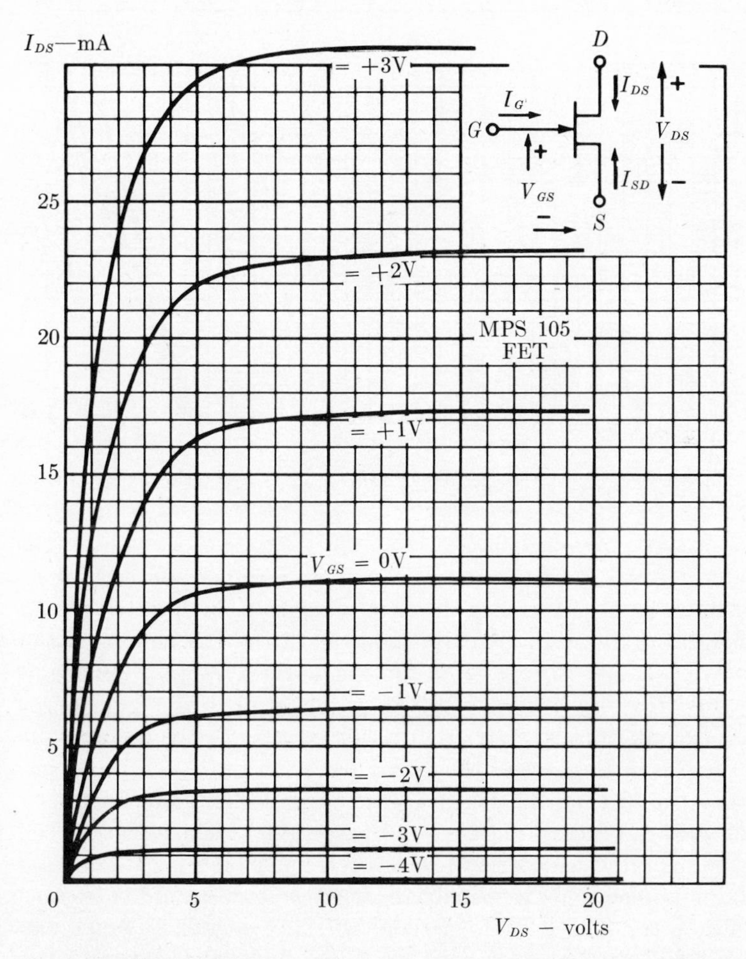

Fig. 3-39 The drain characteristics of a typical n-channel, junction FET, the type MPS 105.

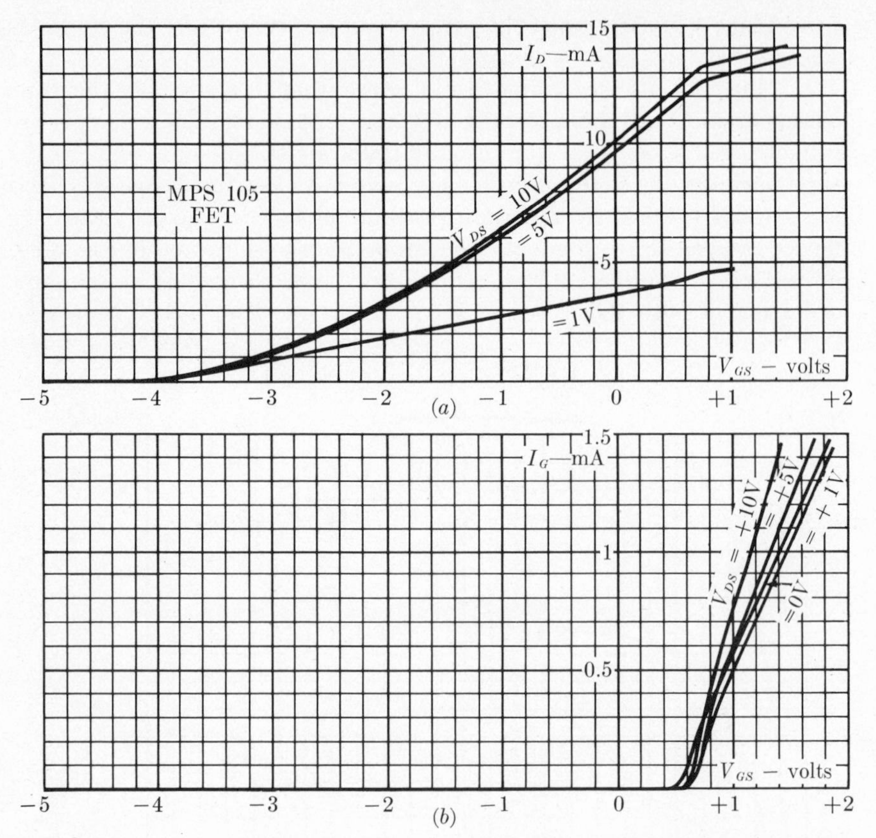

Fig. 3-40 (*a*) The transfer characteristics of the MPS 105. (*b*) The gate characteristics of the MPS 105.

approximate an ideal switch quite well in its OFF condition as long as V_{DS} is not made so negative that the device begins to be forward biased with the normal drain as the effective source. In the ON condition the device acts like a 200-ohm resistor for the current range shown and $V_{GS} = 0$. One advantage of the FET switch is that the curves for $V_{GS} \leqq 0$ go exactly through the origin; there is no offset voltage in the switch as there is in the case of the transistor. Making $V_{GS} > 0$ reduces the ON resistance further, but because gate current now flows, there is a small offset voltage of about 0.04 V for $V_{GS} = +4$ V. The fact that the FET may be operated as a zero offset switch makes it very interesting for modulator applications in which an offset voltage would introduce an error in dc level. The FET is also one of the few devices which can act like a nearly linear resistor with resistance controlled by the gate.

A simple equivalent circuit suitable for the linear region of the FET is shown in Fig. 3-42. The major changes in this circuit from that used to represent the transistor are that an open circuit appears from gate to source, and that the current generator is controlled by the input voltage rather than the input current. The quantities involved are defined in the illustration (Fig. 3-42*a*) but an example of an actual circuit will help clarify their use.

The circuit shown in Fig. 3-43 is a *source follower*, which is analogous to the emitter follower of Fig. 3-28 or the older vacuum-tube cathode follower. The advantage of the FET in this application is that the input impedance is extremely high—greater than 10^8 ohms for a good FET. In this case the FET is driving a capacitive load of 100 pF in parallel with a 200-ohm load resistor. For $t < 0$ the FET is cut off by the input $v_1 = -6$ V; at $t = 0^+$, $V_{GS} = 0$ V since the capacitor was previously

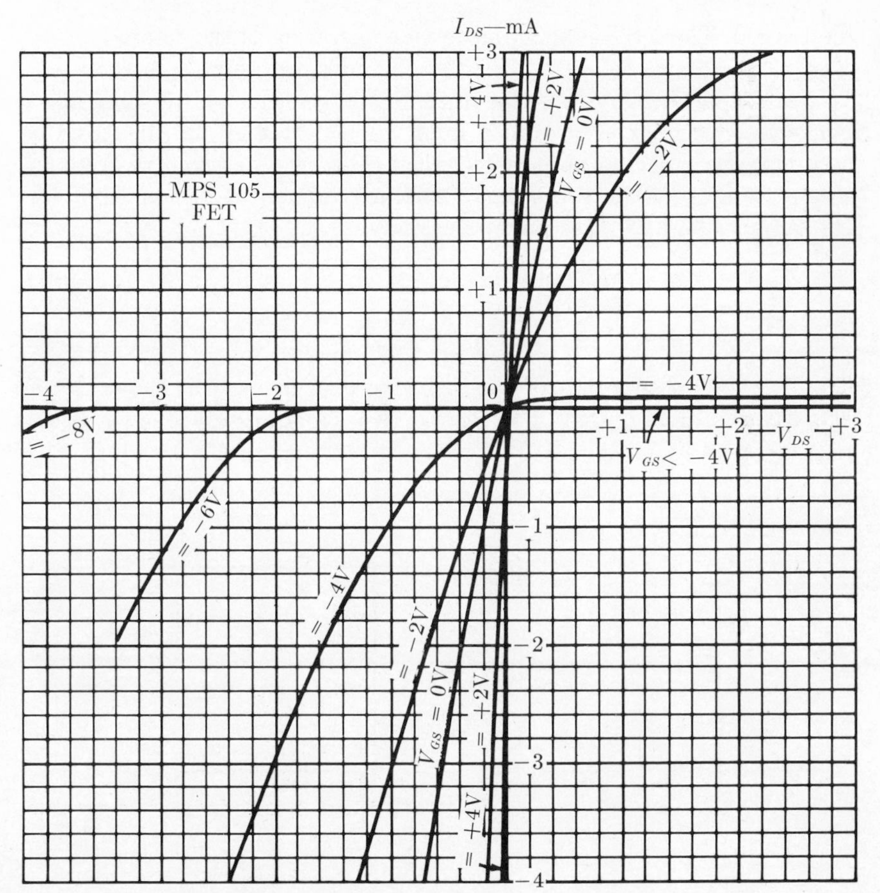

Fig. 3-41 The drain characteristics of the MPS 105 *n*-channel FET magnified about the origin and extended into the third quadrant.

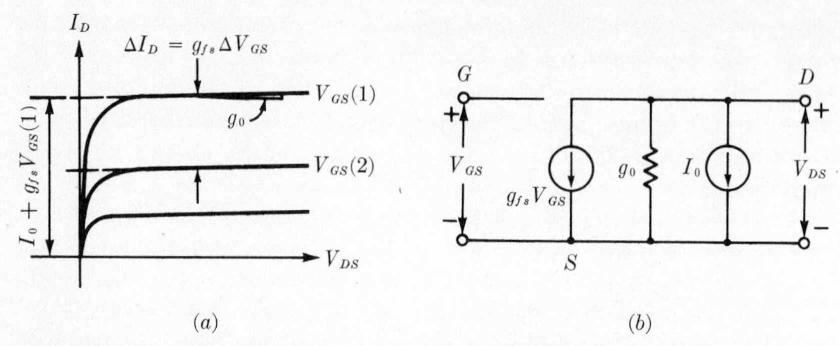

Fig. 3-42 (*a*) A typical set of curves for the *n*-channel FET. (*b*) An equivalent circuit valid in the horizontal region of the drain curves.

discharged. The FET at this time is then operating at the point $V_{GS} = 0$, $V_{DS} = 20$ V, and $I_D = 11.2$ mA by the curves of Fig. 3-39. The total 11.2 mA is flowing into C at the initial instant and none is flowing in R_S. The capacitor charges toward a final value that may be calculated by assuming the capacitor absent. This final value may be found by graphical means from Fig. 3-39. From Fig. 3-43 three different equations involving the currents and voltages may be written

$$V_{DS} = V_{dd} - I_D R_S \qquad (3\text{-}30)$$

$$V_{DS} = V_{dd} - V_1 + V_{GS} \qquad (3\text{-}31)$$

$$v_1 = V_{GS} + I_D R_S \qquad (I_S = -I_D) \qquad (3\text{-}32)$$

The first equation is that of the ordinary load line and is shown in the construction of Fig. 3-44. The second two equations are known as *bias*

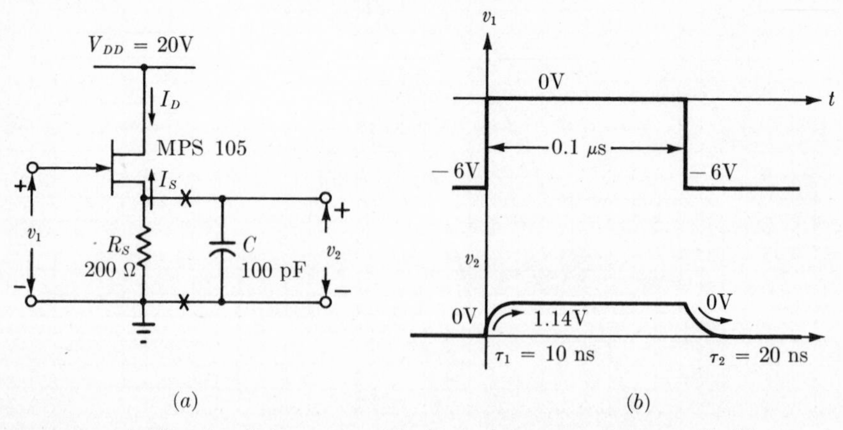

Fig. 3-43 (*a*) The circuit for a capacitively loaded source-follower. (*b*) The input and output waveforms.

lines, but in this particular case only Eq. (3-31) can be usefully plotted as in Fig. 3-44. The intersection of the load line and the bias line (point B) defines the static operating point. $I_D = 5.5$ mA, $V_{DS} = 18.9$ V, and $V_{GS} = -1.1$ V. This V_{GS} is the negative of the final value toward which the capacitor is charging.

To determine the time constant for the capacitor charging an equivalent circuit must be used, but the preceding discussion serves to pinpoint where on the FET characteristics the equivalent circuit should be made. The locus of the operating point begins at point A at $t = 0^+$ (Fig. 3-44) and ultimately ends at point B for $t = \infty$. For this region of operating $g_{fs} = 4.7$ mA/V $= 4700$ $\mu℧$, $I_0 = 11.1$ mA, and $g_0 \approx 0$. (Note that in the case of the FET, I_0 is an important quantity if the total drain current is required.) By placing these quantities in the circuit we arrive at the circuit of Fig. 3-45. This circuit may be easily reduced by splitting it at A-A' and finding the short-circuit current and open-circuit voltage.

$$I_{SC} = g_{fs}V_{GS} + I_0 = g_{fs}v_1 + I_0 \qquad (v_2 = 0) \tag{3-33}$$

$$\left.\begin{aligned} 0 &= V_{GS}g_{fs} + I_0 \\ V_{GS} &= v_1 - v_2 = v_1 - V_{OC} \end{aligned}\right\} \quad (I = 0) \qquad \begin{aligned}(3\text{-}34)\\(3\text{-}35)\end{aligned}$$

$$V_{OC} = \frac{v_1 g_{fs} + I_0}{g_{fs}} = v_1 + \frac{I_0}{g_{fs}} \tag{3-36}$$

$$R_{eq} = \frac{V_{OC}}{I_{SC}} = \frac{1}{g_{fs}} \tag{3-37}$$

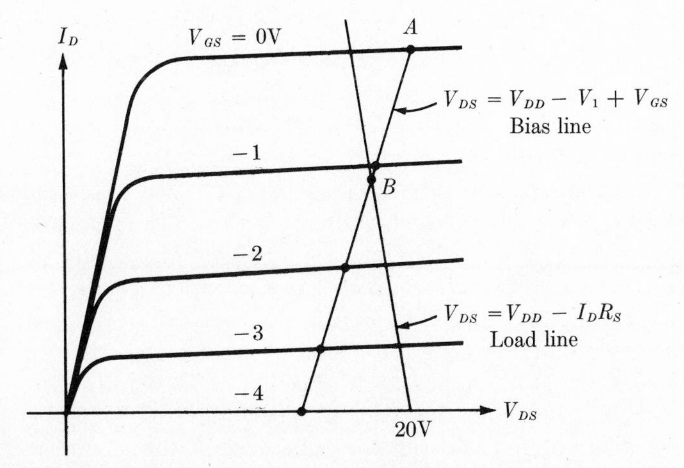

Fig. 3-44 FET curves with bias and load-line constructions shown for the circuit of Fig. 3-43. (The capacitor is assumed either absent or charged.)

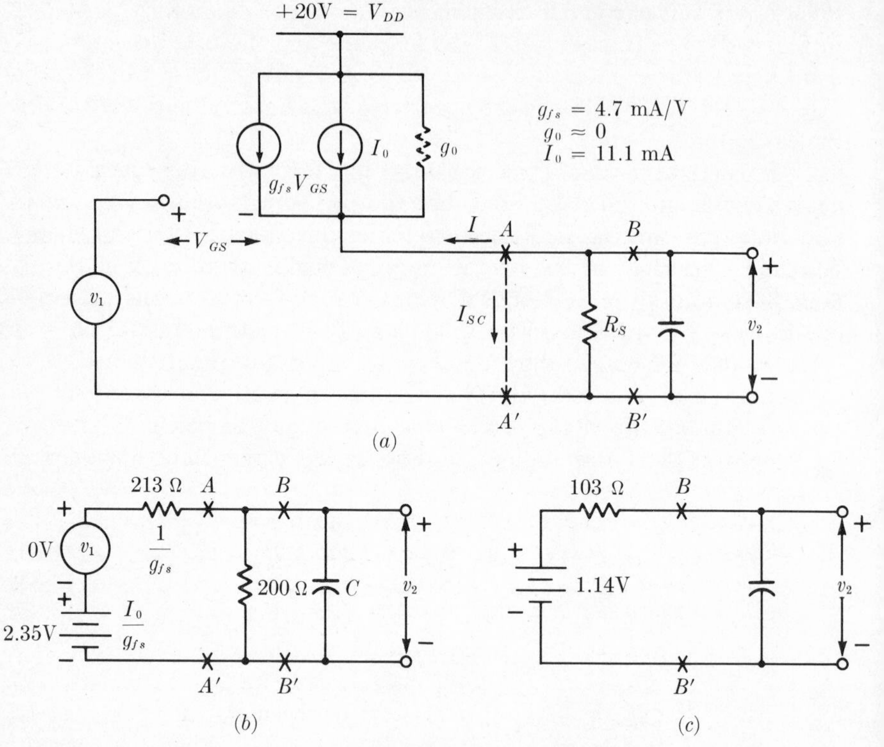

Fig. 3-45 Equivalent circuits for the source follower of Fig. 3-43.

Equations (3-36) and (3-37) for the open-circuit voltage and equivalent output resistance are shown schematically in Fig. 3-45*b*. The source follower is also seen to provide a low-output resistance of about $1/g_{fs}$ in parallel with the load resistance R_S. In Fig. 3-45*c* this paralleling has been done and results in a resistance to drive the capacitor of only 103 ohms or a time constant of about $10^2 \cdot 10^{-10} = 10$ ns. The resulting output waveform is shown in Fig. 3-43*b*, where the leading edge of the output pulse rises toward +1.14 V with $\tau_1 = 10$ ns. At the end of the 0.1-μs input pulse the FET is again cut off with $V_{GS} = -7.14$ V at $t = 0.1^+$ μs, and the capacitor discharges through the 200-ohm resistor toward zero volts with a $\tau_2 = R_SC = 20$ ns. Again, as in the case of the emitter follower driving a capacitative load, the leading and trailing edges of the output pulse have different rise and fall times because of the nonlinear operation of the active device. If both fast rise and fall times are desired, a pair of complementary FET's may be used in a manner much like the transistor complementary pair of Fig. 3-34*a*; however, the FET's will not exhibit the dead zone around zero output from which the transistor pair suffer.

3-14 OTHER TYPES OF FIELD-EFFECT DEVICES

As already shown in Fig. 3-38, there are two basic types of junction FET devices. The one already described in detail is the *n*-channel FET, which has the typical curves shown in Fig. 3-39. The *p*-channel FET is somewhat analogous to the PNP transistor in that the gross characteristics are exactly analogous to those of the *n*-channel device with the signs of all the currents and voltages reversed, i.e., the linear region occurs with V_{DS} negative and V_{GS} positive. Too large a positive V_{GS} will cut off the drain current.

Another type of field-effect device, which appears promising because of the expected ease of making large, complicated, integrated circuit arrays, is the metal-oxide FET shown in cross section in Fig. 3-46. These are abbreviated MOSFET (*m*etal-*o*xide-*s*ilicon FET) or IGFET (*i*nsulated *g*ate FET). In the *n*-channel device, which is rather analogous to the *n*-channel FET previously described, two high-conductivity, *n*-type regions are diffused into a high resistivity *p*-type substrate. Over the intervening region an SiO_2 (quartz) layer is grown which is very thin. On top of this layer an aluminum plate is evaporated, which forms the gate electrode. Application of a voltage between the drain and source results in very little current flow because one or the other of the two *pn* junctions is always reverse biased. The application of a positive voltage

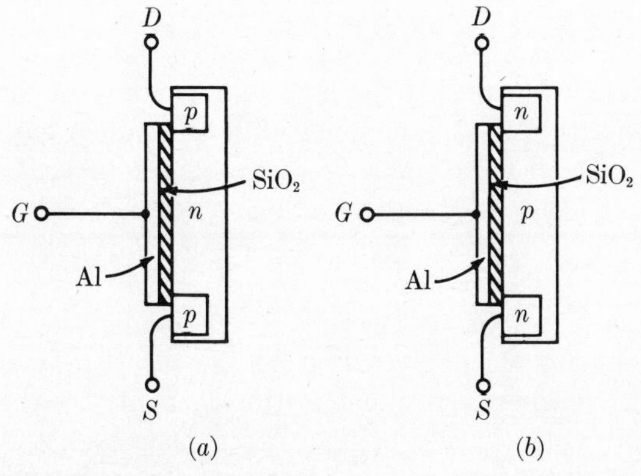

Fig. 3-46 Cross-sectional views of the metal-oxide semiconductor FET. (*a*) *p*-channel device. (*b*) *n*-channel device.

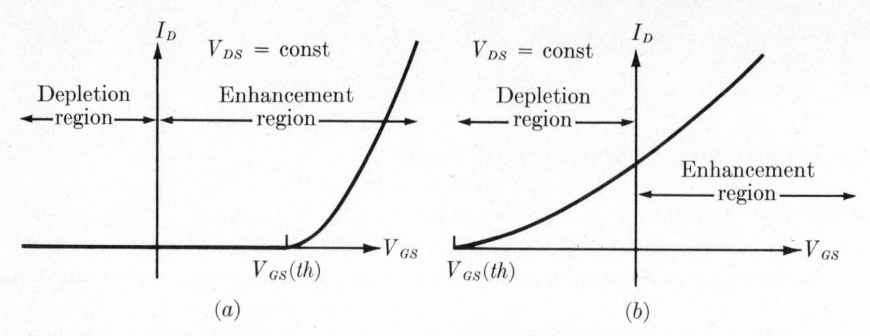

Fig. 3-47 Transfer characteristics of *n*-channel MOSFET's. (*a*) An enhancement mode device. (*b*) A depletion mode device. (Also shown operating in the enhancement mode.)

to the gate terminal establishes a very high field between the gate and *p*-type substrate in the SiO_2 dielectric, which causes the surface of the *p* region to become an *n*-type semiconductor. In the thin *n*-type region or channel thus formed current can flow from the drain to the source. The amount of this current is controlled by the gate voltage, which in turn controls the conductivity of the channel. The major advantage of the device is that the high input impedance is retained for either a positive or negative input voltage, since the gate is insulated from the rest of the structure by the SiO_2 layer. This behavior contrasts with the junction FET where the gate is a reverse-biased junction for one input polarity, but a forward-biased junction for the other input polarity.

Several types of devices are possible with an *n*-channel. In the device described there is virtually zero conduction with zero gate voltage, and the channel is formed by applying a positive gate voltage. Such a mode of operation is known as an *enhancement*-mode device because conduction requires increasing the channel conductivity, i.e., enhancing the conductivity. A second type of device is formed if the layer immediately under the gate is fabricated to be a thin *n*-type channel with no applied gate voltage. The gate voltage applied is negative and reduces the conductivity of the channel by depleting the thin *n*-type region. Such a device is called a *depletion*-mode device. A third type of device is sometimes described which operates in both modes. The easiest way to see the differences between the two basic types is to look at the transfer characteristics of Fig. 3-47. In the enhancement-mode device (Fig. 3-47*a*) the threshold $V_{GS}(th)$ is several volts positive, whereas in the depletion-mode device the threshold voltage is several volts negative. The drain characteristics shown in Fig. 3-48 for both types of device illustrate that the *n*-channel depletion-mode IGFET has characteristics very similar to the *n*-channel FET. The equivalent circuit for the latter device is adequate for

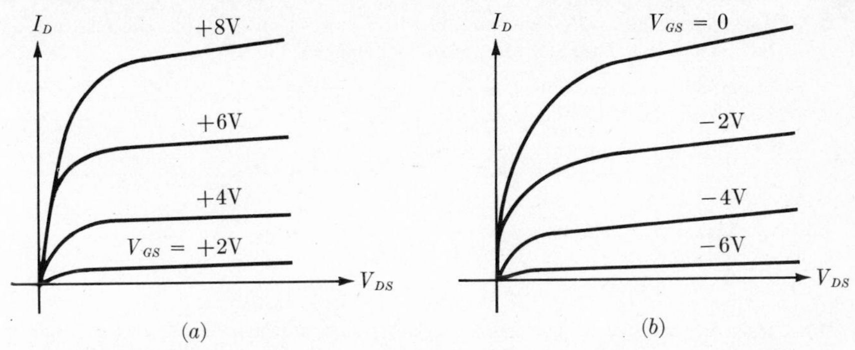

Fig. 3-48 Characteristics of typical n-channel insulated-gate FET's. (a) The enhancement-mode device. (b) The depletion-mode device.

either type of IGFET, providing proper values for the parameters are used.

As in the case of the junction type of FET a complementary type of IGFET also exists. For these devices the channel is of p-type material, and the applied polarities are the reverse of those for an n-channel device.

PROBLEMS

3-1. The waveshaping circuit shown in the figure is supposed to produce the output waveform v_2 from an input triangular waveform. (a) Choose the values of R_1, R_2, V_1, and V_2 to give the desired waveform. (b) Draw a graph giving v_2 as a function of v_1. Also sketch the curve obtained by changing from ideal diodes to the 1N4152 silicon diode.

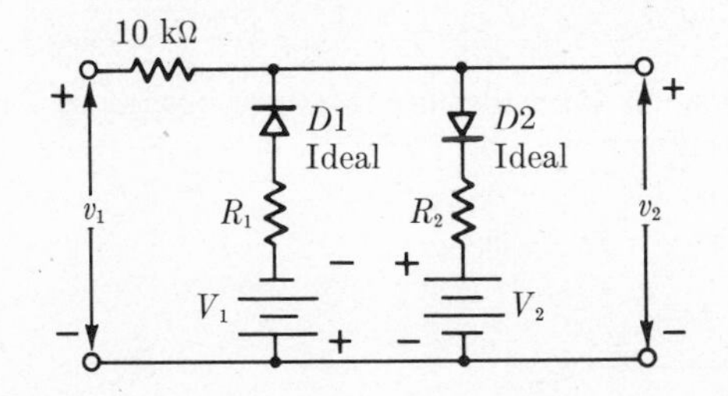

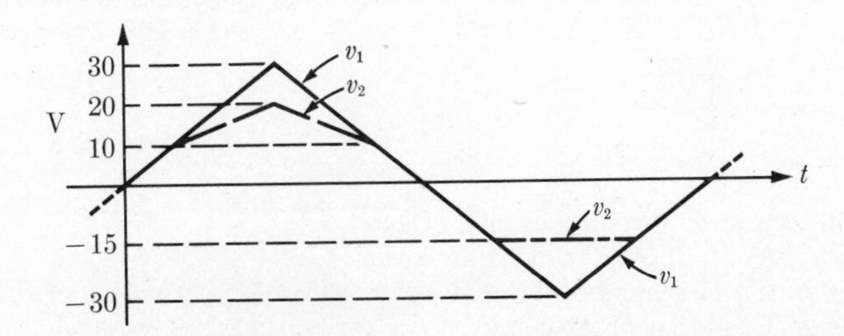

3-2. Two zener diodes are connected in series opposition. What is the relationship between V and I for the pair? Sketch the curve.

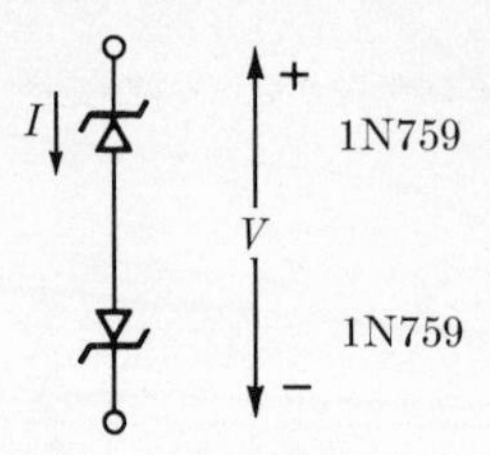

3-3. Plot the transfer characteristic of the circuit shown for $-20 < v_1 < +20$ V. Plot three curves: one for the ideal diode shown, one for a 1N96 diode, and one for the 1N759 diode.

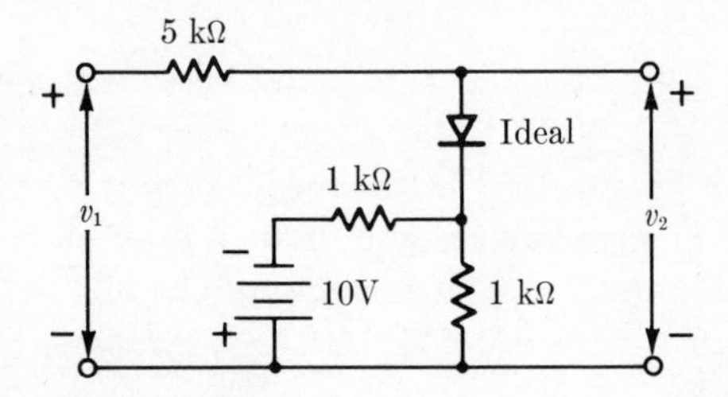

3-4. The transistor amplifier shown in the figure becomes a waveshaping circuit when overdriven. Several different results (two of which will be calculated) are possible depending upon circuit values.

(*a*) Let $R_b = 50$ kΩ and $R_e = 0$. Sketch and carefully label the transfer characteristic (v_2 vs. v_1) for $10 < v_1 < 10$ V. Sketch the output for an input $v_1 = 10 \sin(2\pi)100t$.

(*b*) Let $R_b = 1$ kΩ and $R_e = 2$ kΩ and repeat the preceding.
(NOTE: The fact that emitter-base breakdown occurs for very negative input voltages will not affect the output from the collector. The waveform across R_e is interesting to compute for the (*b*) part because of the breakdown effect.)

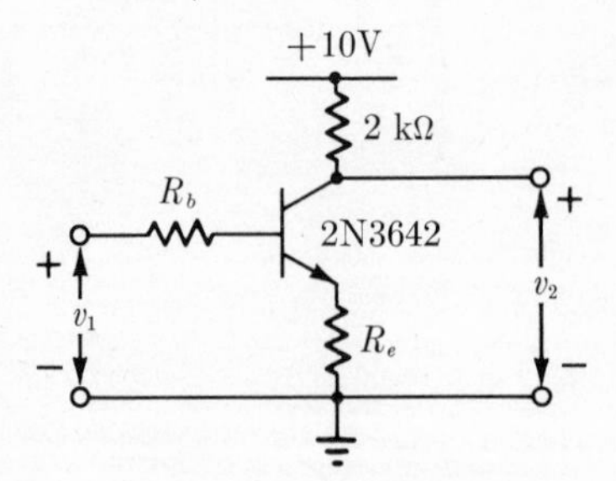

3-5. The circuit is a form of resistor-transistor gate.

(*a*) Sketch the output from the gate as shown.

(*b*) What is the maximum value of R_1 that will cause no output with the presence of either v_1 or v_2 alone?

(c) What is the minimum value of R_1 that will cause the transistor to saturate when a pulse is present on both inputs?

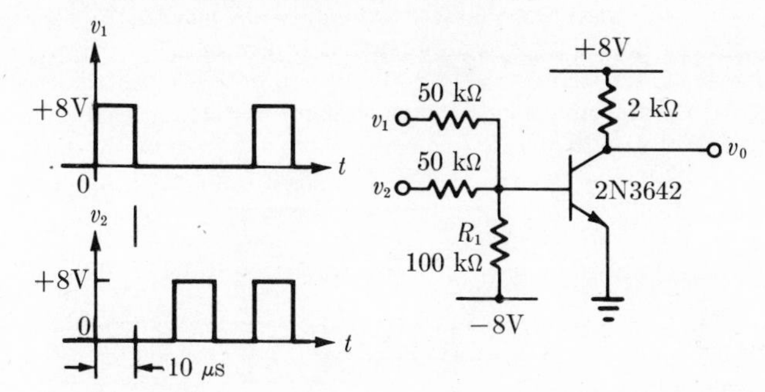

3-6. (a) Solve for $v_2(0^-)$ and $v_2(\infty)$ graphically (without an equivalent circuit).
(b) Find values for I_0, h_{FE}, r_{bII}, and V_{bII} suitable for the operating region found in (a).
(c) Find the Thevenin equivalent looking into the transistor at terminals $a - a'$. Draw the equivalent circuit giving numerical values for the elements. Regard v_1 as a variable source.
(d) What is the voltage gain when the load is connected? (Disregard the capacitor.)
(e) Using the input waveform shown, calculate and sketch the output waveform, $v_2(t)$.
(f) Why is there no discontinuity in slope in $v_2(t)$ when the transistor resumes conducting even though there is a radical change in time constant?

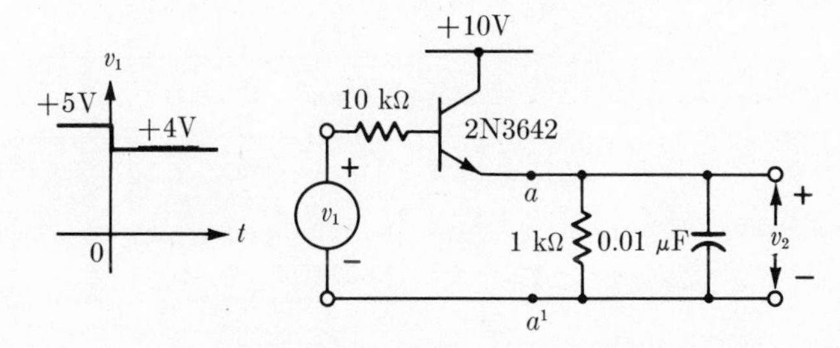

3-7. Draw a circuit using only ideal diodes, resistors, voltage and current sources that will give the V-I diagrams shown. Note that the answer is not unique, and many solutions are possible.

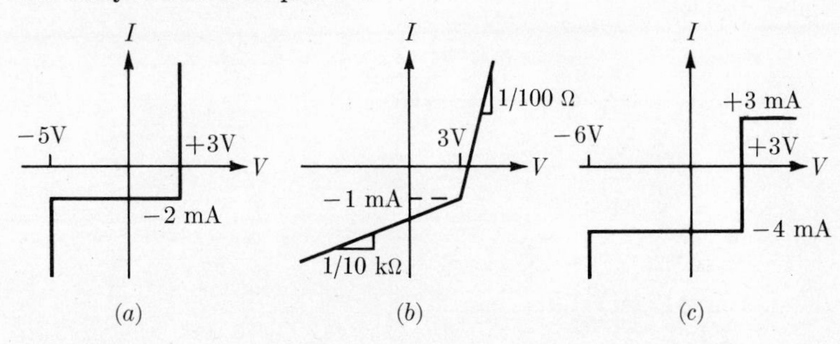

3-8. The transistor circuit shown develops a moderately linear voltage ramp at $t = 0$.
(*a*) Find and sketch the waveform of $v_2(t)$. (The transistor is a 2N3134.)
(*b*) On a sketch of the transistor characteristics draw the locus of the operating point indicating the time for each important transition.
(*c*) What simple circuit change would you make to decrease the value of the recovery time of the circuit by a factor of about 5? (Do not change $v_2(t)$ in the interval $0 < t < 20\ \mu$s.)

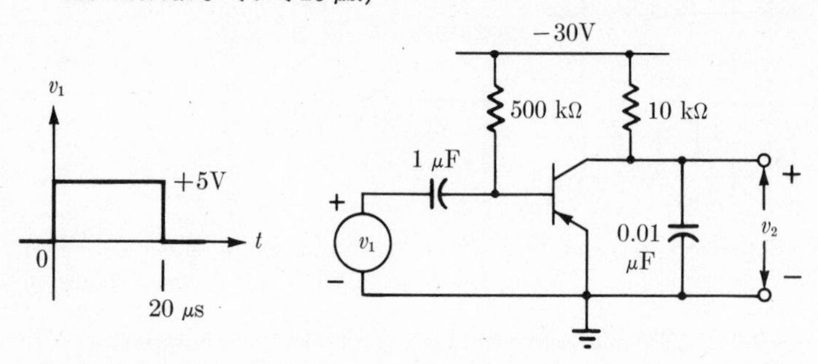

3-9. The function of the circuit is to produce a pulse at the output delayed in time from the input step voltage.
(*a*) First calculate and sketch v_{b1} and v_{c1}. Include an estimate on the fall-time of v_{c1}.
(*b*) Now calculate v_{b2} and the output. When sketching be sure to show the time reference for each waveform to show clearly the delays produced by the circuit.

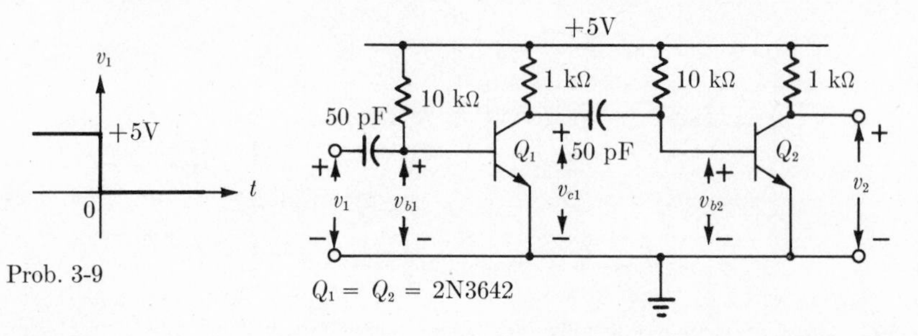

Prob. 3-9

3-10. Find the volt-ampere characteristic at the terminals of the circuit. Sketch and label completely.

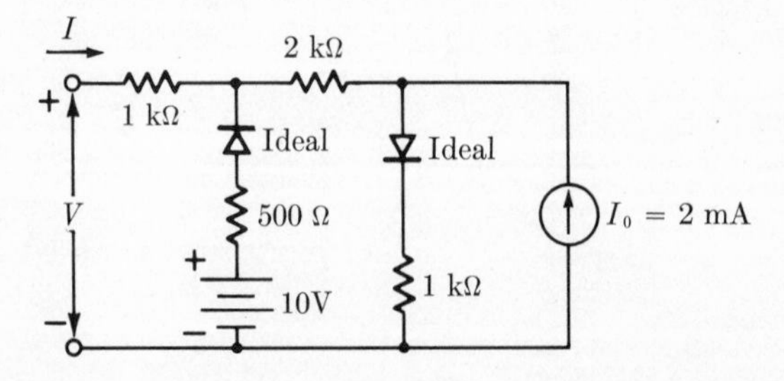

3-11. The circuit shown on the left has the given conditions at $t = 0^-$. At $t = 0$ the circuit is connected to the terminals of the circuit in Prob. 3-7*a*. (NOTE: you do not need to solve for the actual circuit to solve the problem).

(*a*) Find the waveform for the voltage $V(t)$. Sketch and label.

(*b*) On a sketch of the V-I characteristic (Prob. 3-7*a*) show the locus of operation, give the time each inflection point is reached, and the operating point as $t \to \infty$.

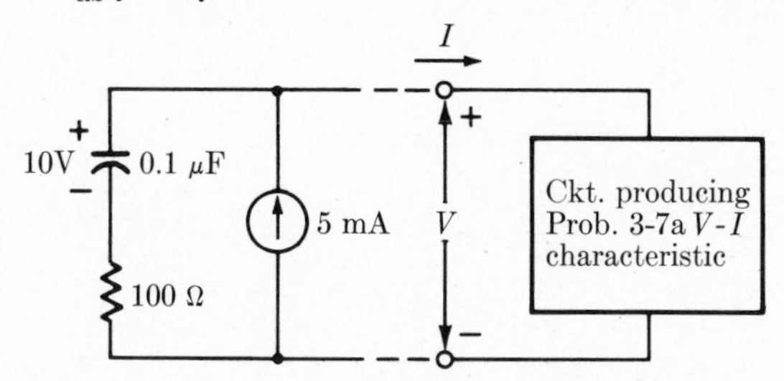

3-12. Plot the transfer characteristic for the circuit shown, i.e., v_2 as a function of v_1.

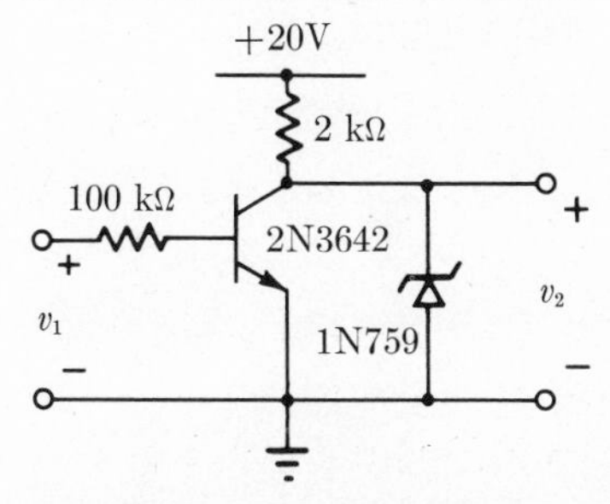

3-13. Sometimes a simple, nearly perfect current source is required. A transistor connected as shown provides this over a limited range of load voltage V.

(*a*) Over what range of V will the current be substantially constant? (This is the same range of V that keeps the transistor in Region II.)

(*b*) Using approximate methods, what is I for $V = 5$ V?

(*c*) To see how good a current source the circuit is, replace the transistor by its Region II, common-emitter equivalent circuit. Then reduce the circuit to the left of (*a*) to a Norton equivalent. (The parallel resistance you find should be of the order of a megohm and the value of the current generator should closely agree with (*b*).)

(*d*) By what percent does I change for V changing from 0 to 10 V?

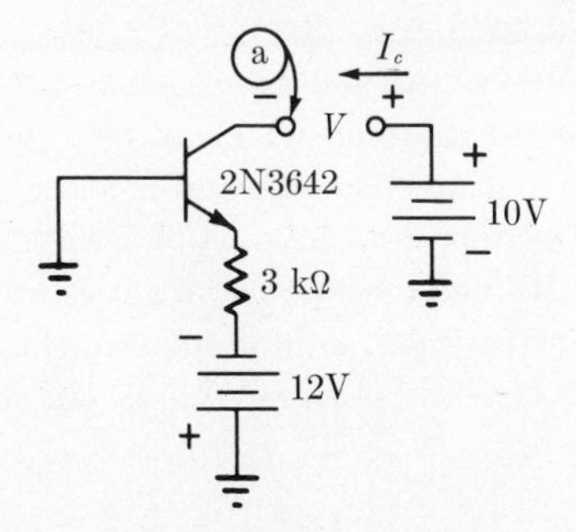

4

Some Examples of *RC* Circuits Incorporating Active Devices

We are now ready to embark upon a series of case studies of circuits that are useful not only in their own right but also because they provide an opportunity to exercise some of the techniques presented in the preceding chapters.

4-1 A CLAMPING OR dc RESTORER CIRCUIT

For some purposes it is desirable to reference one extreme of a waveform to a particular dc level, e.g., a square wave may be desired which has a maximum positive value of zero volts, no matter what amplitude the square wave assumes. A circuit which accomplishes this end is sometimes called a *clamping* circuit because it holds or clamps one part of a waveform to the desired level. The same circuit is utilized on television waveforms to clamp the blanking pulse in the composite video signal to zero, thereby restoring the original dc reference level to the video signal; hence the name for the circuit used in this way: dc restorer.

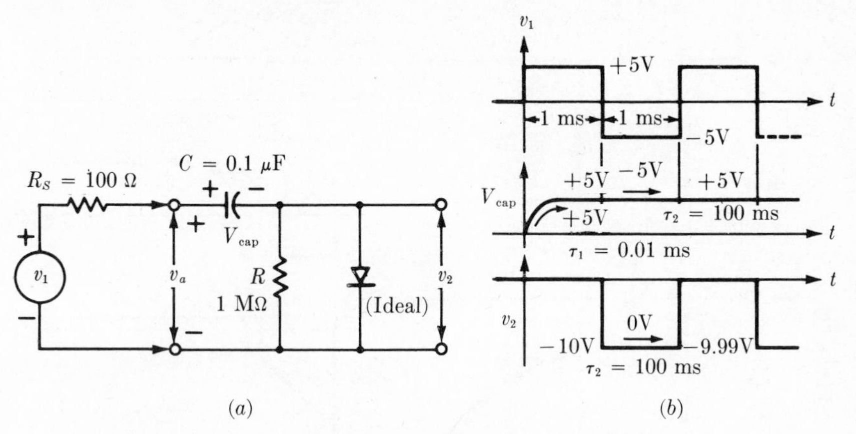

Fig. 4-1 A clamping or dc restorer circuit. (*a*) Circuit. (*b*) Waveforms.

The circuit shown in Fig. 4-1 is similar to an *RC* coupling circuit with a diode added across *R*. The input to the circuit is a square wave starting at $t = 0$ with equal positive and negative portions. (The square wave could have any dc reference level, and the operation of the circuit would be the same.) In the first half cycle of the square wave the diode acts as a short circuit preventing the output from going positive, and charging the capacitor to essentially the full value of v_1 (+5 V) with a time constant $\tau_1 = (100\ \Omega)(0.1\ \mu\text{F}) = 0.01$ ms. At the end of 1 ms both the input and output go negative, and the diode is reverse biased (an open circuit). The capacitor now begins to charge toward −5 V, but with a much longer time constant: $\tau_2 = (1\ \text{M}\Omega)(0.1\ \mu\text{F}) = 100$ ms. The capacitor cannot discharge appreciably in the 1 ms time, so that the output V_2 is maintained almost constant at −10 V (the voltage at $t = 2$ ms actually will be −9.99 V). In the following cycle the small amount of charge lost by the capacitor will be regained in a few microseconds, and the waveforms will essentially repeat those of the first cycle. If the input waveforms were suddenly made larger, the circuit would continue to hold the positive excursion at zero, and the increased amplitude would only increase the negative excursions. On the other hand, a decrease in amplitude leads to an output all negative with respect to ground. The whole waveform then slowly (approximately with time constant τ_2) drifts upwards until the positive portion is again clamped to ground potential.

To study the operation of such circuits in more detail and with non-ideal elements, let us consider the clamping circuit of Fig. 4-2. In this circuit the high-resistance resistor has been replaced by a current source both as a useful circuit element and to give practice with using this relatively

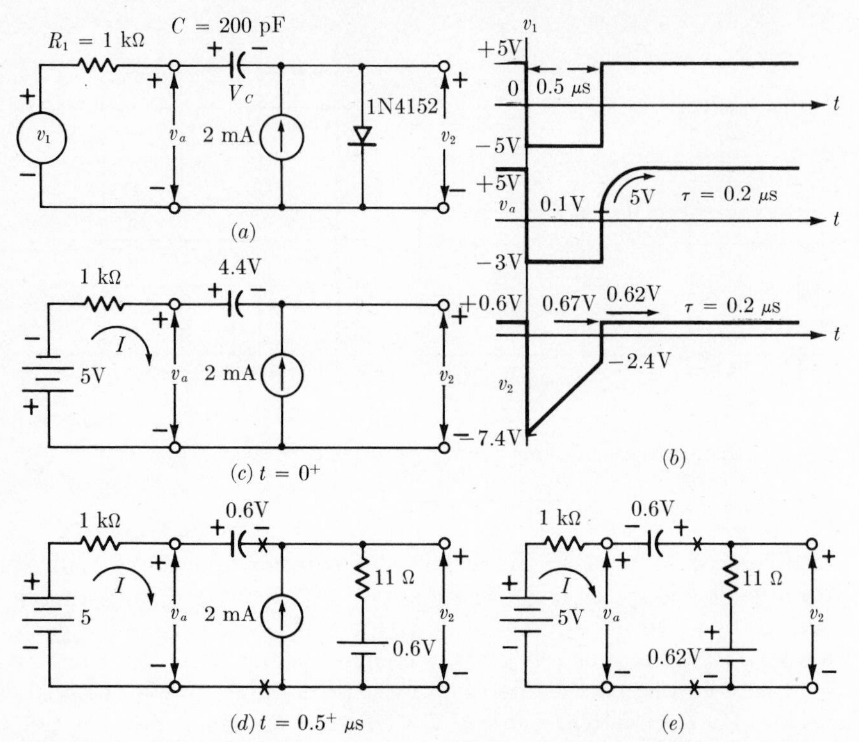

Fig. 4-2 (*a*) A diode clamping circuit. (*b*) The waveforms of the circuit. (*c*), (*d*), (*e*) Equivalent circuits at different instants in time.

unfamiliar element. The time constant has also been shortened so that the effects on the waveform can be more easily seen.

To begin with, note that the circuit has been at rest an infinitely long time so that the capacitor current must be zero; therefore $v_a(0^-)$ is $+5$ V, and the 2 mA is flowing through the diode. From the diode curve $v_2 = 0.6$ V. The capacitor is initially charged to the voltage

$$V_C = v_a - v_2 = 5 - 0.6 = 4.4 \text{ V}$$

These initial voltages are indicated in Fig. 4-2*b*, where the three waveforms of interest are shown one above another so that they may be easily compared and checked. At $t = 0$ the pulse voltage changes from $+5$ to -5 V and the circuit conditions are as shown in Fig. 4-2*c*. The capacitor charge stays constant during the switching, as shown, and the diode current decreases, but as yet we do not know for certain that it turns off completely. We shall assume that it does, and test the assumption. The current shown as I must be equal to the -2 mA through the ideal current generator. Therefore at $t = 0^+$,

$$v_a = -5 - 1\ \text{k}\Omega(-2\ \text{mA}) = -5 + 2 = -3\ \text{V}$$

Note: It is often convenient to calculate voltages as $\text{k}\Omega \times \text{mA} = \text{V}$, and in following calculations kΩ will be indicated and current assumed in mA.

$$v_2 = v_a - 4.4 = -3 - 4.4 = -7.4\ \text{V}$$

The diode is truly turned off as was supposed. Because there is a current source in the series circuit, the time constant is infinite, and other means of finding the circuit behavior must be used. Since $i(t)$ is a constant immediately after $t = 0$, the change in V_C is

$$V_C(t) = V_C(0^-) + \int_0^t \frac{i\,dt}{C} = V_C(0^-) + \frac{it}{C} \tag{4-1}$$

$$\frac{i}{C} = \frac{2 \times 10^{-3}}{2 \times 10^{-10}} = 10^7\ \text{V/s} = 10\ \text{V}/\mu\text{s}$$

From Eq. (4-1) the capacitor voltage is seen to be changing linearly with time, and at $t = 0.5\ \mu$s is $4.4 - (0.5\ \mu\text{s})(10\ \text{V}/\mu\text{s}) = -0.6$ V. The output voltage at this time is

$$v_2 = -5 + 2(1\ \text{k}\Omega) + 0.6 = -2.4\ \text{V} \tag{4-2}$$

For the second phase of the circuit operation for $t > 0.5\ \mu$s the equivalent circuit is as shown in Fig. 4-2d with the equivalent circuit for the conducting diode and the capacitor shown charged as computed for $t = 0.5^-\ \mu$s. In order to make a simple RC series circuit, the portion to the right of x-x is replaced by its Thevenin equivalent, as shown in Fig. 4-2e. From this circuit the current I and the desired voltages may be easily computed.

$$I = \left(\frac{5 + 0.6 - 0.62}{1.011\ \text{k}\Omega}\right)\epsilon^{-(t-0.5)/\tau} \qquad \begin{aligned}\tau &= (1.011\ \text{k}\Omega)C \\ &= 0.202\ \mu\text{s}\end{aligned} \tag{4-3}$$

$$= 4.9\epsilon^{-(t-0.5)/0.2} \qquad (t \text{ in } \mu\text{s}) \tag{4-4}$$

$$\begin{aligned} v_2(t) &= 0.62 + (0.011\ \text{k}\Omega)(4.9\epsilon^{-(t-0.5)/0.2}) \\ v_2(0.5^+) &= 0.62 + 0.054 = 0.67\ \text{V} \\ v_2(\infty) &= 0.62\ \text{V} \end{aligned} \tag{4-5}$$

$$\begin{aligned} v_a(t) &= 5 - (1\ \text{k}\Omega)(4.9\epsilon^{-(t-0.5)/0.2}) \\ v_a(0.5^+) &= 5 - 4.9 = 0.1\ \text{V} \\ v_a(\infty) &= 5\ \text{V} \end{aligned} \tag{4-6}$$

[The notation $v(\infty)$ denotes the voltage approached for $t \gg \tau$.] These values are used to plot the waveform in Fig. 4-2b. Note that three important values are given for each exponential portion of the waveform: the initial value, the final value, and the time constant. With these three items of information no additional points need be indicated. The output of the circuit v_2 is also a pulse but with the baseline fixed in the region of 0.6 V. The pulse "top" has been distorted by the discharge of the capacitor, but could be made as flat as desired by increasing C. The circuit "recovers" or returns to its original set of conditions in about 5τ or one microsecond after the end of the pulse. At this time another pulse could be applied, and an essentially identical output waveform to that obtained for the first pulse would be obtained. Note that $v_2(0^-) = 0.6$ V and $v_2(\infty) = 0.62$ V. These values should be identical and differ because the first value was obtained directly from the diode curve, whereas the second came from the equivalent circuit chosen. The error is negligible, but can be eliminated by picking an equivalent representation for the diode that intersects the diode characteristic at the static operating point ($I_d = 2$ mA).

An important point to notice here is that perfectly satisfactory results could have been obtained with the diode resistance chosen to be zero as in Fig. 3-2*c*. The only noticeable difference would be the omission of the exponential in v_2 going from 0.67 to 0.62 V. The time constant, recovery time, and v_a waveform would be nearly identical, but the analysis would be somewhat simpler.

4-2 DECOUPLING FILTER

The first of these circuits is the so-called "decoupling filter," an *RC* circuit which provides the dc voltage to the collector of an amplifier or switching transistor, for instance, in such a way that the voltage fluctuations of the collector of this transistor do not produce variations in other portions of a circuit supplied from the same dc source. Such variations in other parts of the circuit occur, of course, because the power source has a nonzero internal resistance (impedance in the ac case). An example of such a circuit is shown in Fig. 4-3. The power supply-internal resistance is considered to be lumped in with a series resistor, the combined value being 3 kΩ. The objective of the circuit is to keep the voltage v_2 as nearly constant as possible. The collector voltage V_{CE} will swing up and down in accordance with the signal, but these swings should not be transmitted back to the supply to interfere with other circuits.

Presumably the circuit has been designed to do its job well. For our first approximation, let us assume it does the job perfectly, i.e., that the capacitor voltage v_2 does not change during the cycle of operation once the steady-state values of voltage and current are reached. We shall assume the input is a periodic pulse waveform which was applied long

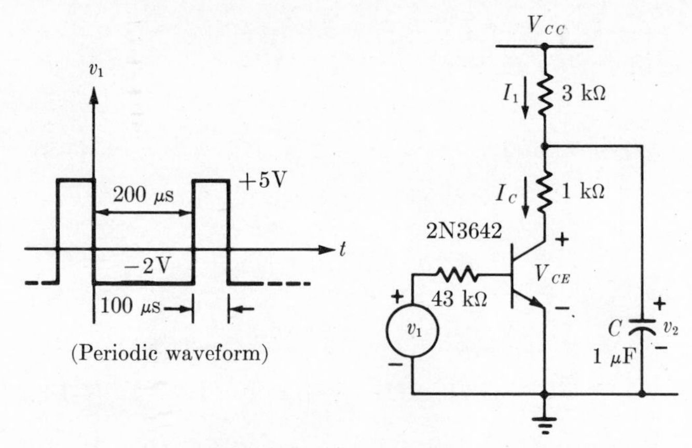

Fig. 4-3 A decoupling filter (the 3-kΩ resistor and capacitor) driven by a periodically operated transistor switch.

before $t = 0$. For the approximation v_2 = const to be good the time constants associated with charging and discharging the capacitor must be very large compared with the periods of charging and discharging. The voltage v_2 becomes V_2, a constant, and there is a constant current flowing through the 3 kΩ resistor, which we shall call $I_1 = (V_{CC} - V_2)/3$ kΩ. During the cycle the circuits of Fig. 4-4 are appropriate.

None of the currents or voltages is known. We are primarily interested in V_2 and V_{CE}. The principle that we shall apply in solving for the circuit values is that the charge flowing onto the capacitor during the charging interval must equal the charge flowing off during the discharge time. If this were not true, the voltage V_2 would be different at the end of the cycle from what it was at the beginning. Another way of stating this

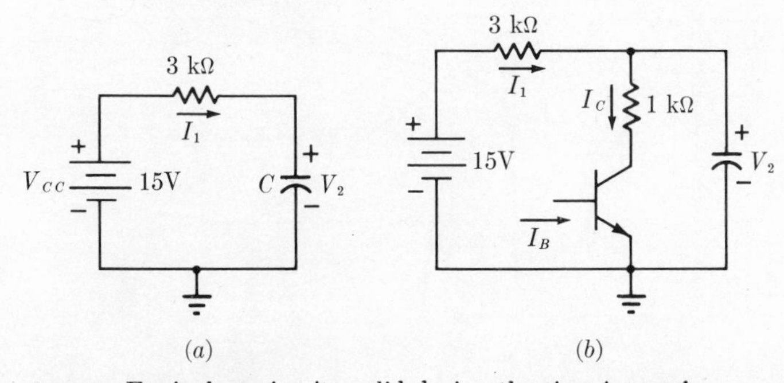

Fig. 4-4 Equivalent circuits valid during the time intervals (*a*) when the transistor is OFF and (*b*) when the transistor is ON.

principle is that when steady conditions have been established in a circuit, the average current through a capacitor is zero.

$$Q_{\text{charge}} = Q_{\text{discharge}} \tag{4-7}$$

$$I_1 \times 200\ \mu\text{s} = (I_C - I_1)100\ \mu\text{s} \tag{4-8}$$

From which

$$I_C = 3I_1 \tag{4-9}$$

The collector voltage V_{CE} is then

$$V_{CE} = V_{CC} - 3\ \text{k}\Omega(I_1) - 1\ \text{k}\Omega(I_C)$$

$$= V_{CC} - \frac{3\ \text{k}\Omega(I_C)}{3} - 1\ \text{k}\Omega(I_C) = V_{CC} - 2\ \text{k}\Omega(I_C) \tag{4-10}$$

Equation (4-10) is shown schematically in Fig. 4-5*a*, where the collector voltage is shown to be the same as found by considering I_C to be the current flowing through 2 kΩ with no capacitor present. This is, of course, true only when the transistor is conducting. The operating point is then on the 2 kΩ load line as shown in Fig. 4-5*b*, and the base current will determine where on this line the actual operating point is. The base current is given by

$$I_B = \frac{+5 - V_{BE}}{43\ \text{k}\Omega} = 100\mu\text{A} \tag{4-11}$$

and the actual operating point during the 100 μs conduction period is $I_C = 5.1$ mA and $V_{CE} = 4.75$ V.

The capacitor voltage may be found in two ways:

$$V_2 = V_{CC} - \frac{I_C(3\ \text{k}\Omega)}{3} \quad \text{or} \quad V_2 = V_{CE} + I_C(1\ \text{k}\Omega) = 9.9\ \text{V} \tag{4-12}$$

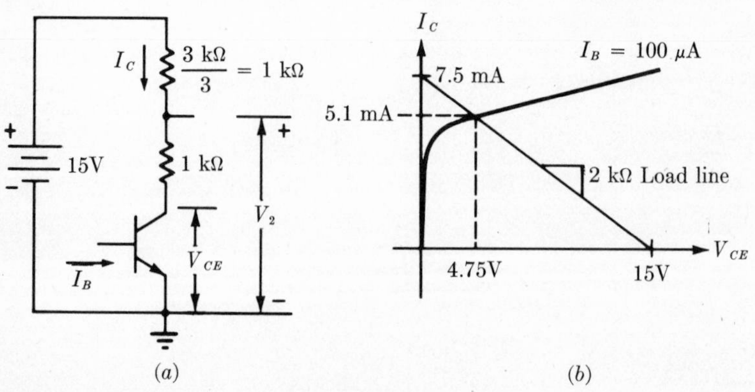

Fig. 4-5 (*a*) Equivalent circuit valid for Fig. 4-1 when the transistor is conducting. (*b*) Load line for the circuit of Fig. 4-3*a*.

In actual fact the capacitor voltage does vary a slight amount during the cycle of operation, but this amount can be made as small as desired by choosing the size of the capacitor properly. In this case the variation of the capacitor voltage is easily found by dividing the change in charge during a part of the cycle by the capacitance such that

$$\begin{aligned}\Delta V_2 &= \frac{\Delta Q}{C} = \left(\frac{I_C}{3}\right)\left(\frac{200\ \mu\text{s}}{C}\right) = (I_C - I_1)\left(\frac{100\ \mu\text{s}}{C}\right)\\ &= \frac{(5.1 \times 10^{-3})(2 \times 10^{-4})}{(3)(10^{-6})} = \frac{(5.1 - 1.7)(10^{-3})(10^{-4})}{10^{-6}}\\ &= 0.34\ \text{V} \end{aligned} \tag{4-13}$$

The maximum and minimum values of V_2 are then

$$V_2 \quad \min = 9.9 - 0.34/2 = 9.73\ \text{V}$$
$$V_2 \quad \max = 9.9 + 0.34/2 = 10.07\ \text{V}$$

The collector voltage is approximately $V_2 - (I_C)(1\ \text{k}\Omega)$ and has about the same variation in voltage during the pulse as occurs on the capacitor. Since the variation in capacitor voltage during the pulse is relatively small, the current I_1 taken from the power supply is nearly constant as was desired.

This example is representative of a class of capacitor charge and discharge problems in which the capacitor is never really completely charged or discharged because the times allowed are insufficient compared to the time constants involved. A more exact solution than the preceding may be made using equivalent circuits and the generalized waveform of Fig. 4-6. In this figure the waveform sought is the combination of two exponentials with time constants τ_1 and τ_2 as shown. The time T_1 (or T_2)

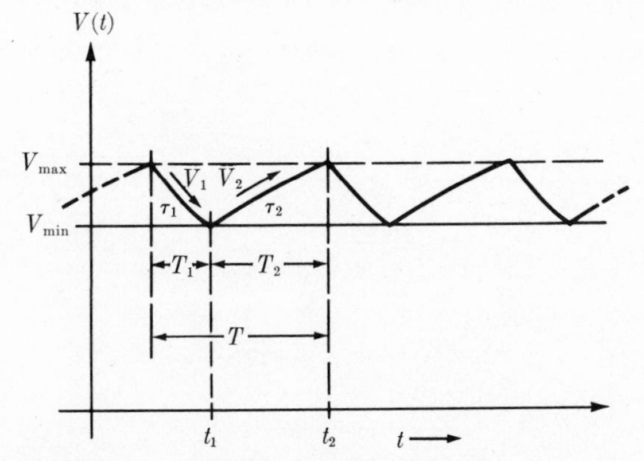

Fig 4-6 A periodic waveform caused by alternately charging and discharging a capacitor. Both $\tau_1 > T_1$ and $\tau_2 > T_2$ so that $V(t)$ never reaches the final values V_1 and V_2, respectively.

may be so long compared to τ_1 that the waveform is appreciably curved, or as in the case of the previous example, T_1 may be short compared to τ_1 so that the waveform is nearly straight. The voltage waveform in terms of the constants shown in Fig. 4-6 for the discharging portion of the cycle T_1 is

$$V(t) = V_1 + (V_{\max} - V_1)\epsilon^{-t/\tau_1} \tag{4-14}$$

At the time t_1 the voltage $V(t)$ will reach the minimum value

$$V(t_1) = V_1 + (V_{\max} - V_1)\epsilon^{-T_1/\tau_1} = V_{\min} \tag{4-15}$$

A similar equation for $V(t)$ may be written, which is valid during the charging portion of the period T_2, and again this voltage can be equated to the maximum voltage $V_{\max}$

$$V(t_2) = V_2 + (V_{\min} - V_2)\epsilon^{-T_2/\tau_2} = V_{\max} \tag{4-16}$$

Equations (4-15) and (4-16) are simultaneous equations for the voltage $V(t)$ and contain as unknowns only the minimum and maximum values, $V_{\min}$ and $V_{\max}$. Solving the two equations for $V_{\min}$ and $V_{\max}$ gives

$$V_{\max} = \frac{V_1\epsilon^{-T_2/\tau_2}(1 - \epsilon^{-T_1/\tau_1}) + V_2(1 - \epsilon^{-T_2/\tau_2})}{1 - \epsilon^{-T_1/\tau_1}\epsilon^{-T_2/\tau_2}} \tag{4-17}$$

$$V_{\min} = \frac{V_2\epsilon^{-T_1/\tau_1}(1 - \epsilon^{-T_2/\tau_2}) + V_1(1 - \epsilon^{-T_1/\tau_1})}{1 - \epsilon^{-T_1/\tau_1}\epsilon^{-T_2/\tau_2}} \tag{4-18}$$

The equivalent circuits of Fig. 4-7 may be used to determine the necessary final values V_1 and V_2 and the time constants τ_1 and τ_2. The

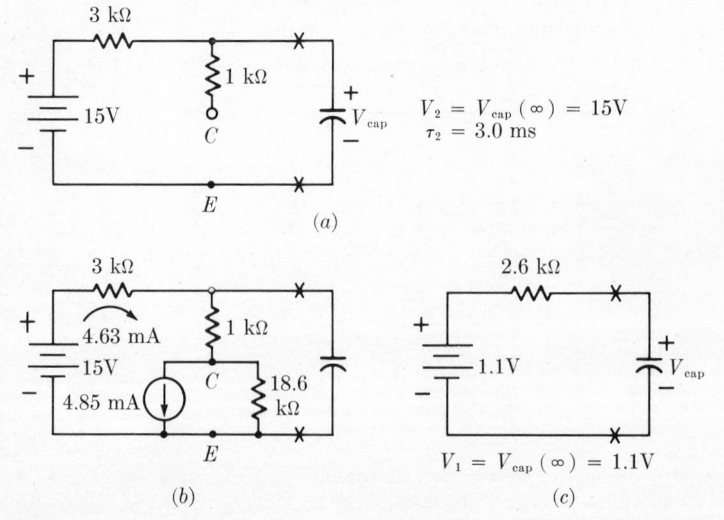

Fig. 4-7 Equivalent circuits for the decoupling circuit. (*a*) Circuit valid during T_2. (*b*) Circuit valid during T_1. (*c*) The Thevenin equivalent to (*b*).

circuit of Fig. 4-7*a* represents the conditions in Fig. 4-3 when the transistor is nonconducting and the capacitor is charging. The necessary values $V_2 = 15$ V and $\tau_2 = 3$ ms may be obtained by inspection from the equivalent circuit. The circuit of Fig. 4-7*b* represents conditions during the capacitor discharge time, and the appropriate values for the transistor are found from the knowledge of the operating point obtained from the previous solution. The equivalent circuit must be reduced to the Thevenin equivalent of Fig. 4-7*c* from which the values $V_1 = 1.1$ V and $\tau_1 = 2.6$ ms are obtained. These values may then be substituted into Eqs. (4-17) and (4-18) to give $V_{max} = 10.08$ V and $V_{min} = 9.71$ V. These values agree closely with those obtained from the previous approximate solution. To get this degree of agreement, however, the *exact* calculations had to be carried out to five significant figures to keep the small differences $(1 - \epsilon^{-t/\tau})$, and so on, accurate enough for a two or three significant figure final answer. Therefore, the *approximate* approach is much more satisfactory if the fractions of the time constants used are 0.1 or 0.2 or less.

The exact method of solving two exponentials just made will work in any similar *RC*-circuit problem involving steady-state conditions as long as the necessary equivalent circuits can be obtained, and if only two exponentials are involved. In the case of the problem just solved, the solution could not have been obtained in the previous manner if the capacitor was made so small that the large excursions in collector voltage caused the transistor to go into saturation for a portion of t_1. The type of solution just used we shall call *Case I*. This solution is necessary if neither exponential continues to essentially its final value. Whether the approximate or exact method of solution is used for Case I depends upon the proportion of the time constant used. For situations using only 10 or 20 percent of the time constant, the approximate solution is much preferred.

Two other cases of periodic operation of *RC* circuits arise and can be solved much more easily. *Case II* is the situation where one of the time constants is much shorter than the period the time constant is effective. Such a case can be easily solved by starting the solution at the end of such a period, because the initial conditions for the subsequent period will be known. *Case III* is the situation where both time constants are shorter than the appropriate periods, and the solution can be begun at the beginning of either period.

4-3 *RC* COUPLING CIRCUIT

An example which can illustrate all three cases is Fig. 4-8. Here is a simple *RC* coupling circuit driven by a transistor switch, in turn driven by a 1-kHz square wave. Because of the sizes of the resistors associated with the transistor, a very good approximation is to consider the transistor an ideal

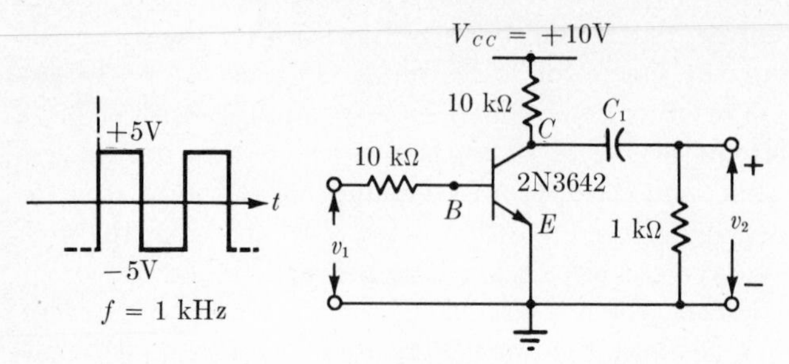

Fig. 4-8 Periodically driven transistor switch driving an RC coupling circuit.

switch. The equivalent circuits for the conducting and nonconducting periods are shown in Fig. 4-9*a* and *b*, respectively. The different cases of interest are obtained by using an appropriate value for the capacitor. The three cases will be treated in reverse order starting with the simplest.

Case III. For this case both τ_1 and τ_2 need to be small compared with the 500-μs half period. Let $C_1 = 0.005$ μF giving $\tau_1 = 5$ μs, and $\tau_2 = 55$ μs. The solution can begin at either $t = 0$ or $t = 500$μs, since the initial conditions at either time can be easily found. At $t = 0^-$ the circuit conditions are as in Fig. 4-9*b* with the capacitor completely charged to $V_C = 10$ V since $500/55 = 9.1$ time constants have been allowed for it to charge. The

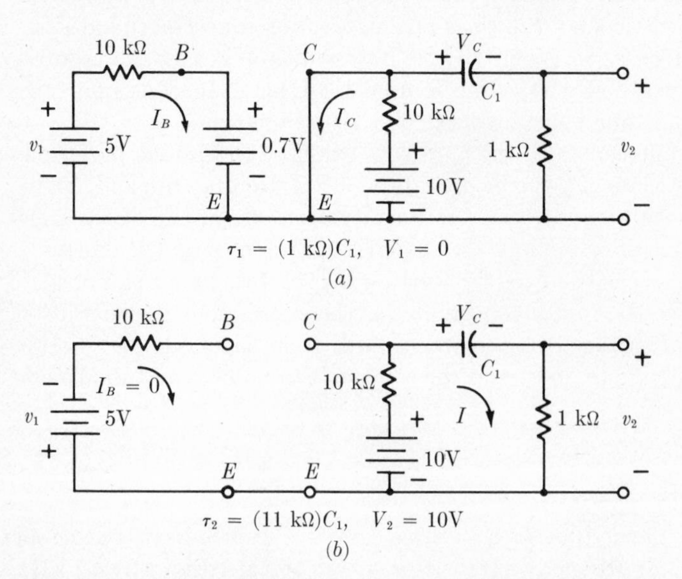

Fig. 4-9 Equivalent circuits for the two halves of a typical period: V_1 and V_2 are the final values for V_C. (*a*) Switch ON: $0 < t < 500$ μs. (*b*) Switch OFF: $500 < t < 1000$ μs.

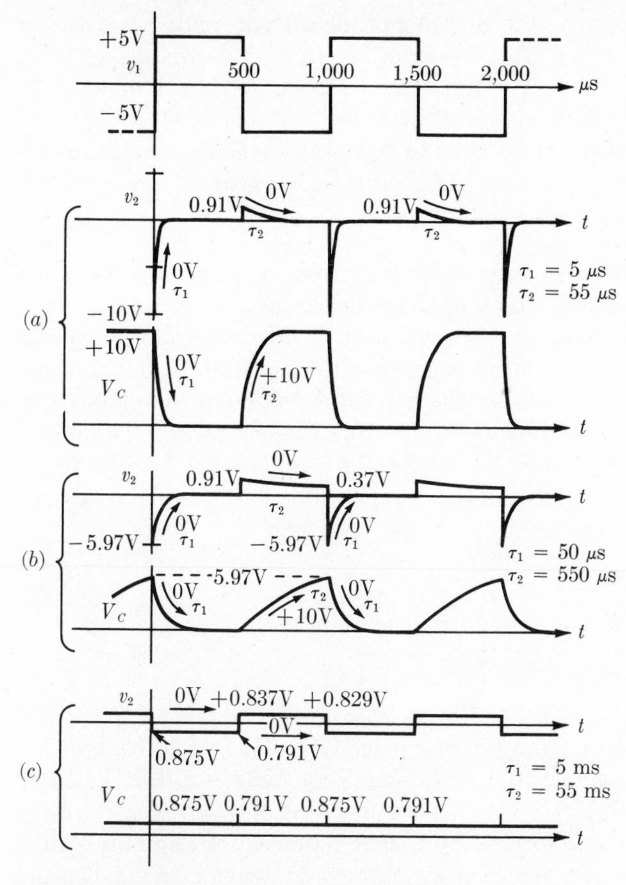

Fig. 4-10 The circuit of Fig. 4-8 solved for three widely varying time constants. (*a*) Case III, $C_1 = 0.005\ \mu F$. (*b*) Case II, $C_1 = 0.05\ \mu F$. (*c*) Case I, $C_1 = 5\ \mu F$.

conditions at $t = 0^+$ are those of Fig. 4-9*a* with $V_C = 10$ V. The output $v_2 = -10$ V then, since the other end of the capacitor is substantially grounded through the conducting transistor. The capacitor and output voltages decay to zero with a 5-μs time constant as shown in Fig. 4-10*a*. At the end of the first 500 μs, 100 time constants have elapsed, and both voltages are zero, giving $V_C = 0$ at $t = 500^+$ μs. The current I which charges the capacitor is

$$I = 10/10\ \text{k}\Omega + 1\ \text{k}\Omega = 0.91\ \text{mA} \qquad (t = 500^+\ \mu\text{s}) \tag{4-19}$$

and the output voltage is

$$v_2 = (1\ \text{k}\Omega)(0.91) = 0.91\ \text{V} \qquad (t = 500^+\ \mu\text{s}) \tag{4-20}$$

The time constant for both the charging of the capacitor and the decay of the output voltage is $\tau_2 = 55\ \mu s$. These values are also shown in Fig. 4-10*a* and complete the solution. (The transistor collector current should also be computed to insure that sufficient base drive is supplied to actually saturate the transistor.) The solution to this problem could also be accomplished by using Eqs. (4-17) and (4-18), but the complexity is unwarranted, and no increase in accuracy would result.

Case II. For this case only the shorter time constant need be much shorter than 500 μs so let us pick 10 times the former capacitor $C_1 = 0.05\ \mu F$. Then $\tau_1 = 50\ \mu s$, and $\tau_2 = 550\ \mu s$. In this case the solution cannot start at $t = 0$ because the initial conditions are not known, but solution can begin at $t = 500\ \mu s$, because the capacitor will have completely discharged in the previous time ($500/50 = 10$ time constants). The initial condition for Fig. 4-9*b* is then $V_C = 0$; $I(500^+)$ and $v_2(500^+)$ are as found for Case III, but $\tau_2 = 550\ \mu s$ and the capacitor does not completely charge as before. At $t = 1{,}000^-\ \mu s$ the capacitor voltage is

$$V_C(1{,}000^-) = 10(1 - \epsilon^{-500/550}) = 5.97\text{ V} \tag{4-21}$$

and the output voltage is

$$v_2(1{,}000^-) = \frac{(10 - 5.97)(1\text{ k}\Omega)}{(10\text{ k}\Omega + 1\text{ k}\Omega)} = 0.37\text{ V} \tag{4-22}$$

The values are shown on the waveform for Case II, Fig. 4-10*b*, and the capacitor voltage is the desired initial condition for $t = 1{,}000^+$ and also $t = 0^+$. Placing $V_C = 5.97$ V into the circuit of Fig. 4-9*a* gives $v_2 = -5.97$ V at $t = 0^+$, and the waveforms may be completed as shown in Fig. 4-10*b*. Again the solution could have been obtained by using Eqs. (4-17) and (4-18), but the additional difficulty would in no way increase the accuracy of the solution.

Case I. For this case neither time constant may be short compared with the 500-μs half period. If we again increased C_1 by 10 times to 0.5 μF, the two time constants would be $\tau_1 = 500\ \mu s$ and $\tau_2 = 5000\ \mu s$, and the only satisfactory way to solve the circuit is to use Eqs. (4-17) and (4-18). Since this involves only straight substitution into the equations, let us increase C_1 to 5.0 μF and solve the circuit using the previous approximation that V_C is almost constant. Writing the equations for charging and discharging we obtain

$$Q_{\text{ch}} = \left(\frac{10 - V_C}{11\text{ k}\Omega}\right) 500\ \mu\text{s} \tag{4-23}$$

$$Q_{\text{dis}} = \left(\frac{V_C}{1\text{ k}\Omega}\right) 500\ \mu\text{s} \tag{4-24}$$

By equating the charge to the discharge, and solving for V_C we obtain

$$V_C = \frac{(10)(1\ \mathrm{k\Omega})}{12\ \mathrm{k\Omega}} = 0.833\ \mathrm{V} \qquad (4\text{-}25)$$

By placing this value for V_C in the circuit of Fig. 4-9*b*, we get for the output v_2 during the charging portion of the cycle,

$$v_2 = \frac{(10 - 0.83)(1\ \mathrm{k\Omega})}{11\ \mathrm{k\Omega}} = 0.833\ \mathrm{V} \qquad (4\text{-}26)$$

During the discharging portion of the cycle $v_2 = -V_C = -0.833$ V. If the capacitor were infinitely large, these values of v_2 would be exactly correct, but the finite size of C_1 causes v_2 to vary somewhat during the cycle. The amount is easily found by computing ΔQ.

$$\Delta V_C = \frac{\Delta Q}{C} = \frac{It}{C} = \left[\frac{(10 - 0.833)}{(11\ \mathrm{k\Omega})(5 \times 10^{-6})}\right] 500 \times 10^{-6} = 0.0833\ \mathrm{V} \qquad (4\text{-}27)$$

This variation is also the variation in v_2 during the discharging portion of the cycle, but to find the variation during the charging portion of the cycle, the value of V_C must be put into the circuit of Fig. 4-9*b* and the circuit solved for v_2. The resulting values are shown on the waveforms of Fig. 4-10*c*. Note that the average value of both the positive and negative portions of v_2 is 0.083 V as computed in Eq. (4-26) giving, of course, a whole period average for v_2 of zero.

Other examples of the use of these techniques for solving *RC* circuit problems are found in the following problems and chapters, where the *RC* circuit is used as an essential timing and coupling element. As in all problems where a variety of solution methods are possible, one may have to guess at a proper method initially and make a trial solution. If the solution results in a contradiction, another method will have to be tried. Usually an approximate solution is a good one to try first to get a feel for the proper final solution.

PROBLEMS

4-1. Assume that the input to the clamping circuit of Fig. 4.1 is reduced from a peak value of 5 V to a peak value of 2.5 V at $t = 2$ ms. Sketch and carefully label the output waveform for $0 < t < 5$ ms. At what approximate time would you estimate the clamping action to recommence?

4-2. The circuit illustrated is sometimes called a step charging rectifier because the output waveform will resemble a staircase. (The resemblance would be even better if the series capacitor were smaller.) The input is a square wave which is turned on at $t = 0$, and the capacitors are initially discharged.

Calculate the output waveform and carefully sketch it for $0 < t < 20\ \mu s$. What will the output v_0 approach as $t \to \infty$?

If the same circuit were used with a sine wave input (e.g., 60 Hz) the circuit would be called a voltage doubling rectifier. Explain why.

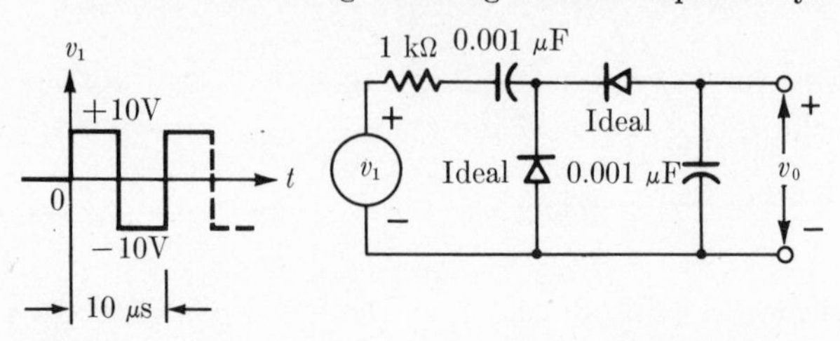

For $t < 0$, $v_1 = 0$

4-3. Draw the circuit including reasonable element values for a clamping circuit which will clamp the *negative* extreme of the output to +10 V. Assume the input is a 1-kHz square wave and the diode is a 1N4152.

4-4. The circuit shown is to generate a sawtooth waveform that starts at ≈ 0 volts and has a peak value of +6 V. What is the maximum value of R_b that will insure that the output is clamped to zero, and what is the value of C?

If the input frequency is doubled, does R_b need to be changed to keep the output clamped at 0V? (Leave C unchanged.)

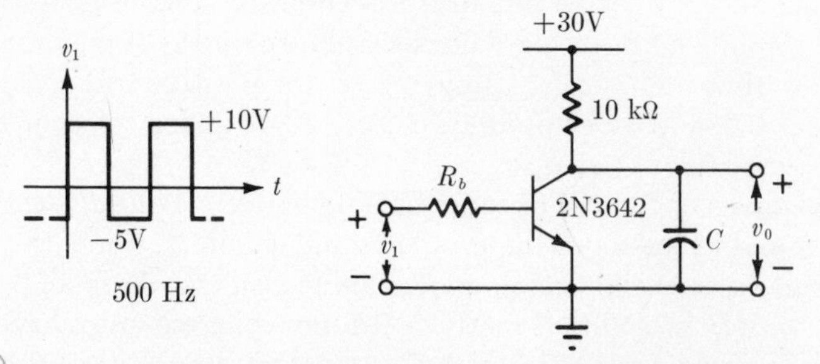

4-5. The circuit shown illustrates a rather extreme form of pulse distortion due to a coupling capacitor. Find and sketch v_0 as a function of time. Determining v_b and i_b as preliminary steps is helpful.

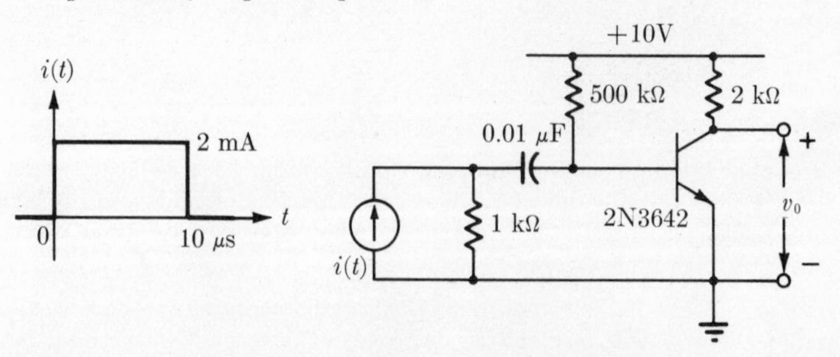

4-6. The circuit is operated in steady-state with a 10 μs pulse occurring once every millisecond. Calculate and sketch the waveform across C for $C = 0.001\ \mu F$,

0.01 μF, 1.0 μF. (A little thought on where to begin the calculations may save much work.)

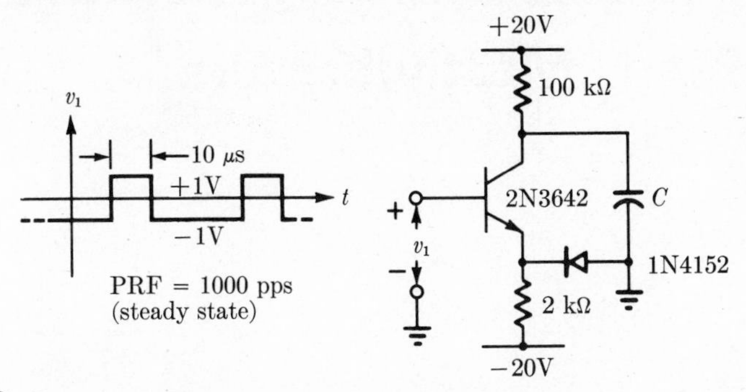

4-7. The circuit shown can act as a crude discriminator in that the dc capacitor voltage is roughly proportional to frequency over a limited frequency range. First find C so that the peak-to-peak ripple in v_2 is 1 percent of the average value of v_2 for the input shown. Then calculate v_2 for one-half and one-fourth of the frequency shown. Sketch a curve of v_2 versus f using the three calculated points.

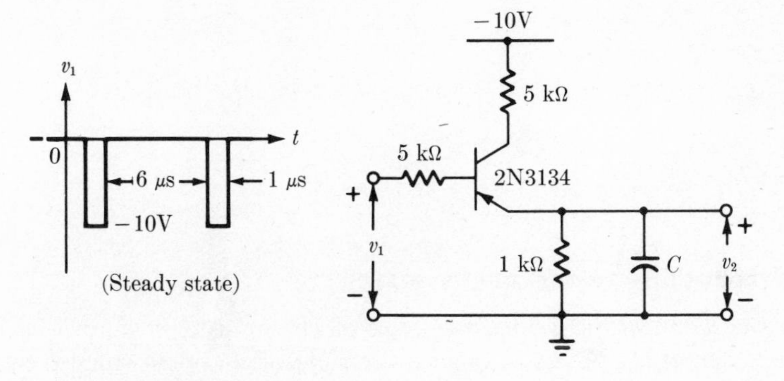

5

Stable States and Regenerative Switching. Multivibrators

5-1 INTRODUCTION TO THE MULTIVIBRATOR

In this chapter we introduce the class of circuit usually known as the *multivibrator*. It is an extremely versatile and useful group, which is called upon to do a great variety of functions. Among the important ones are storage of binary information (memory), generation of square waves and pulses, generation of time delays, the division of frequency, counting, waveform squaring, and the generation of timing waveforms. Three general types of multivibrators can be treated quite separately:

1. A bistable variety which has two states either of which remains until a stimulus is applied to make it change states.
2. An astable variety which will remain in neither state more than a predetermined length of time, and continuously oscillates back and forth between states.
3. A monostable form which has one state it will remain in until stimulated into the second state where it stays only a predetermined length of time.

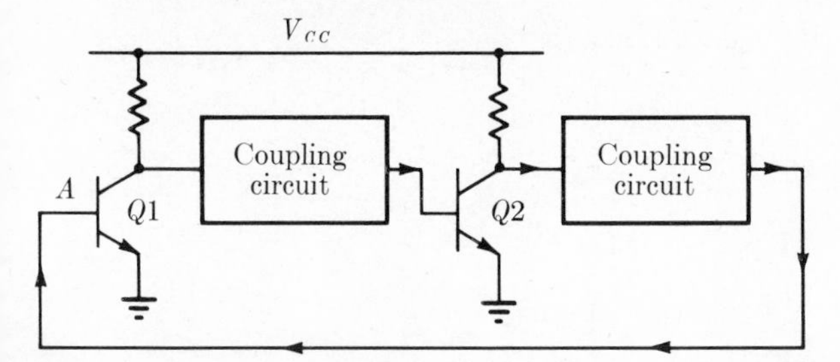

Fig. 5-1 Basic multivibrator circuit shown as a two-stage amplifier with the output coupled back to the input.

Which variety a given multivibrator is depends upon both the circuit configuration and the values of the elements in the circuit. For example, a given circuit configuration can be either an astable or a monostable multivibrator depending upon the values of the circuit elements.

A typical multivibrator may be thought of as a two-stage amplifier with its output connected back to its input, as shown in Fig. 5-1. The boxes shown as coupling circuits may contain either an ac coupling (series capacitor) or dc coupling (series resistor) and determine what type of multivibrator is obtained. Any disturbance occurring at point A will be amplified by $Q1$, coupled to the input of $Q2$, and again amplified. The output of $Q2$ will be in phase (or polarity) with the initial signal at point A, and will further reinforce the initial disturbance. Because of this reinforcement the initial disturbance will grow or *regenerate* as long as the combined gain of the two stages is greater than unity. As the signal grows, one or both of the stages will be driven into a limiting condition—e.g., cutoff or saturation—and further increase in the signal is prevented. If the circuit is directly coupled, the stages will stay in the limiting condition, as will be seen in the following bistable example. If the circuit is ac coupled, the signal conditions necessary to keep the transistors in the limiting condition cannot be permanently sustained; later the stages will become regenerative again, and a new regenerative cycle will begin. Such a situation obtains in the astable case discussed in Sec. 5-9.

5-2 THE BISTABLE MULTIVIBRATOR

Figure 5-2 shows the most elementary form of bistable multivibrator or binary possible with two similar transistors. (A slight further simplication would be to make $R_{B1} = R_{B2} = 0$, but the circuit is not quite as useful.) As is usual with such circuits, there are two stable states in which the circuit will rest. The first such state is with $Q1$ saturated and $Q2$ nonconducting; the second is with the conditions of $Q1$ and $Q2$ reversed. These conditions

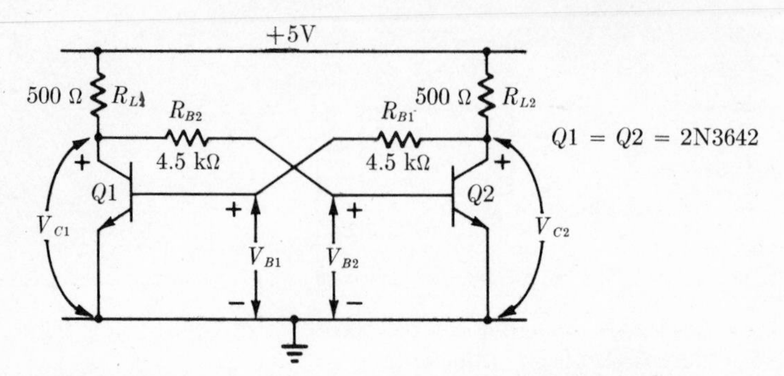

Fig. 5-2 Simple form of collector-coupled bistable multivibrator.

in the bistable circuit can be portrayed by the crude analogy of a mechanical switch as in Fig. 5-3. There are two positions of the switch, 1 and 2, corresponding to the two stable states of the circuit, together with a spring whose functioning corresponds roughly to the regenerative action of the circuit. Because of the spring the switch is held stable in either extreme position. It would be stable in the exact center position, except that any slight displacement, however small, would quickly result in a transition to one of the end positions. Finally, if the switch is to be moved from one position to the other, it is not necessary to manually provide the complete displacement; the switch need be carried only part way, at least to the center, and the spring will complete the action. All of these features have their counterpart in the electrical behavior of the multivibrator, as becomes apparent in the discussion following.

The useful outputs of the bistable circuit are the voltages V_{C1} and V_{C2} appearing at the collectors of the two transistors. The collector of the conducting transistor will have very nearly 0 V, whereas the collector of the nonconducting transistor will have nearly 5 V.

To show the exact circuit conditions and prove that the two stable states do indeed exist, begin by assuming a state and see if all conditions are consistent with the assumption. Let us assume $Q1$ is ON and saturated, and that $Q2$ is completely OFF. Two loops must always be solved in a multivibrator as shown in Fig. 5-4. One loop contains the base of $Q1$ and

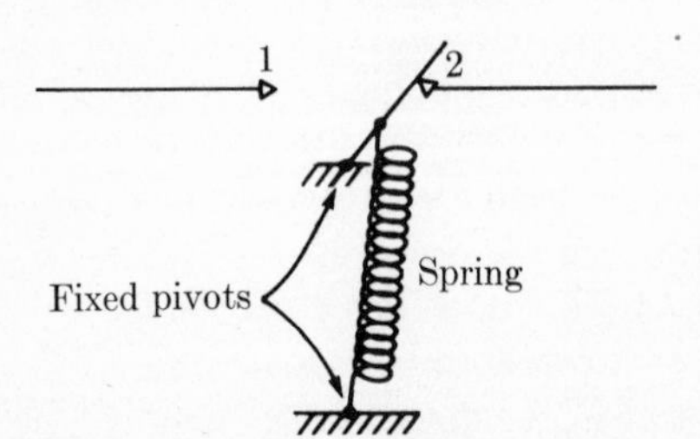

Fig. 5-3 Crude mechanical model to illustrate bistable multivibrator.

the collector of $Q2$ (Fig. 5-4a), and the other loop contains the collector of $Q1$ and the base of $Q2$. Equivalents appropriate to the assumptions have been shown for the transistor. The base current from Fig. 5-4a is $I_{B1} = (5 - V_{B1})/5\ \text{k}\Omega = 0.86$ mA. Using a load line for $V_{CC} = 5$ V and $R_L = 500$ ohms on the transistor curves of Fig. 3-14 shows that $V_{C1} = 0.1$ V, and the transistor is heavily saturated (much more base current than necessary for saturation is flowing). From Fig. 5-4b, $V_{B2} = V_{C1} = 0.1$ V. With such a low base voltage $I_{B2} \approx 0$, $Q2$ is OFF,

$$V_{C2} = 5 - (0.5\ \text{k}\Omega)I_{B1} = +4.57\ \text{V} \tag{5-1}$$

and the initial assumptions are verified.

The circuit will stay in this state until something disturbs it sufficiently to make it change states. A negative signal coupled to the base of $Q1$, for example, would decrease the current flow in $Q1$ causing its collector voltage to rise. This rise coupled to $Q2$ through R_{B2} turns $Q2$ ON and makes its collector voltage fall. The fall coupled through R_{B1} tends to turn $Q1$ OFF as did the originating signal. The original negative signal therefore initiates the regenerative process, and if the signal is big enough, it will cause $Q1$ to go completely OFF and $Q2$ to saturate. The circuit conditions are then exactly as before, with the subscripts 1 and 2 reversed on the currents and voltages.

Because of the symmetry of the circuit it might well be asked why another stable state would not occur when both transistors are carrying equal currents and both base voltages are identical. The conditions for such a situation can be found mathematically, and are approximately $V_{CE} = 0.8$ V and $V_{BE} = 0.7$ V. However, the circuit is analogous to the spring-actuated switch in dead center; any slight disturbance quickly unleashes a driving force which sends the device toward one of its truly stable states. As pointed out previously, the transistors are connected essentially in cascade with a return connection to provide regeneration or positive feedback. The gain of each stage A_1 at the unstable operating point $V_{CE} = 0.8$ V is much greater than unity, and the overall gain is A_1^2,

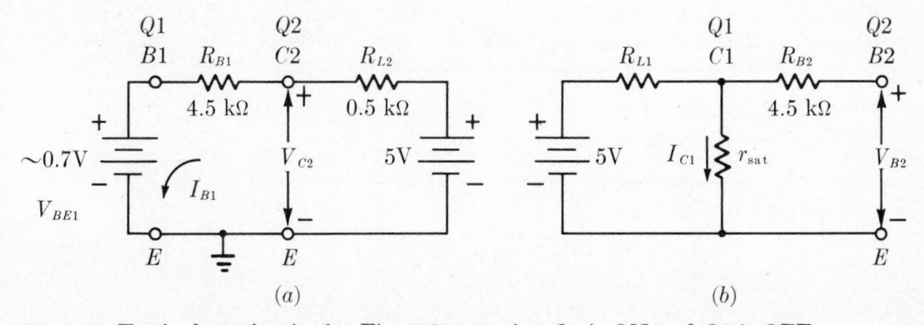

Fig. 5-4 Equivalent circuits for Fig. 5-2 assuming $Q1$ is ON and $Q2$ is OFF.

which is therefore greater than unity. Any small disturbance such as thermal noise will cause the circuit to move rapidly to one of the stable states. The stable state achieved by the circuit depends upon the nature of the disturbance initiating the change, since both stable states are equally probable at the unstable operating point above.

5-3 THE SET-RESET BINARY

To make the circuit switch reliably from one state to another, some form of coupling from the initiating signal (trigger) to the binary is required. Two transistors are added to the binary in Fig. 5-5*a* to apply the trigger signals called in this case SET and RESET. Both trigger lines, v_1 and v_2, are assumed to be at zero volts in the absence of a trigger pulse so that both added transistors $Q3$ and $Q4$ are OFF, and the binary is unaffected by their presence. Either binary state is equally probable, but assume $Q1$ is OFF, $Q2$—ON. V_{C1} is high (+4.6 V) or in the ONE state. The arrival of the SET pulse turns $Q3$ ON, and if the pulse v_1 is large enough, Q_2 is

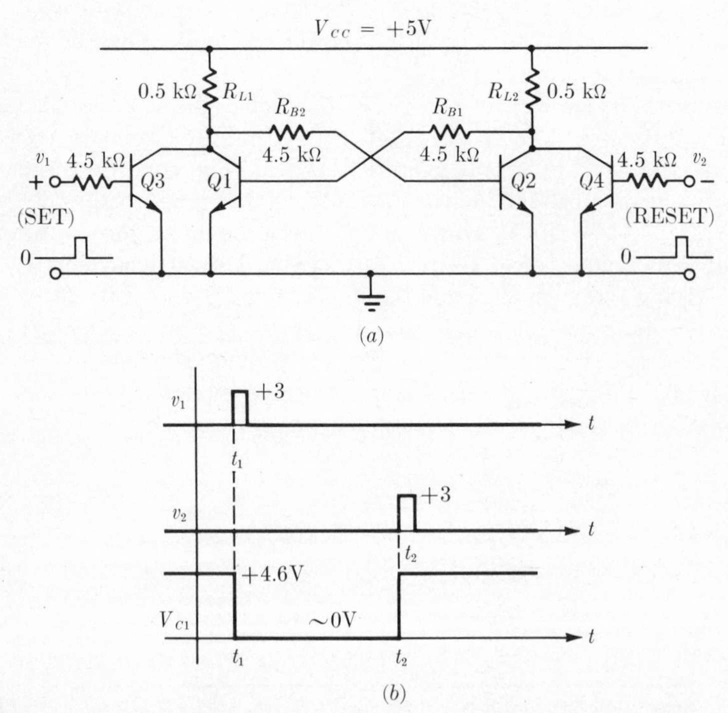

Fig. 5-5 (*a*) Flip-flop of Fig. 5-2 with gates added to apply SET and RESET pulses. (*b*) Waveforms from the circuit assuming the circuit to be initially RESET.

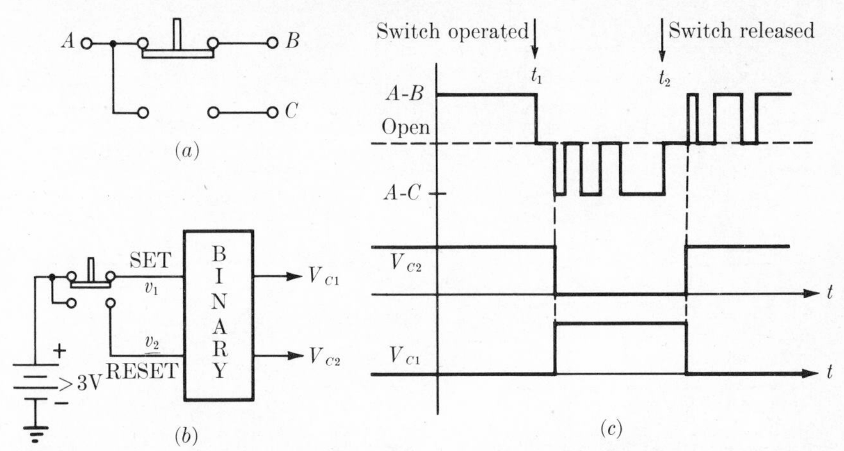

Fig. 5-6 (*a*) A single-pole double-throw push-button switch with normally-closed (NC) contacts AB and normally-open contacts (NO) AC. (*b*) Switch connected to SET-RESET binary. (*c*) Switch behavior and waveforms from binary.

turned OFF. Its collector voltage rises and turns $Q1$ ON holding V_{C1} at nearly zero volts. The pulse v_1 may now be removed, and the circuit will stay in the state $Q1$—ON, $Q2$—OFF, which is the SET state. At some later time a pulse may be applied to the RESET line turning $Q4$ ON. This will cause $Q1$ to go OFF and $Q2$ to go ON, and the circuit will be in its original RESET state. Since the circuit will remain in the SET state after the application of the SET pulse, it may be said to "remember" whether or not a set pulse was received.

The SET-RESET binary may be realized by using two of the NOR circuits of Fig. 3-16. Transistors $Q1$–$Q3$ are provided by one NOR circuit and $Q2$–$Q4$ by the other. This realization of the binary is easily accomplished using integrated circuits for the NOR function.

Another simple but useful application of the SET-RESET binary is in filtering or smoothing the action of mechanical switches. Often one would like to trigger a circuit at the push of a button. A mechanical switch rarely makes only a single contact when you actuate it, but instead is likely to bounce a number of times before the contact becomes settled in its new position. Figure 5-6*a* shows a schematic diagram for a single-pole, double-throw push button switch, and Fig. 5-6*b* shows the behavior of the contacts when the switch is pushed at t_1. The NC (normally closed) contact almost immediately opens, but a moment elapses before the first contact is made on the NO (normally open) side. The switch then bounces a few times before finally settling down with the NO contact closed. When the button is released, approximately the same sequence is repeated in the reverse order. If some precaution were not taken, the circuit to be

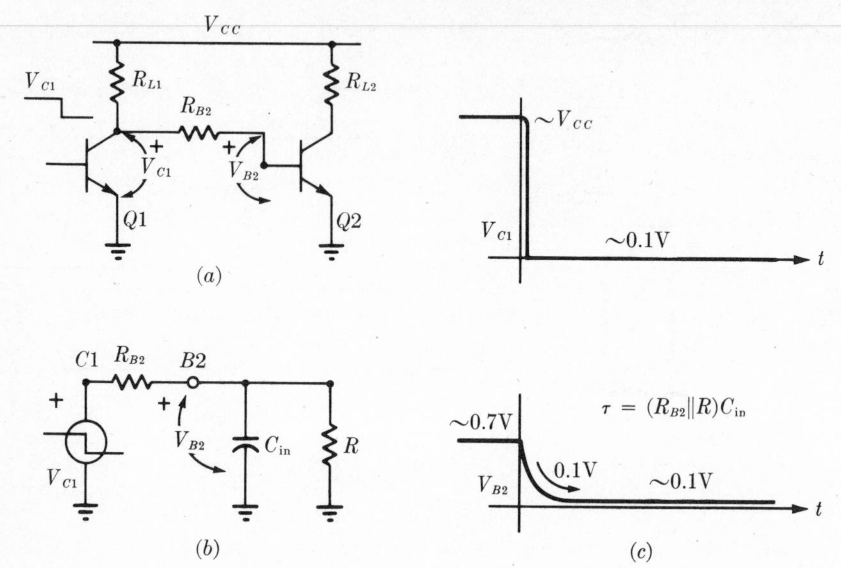

Fig. 5-7 (*a*) Circuit with collector voltage being switched off (as by $Q3$ in Fig. 5-5) to initiate a change of states. (*b*) Equivalent circuit to find change of base voltage in stage being turned off. (*c*) Waveforms at collector $C1$ and base $B2$.

operated by the switch might operate once for each bounce of the switch rather than once for each push, as desired. In Fig. 5-6*c* the offending switch is connected to the binary of Fig. 5-5*a*, as shown. The first contact closure occurring after t_1 provides a SET pulse, and the succeeding SET pulses caused by the switch bounces have no effect because the binary is not RESET between the pulses. Therefore V_{C1} is a negative-going step occurring at the first SET pulse. A similar action occurs upon release of the button at t_2, and the binary is RESET by the first RESET pulse to occur with no effect from succeeding ones. The circuit to be actuated can be connected to either V_{C1} or V_{C2}, and is triggered once for each complete cycle of pushing and releasing the switch.

5-4 SWITCHING SPEED IN THE BINARY

Although the actual calculation of switching speed is beyond the confines of this text, a discussion of the elements which limit the speed and the ways of speeding up the transition is pertinent. It is probably obvious that the active devices (transistors, for the preceding examples) are the ultimate limitation to the switching speed for the ordinary ranges of circuit parameters. To minimize the effect of the transistor, units with high cutoff frequencies, low storage time, and small capacitances are chosen. In addition, the circuit may be designed so that transistor saturation does not occur, thereby eliminating the problem of transistor storage time.

The nature of the switching problem can be understood with reference to Fig. 5-7. In Fig. 5-7*a* the collector voltage waveform of $Q1$ is shown being switched from V_{CC} to ground (as would occur if a SET pulse were applied at $t = 0$ to the circuit of Fig. 5-5). This step is applied through the coupling resistor R_{B2} to the base of $Q2$, the conducting transistor. Ideally $Q2$ would immediately turn off, but a finite charge must be removed from the base of $Q2$ before it will turn off. This situation is shown by the crude equivalent circuit for the base of $Q2$, the resistor R in parallel with C_{in}, in Fig. 5-7*b*. The waveforms at the collector of $Q1$ and the base of $Q2$ are shown in Fig. 5-7*c*. The base waveform is an exponential with a time constant which can be decreased by decreasing the value of R_{B2}. In practice, however, decreasing R_{B2} does not materially improve the switching time, because the increasing base current resulting causes the charge storage in $Q2$ to increase, negating to some degree the effect of the reduced time constant.

The circuit may be improved to reduce the switching time by including a small capacitor known as a speed-up capacitor across R_{B2} (and R_{B1}), as shown in Fig. 5-8. This capacitor couples the high-frequency com-

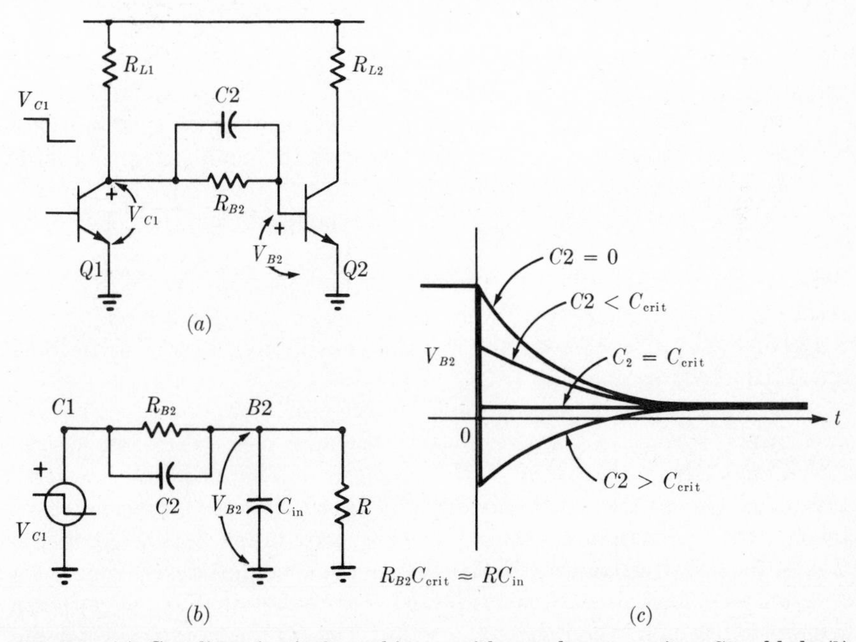

Fig. 5-8 (*a*) Coupling circuit for a binary with speed-up capacitor $C2$ added. (*b*) Equivalent circuit of (*a*) useful during the transition time. (*c*) Waveforms occurring on the base of $Q2$, the ON transistor, assuming a step change at the collector of $Q1$.

ponents in the fast fall of V_{C1} to the base of Q_2. The circuit is identical to the compensated voltage divider of Fig. 2-13. There is a critical value of $C2 = C_{\text{crit}}$ which makes the voltage V_{B2} jump to its final value without any exponentials. Increase of $C2$ beyond this value will not improve the operation of the circuit but will increase the capacitive load driven by $Q1$ or $Q3$. Note that now the waveform V_{C1} cannot be a perfect step without infinite current flowing in the driving transistor at $t = 0$.

The speed-up capacitors have another function in that the state of the circuit is stored or "remembered" in these capacitors for a brief moment during a change of state in the binary. If $C1$ and $C2$ are added across R_{B1} and R_{B2}, respectively, and the binary is in the state $Q1$—ON and $Q2$—OFF, then $C2$ will be discharged and $C1$ charged (to about 4 V). During a change of state the charges on these capacitors may be used to determine which way the circuit switches. In the counting binary which follows, this particular mode of operation is essential.

5-5 COUNTING CIRCUITS USING BINARIES

If a binary is driven from a source applied to both halves simultaneously, the circuit can be used as a counter or frequency divider. Many different trigger methods may be used, and three possible arrangements are shown in Fig. 5-9. Although the trigger signal is applied in different places in the three circuits, it must ultimately appear on the base of the conducting transistor for the negative-going trigger signal shown.

In Fig. 5-9*a* the trigger is applied equally to the two collector load resistors, but little appears on the collector of the conducting transistor because of its low saturation resistance. Almost the entire trigger appears at the collector of the OFF transistor, and is coupled from the collector to the base of the ON transistor by the speed-up capacitor. The ON transistor is thus switched OFF, thereby turning on the previously OFF transistor. Such a scheme will work, but the values of the three capacitors together with the magnitude and rise time of the trigger signal are critical for satisfactory switching.

In the second circuit, Fig. 5-9*b*, the steering to the proper collector, and ultimately the ON transistor base, is accomplished by steering diodes. Assume $Q1$ ON and $Q2$ OFF. With no trigger pulse applied, $D1$ is reverse biased by almost the entire amount of V_{CC}, whereas $D2$ is reverse biased by only the drop across R_{L2} caused by base current flowing to $Q1$. Application of the negative trigger (which should have a magnitude less than V_{CC}) will cause $D2$ to conduct, but not $D1$. Therefore the negative trigger pulse is coupled through C and $D2$ to the collector of $Q2$. From the collector the pulse is coupled via $C1$ to the base of $Q1$, which is then turned off and starts the regenerative process necessary to cause the states to change.

(This type of binary is often called a *flip-flop,* and the triggering action described has just made the circuit flip.) A subsequent pulse applied at the trigger input will cause the circuit to change states back to the original condition (or flop). The triggering path for the second operation is via C, $D1$, and $C2$.

The third circuit, Fig. 5-9c, uses steering diodes in the bases of the transistors, but the steering information is derived from the collector circuits. The circuit is also shown with the additional resistors R_{B3} and R_{B4}. The function of these resistors is to hold the base of the OFF transistor negative and to provide a margin of safety when base leakage currents are

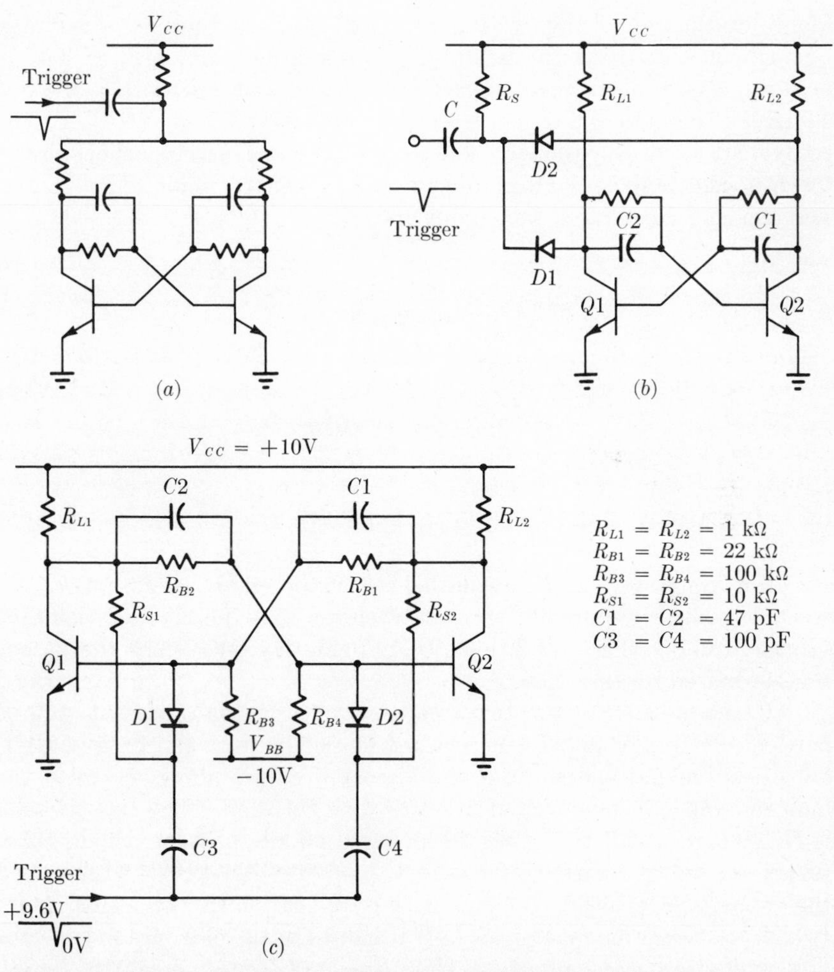

Fig. 5-9 (a) Binary with symmetrical collector triggering. (b) Binary with steering diodes for collector triggering. (c) Binary with steering diodes to the bases.

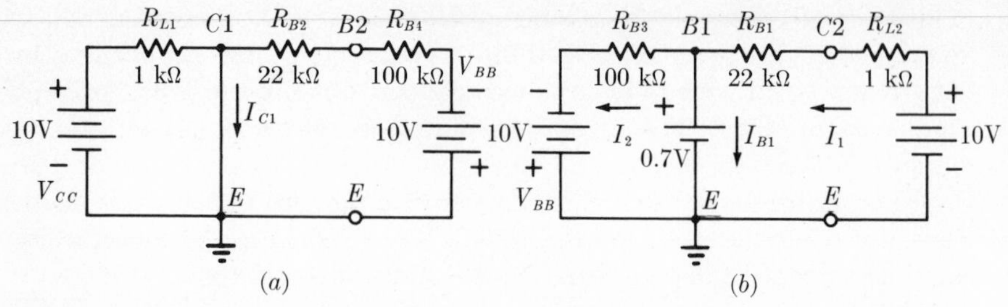

Fig. 5-10 Equivalent circuits to determine the static conditions in the binary of Fig. 5-9*c*.

a problem, as they always are if germanium transistors are used, and may be if silicon transistors are used at higher operating temperatures. The dc or static conditions in the circuit may be found with reference to the two equivalent circuits of Fig. 5-10, which assume $Q1$ ON and $Q2$ OFF. The equivalent circuit for the ON transistor is taken as merely a short circuit for R_{sat} and a fixed voltage for the base-emitter circuit. The various voltages and currents in the circuit are:

$$V_{C1} = 0 \text{ V} \tag{5-2}$$

$$V_{B2} = -\frac{(10)(22 \text{ k}\Omega)}{100 \text{ k}\Omega + 22 \text{ k}\Omega} = -1.8 \text{ V} \tag{5-3}$$

$$V_{C2} = 0.7 + \frac{(9.3)(22 \text{ k}\Omega)}{1 \text{ k}\Omega + 22 \text{ k}\Omega} = 9.6 \text{ V} \tag{5-4}$$

$$I_{B1} = I_1 - I_2 = \frac{9.3}{1 \text{ k}\Omega + 22 \text{ k}\Omega} - \frac{10.7}{100 \text{ k}\Omega} = 0.3 \text{ mA} \tag{5-5}$$

$$I_{C1} = \frac{10}{1 \text{ k}\Omega} - \frac{10}{100 \text{ k}\Omega + 22 \text{ k}\Omega} = 9.9 \text{ mA} \tag{5-6}$$

For the transistor to be saturated as supposed, a minimum h_{FE} of $9.9/0.3 = 33$ is required. Figure 3-14 shows that the 2N3642 would be saturated under these conditions. $Q2$ is definitely OFF since the base is reverse biased by the -1.8 V.

The condition of the two steering diodes in the quiescent state is particularly important: $D1$ is forward biased because the base of $Q1$ is more positive than the collector, but almost no current flows because of the high resistance R_{S1} connected in series with the diode. The capacitor $C3$ is charged to about 9 V with the positive on the bottom. Diode $D2$ is reverse biased by an amount $V_{C2} - V_{B2} = 9.6 - (-1.8) = 11.4$ V, and the capacitor $C4$ has zero charge. The application of the trigger pulse (which has the same magnitude as if it had been taken from the collector or a similar binary stage) causes the cathode of $D1$ to momentarily go to -9.6 V and the cathode of $D2$ to go to zero volts. Therefore $D1$ conducts and makes the base of $Q1$ go negative. However, $D2$ does not conduct because the cathode is not made more negative than the anode, and $Q2$ is unaffected directly by the trigger signal. The effect of the trigger is therefore to turn off $Q1$, which was previously conducting, and turn on $Q2$ because of the rising voltage appearing at the collector of $Q1$. The trigger reverses the states in the binary (causes it to flip) because it is steered to the transistor requiring a turn-off signal. A succeeding trigger signal would cause the circuit to return back to its original state; therefore, the binary goes through *one* complete cycle of operation for *two* complete cycles of the input pulse. The input signal or trigger need not be a pulse, but could instead be a square wave with a peak-to-peak amplitude of 9 V or less (too large a signal would let some of the trigger through the wrong steering diode). The output is a square wave with half the frequency of the input wave.

A number of such binary *frequency dividers* may be connected in cascade to divide an input frequency by 2^n, where n is the number of binaries used. Such a situation is shown in Fig. 5-11 where four dividers are connected in cascade to provide division by $2^4 = 16$. The input could be a pulse or a square wave with a fast rise and fall time and appropriate amplitude at the frequency f_1. The output is a square wave with exactly equal on and off times at the frequency $f_1/16$. The information as to the state of the binary is usually obtained by sensing the collector voltage. If the high collector voltage is defined as a ONE, and the low collector voltage is defined as a ZERO, and if the collector of $Q2$ in each binary is used to provide the output information, then the possible states for the cascade of four dividers are given in the state table (Table 5-1). If the operation is considered to start with $Q2$ in each binary conducting, then all the outputs (the collector of each $Q2$) are at zero, as shown for $N = 0$. One input pulse flips the first binary causing the output to rise to the ONE state. The second binary is not affected by the *rise* in its input voltage, and

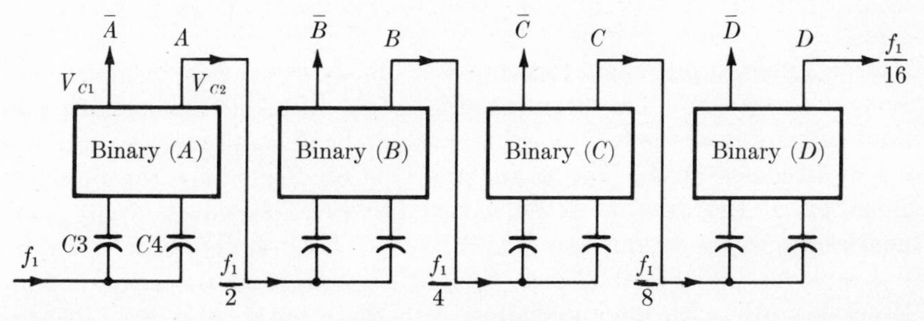

Fig. 5-11 Binary circuits as in Fig. 5-9*c* connected to give frequency division. The input to each divider is from one of the two collectors of the preceding divider. The output frequency of each divider is one-half the input frequency.

Table 5-1 The states of a chain containing four binaries. (*a*) The binaries are in cascade with no feedback. (*b*) The binaries are in a chain but with feedback to provide binary-coded decimal counting

	(*a*) *Binary*					(*b*) *Binary-coded decimal*				
Output Terminal		*A*	*B*	*C*	*D*		*A*	*B*	*C*	*D*
Weight of Output		1	2	4	8		1	2	4	8
Number of Pulses	*N*					*N*				
	0	0	0	0	0	0	0	0	0	0
	1	1	0	0	0	1	1	0	0	0
	2	0	1	0	0	2	0	1	0	0
	3	1	1	0	0	3	1	1	0	0
	4	0	0	1	0	4	0	0	1	0
	5	1	0	1	0	5	1	0	1	0
	6	0	1	1	0	6	0	1	1	0
	7	1	1	1	0	7	1	1	1	0
	8	0	0	0	1	8	0	0	0	1
	9	1	0	0	1	9	1	0	0	1
	10	0	1	0	1	10	0	0	0	0
	11	1	1	0	1			. . .		
	12	0	0	1	1			. . .		
	13	1	0	1	1			. . .		
	14	0	1	1	1			. . .		
	15	1	1	1	1			. . .		
	16	0	0	0	0			. . .		

Note: The value of the count is $A \cdot 1 + B \cdot 2 + C \cdot 4 + D \cdot 8$ where A, B, C, D are either 0 or 1, as indicated above.

therefore stays in the ZERO state. The value of the number stored in the binary chain is now

$$1 \cdot 2^0 + 0 \cdot 2^1 + 0 \cdot 2^2 + 0 \cdot 2^3 = 1 \cdot 1 + 0 \cdot 2 + 0 \cdot 4 + 0 \cdot 8 = 1$$

A second pulse input again flips the first binary causing its output to go from ONE to ZERO. This negative output transition differentiated by the input capacitors of the following binary flips the second binary so that its output becomes ONE. The third binary is unaffected by the positive transition at the output of the second binary. The states of the four binaries are as shown for $N = 2$ in Table 5-1. The number now stored is $0 \cdot 1 + 1 \cdot 2 + 0 \cdot 4 + 0 \cdot 8 = 2$. Succeeding pulses continue the action described until, with the application of the 15th pulse, all of the binaries are in the ONE state. The 16th pulse then causes all of the binaries to flip to the ZERO state, which was the original starting point. The chain can therefore be used to store a number between 0 and 15 inclusive, or there are sixteen possible states. This storage principle is made use of in counters where the input to the binary chain is the pulse group to be counted, and the "dc" outputs from the collectors give the number of pulses applied to the input of the first counter (binary).

To do the counting effectively the state of the counter (group of binaries) must be set before the unknown pulses are applied. This preliminary setting of the counters or RESETTING (to zero) is easily accomplished if a pulse is generated prior to the desired counting interval and applied to the binary, as shown in Fig. 5-12. The positive RESET pulse causes $Q2$ to conduct regardless of the previous state of the binary, and the output of the binary is forced to ZERO. One such resistor is added to each binary in the counter, and a common RESET line serves to place all the binaries in the ZERO state. The resistor R is chosen to provide sufficient base drive to turn on $Q2$ even though the previous state of the binary is such as to try to keep $Q2$ off. Some of the base drive provided by the cross-coupling resistors is removed from the base by R when the RESET signal is absent (at 0 V); therefore the size of the cross-coupling resistors may have to be slightly reduced to keep sufficient base drive.

Although in many situations counting in powers of two is satisfactory, it is preferable to have the output in decimal form, that is, in powers of 10. To do this requires changing the basic group of four binaries (which count 0 through 15) to a group which resets to zero on the application of the tenth pulse. A typical counting sequence for a decade counter is

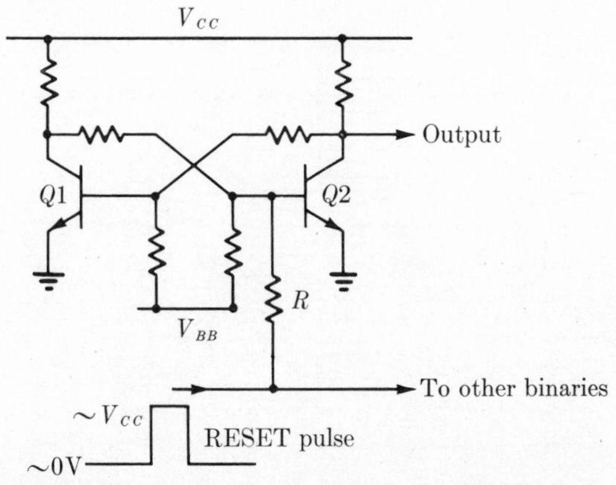

Fig. 5-12 A method of applying a RESET pulse to a binary. Only the dc circuit of the binary is shown. Application of the positive RESET pulse causes $Q2$ to conduct and the output to go to the ZERO state.

given in Table 5-1. Comparison of the two parts of the table shows that the decade counter and the binary counter have identical patterns through the count of nine. The difference occurs at the tenth count, where the second counter in the decade is inhibited from changing from 0 to 1 as it would normally do, and the fourth counter is reset to zero. This may be accomplished by changing the basic binary chain of Fig. 5-11 to the connection shown in Fig. 5-13*a*. Here an AND gate has been added between the first and second stages, and the inputs to the last stage have been separated to steer pulses directly to each side of the binary. If the counter is initially reset, all the outputs $A\ B\ C\ D$ are at *0*, and $\overline{D}$ is in the *1* state.

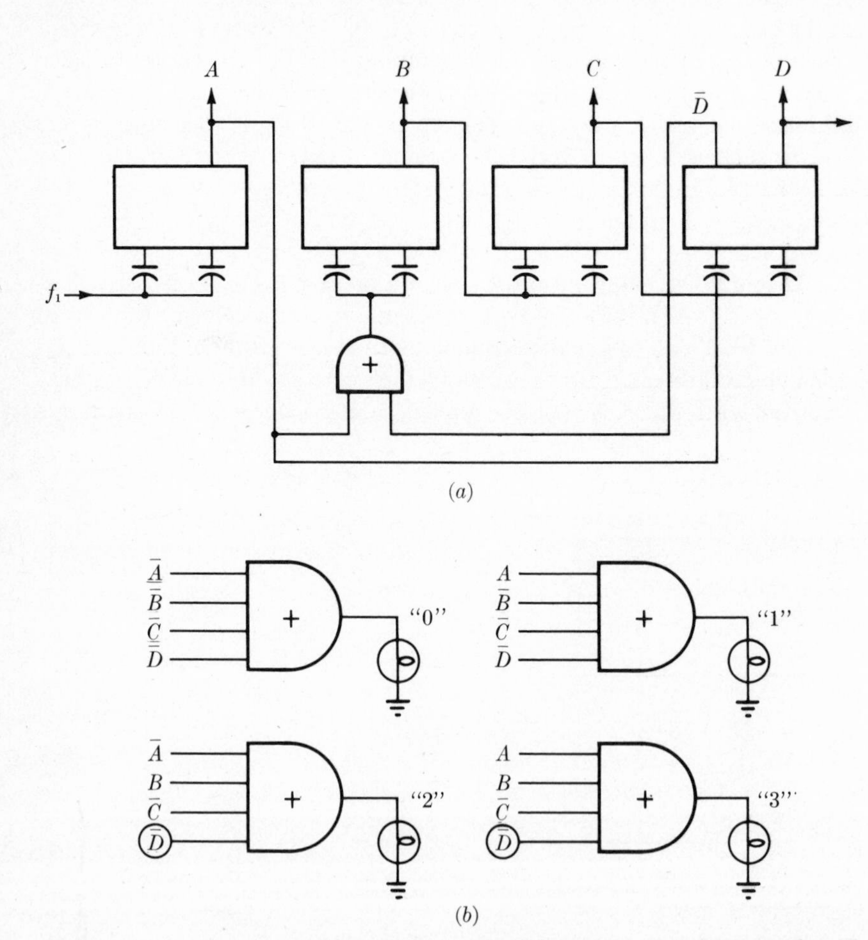

Fig. 5-13 (*a*) A decade counter made up of four binaries connected in cascade with feedback to reset at 10 rather than 16. (*b*) An arrangement to give visual indication for the first four digits. The indicated light bulb might be just that or one cathode of a multinumeral indicator tube.

The output of the AND gate is also *0* because only one input is at *1*. The application of one input pulse causes A to go to the logic *1* state, and the output of the AND gate to go to the logic *1* state. At the second pulse both A and the output of the AND gate go to the logic *0* state so that the second counter is flipped. A pulse is also delivered to the left side of the fourth counter, but it is already in the reset state ($\overline{D}$ high) and no change occurs. The counter continues to count in the normal manner until the eighth pulse is applied, at which time the third counter in going from logic *1* to *0* flips the last counter. This causes $\overline{D}$ to go to zero and therefore the right input of the AND gate. The application of the ninth pulse does nothing but change the output of the first binary back to *1*. The tenth pulse, however, causes A to go from *1* to *0*, a change which is *not* transmitted through the AND gate because its right-hand side is still at zero. The change in A is transmitted directly to the fourth binary and flips it back to the logic *0* state. The decade chain is thus placed back into its starting condition and is ready to continue counting.

The preceding scheme for converting the four binaries into a decade is not the only possible one, but is a commonly accepted method. A complete counter would normally consist of a cascade connection of as many such decades as digits are required in the answer. The answer may be indicated by using the outputs of the decade and suitable logic to run an indicator such as numbered lights or a visual readout tube containing the numbers 0 to 9. Figure 5-13*b* shows the required logic for the numbers 0 through 3. The circuit is a simple AND gate operating a light. In the case of the "2," for example, the presence of a logic *1* at $\overline{A}$, B, $\overline{C}$, and $\overline{D}$ gives a logic *1* at the output of the gate which lights the light. The output from the fourth binary is circled to indicate it could be omitted from a decade arrangement, but would be needed in the basic binary chain to distinguish between "2" and "10."

5-6 THE EMITTER-COUPLED BINARY

The binary circuits discussed thus far are collector coupled, i.e., the coupling is from the collector of a driving stage to the base of the driven stage. Other coupling modes may be employed, and a particularly useful one is to make one of the two couplings from emitter to emitter as shown in Fig. 5-14.[1] Such a circuit may be used as the preceding binaries have been used, and we shall show that it does indeed have two stable states like the preceding binaries. A more interesting use of this circuit, however, is for

[1] The vacuum-tube counterpart of this circuit has been known as a Schmitt Trigger (see O. H. Schmitt, A Thermionic Trigger, *Journ. Sci. Inst.*, **15**: 24–26, 1938), and the term is sometimes applied to the semiconductor versions.

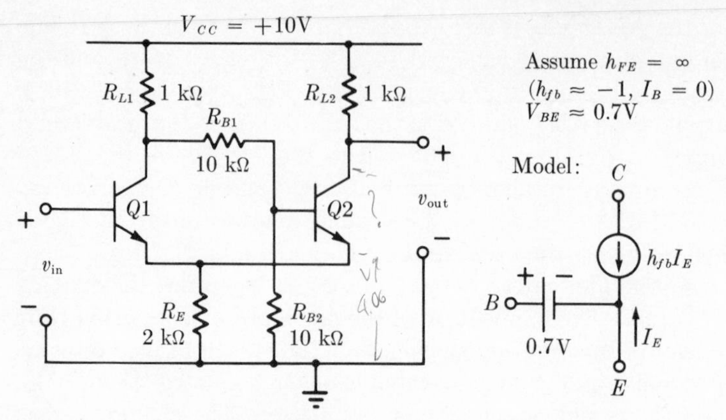

Fig. 5-14 An emitter coupled binary with an idealized transistor model. The model is valid as long as the drop due to I_B in R_{B1} and R_{B2} is small.

waveform shaping or amplitude comparison. The preceding binaries also could be used for these functions, but there is no convenient, independent input terminal, all the terminals of the active devices are connected to important parts of the binary. In the emitter-coupled form, by contrast, there is a convenient input terminal (the base of $Q1$) and a convenient output terminal (the collector of $Q2$) which can be loaded without affecting the binary.

To show that the circuit is probably regenerative, consider what occurs with a small positive step applied to the input terminals as v_{in}. The collector voltage of $Q1$ begins to fall, carrying the base of $Q2$ in the negative direction as well, since the voltage divider R_{B1}-R_{B2} couples the collector of $Q1$ to the base of $Q2$. The negative-going voltage on the base of $Q2$ causes the current in $Q2$ to decrease, thereby decreasing the emitter voltage of $Q2$ and also $Q1$. The decreasing emitter voltage on both $Q1$ and $Q2$ is a change in the direction to increase the current flowing in $Q1$, as was the originally hypothesized signal. The circuit responds to the original input stimulus in a direction to increase the effect of the stimulus, and is therefore regenerative. The action just described will continue to increase the current in $Q1$ and decrease it in $Q2$ until either $Q1$ is saturated, or $Q2$ is cut off, or both.

To analyze the action of the circuit let us assume the simplest possible transistor equivalent circuit as shown in Fig. 5-14. This circuit will give the essential circuit operational details without making the computations difficult. Again the best way to begin the analysis is to assume a state in the binary and see what happens. Let us assume that $Q1$ is OFF and $Q2$ is ON, but not necessarily saturated. Since I_{B2} has been assumed to be negligible or zero, the collector voltage of $Q1$, V_{CN1}, and the base voltage of $Q2$,

V_{BN2}, are easily calculated. (The use of N in a subscript means that voltage is referred to ground, that is, V_{BN1} is the base-to-ground voltage on $Q1$ or v_{in}.)

$$V_{CN1} = \frac{V_{CC}(R_{B1} + R_{B2})}{R_{B1} + R_{B2} + R_{L1}} = \frac{(10)(20\text{ k}\Omega)}{21\text{ k}\Omega} = 9.53\text{ V} \tag{5-7}$$

$$V_{BN2} = \frac{V_{CC}R_{B2}}{R_{B1} + R_{B2} + R_{L1}} = \frac{(10)(10\text{ k}\Omega)}{21\text{ k}\Omega} = 4.76\text{ V} \tag{5-8}$$

The emitter voltage is the base voltage of $Q2$ less 0.7 V or

$$V_{EN1} = V_{EN2} = 4.06\text{ V}$$

For the model chosen for the transistor, the emitter and collector currents are equal, so the collector voltage of $Q2$ is

$$V_{CN2} = 10 - \frac{4.06 \cdot 1\text{ k}\Omega}{2\text{ k}\Omega} = 7.97\text{ V}$$

Transistor $Q2$ is therefore ON as supposed, but not saturated (note $V_{CE2} = 7.97 - 4.06 = 3.91$ V). The input voltage must be low enough to keep $Q1$ OFF, as assumed. From the curve giving a typical base characteristic, as in Fig. 3-14, it is seen that I_B is virtually zero for $V_{BE} < 0.5$ V; therefore $Q1$ will be OFF, as assumed if $v_{\text{in}} < 4.06 + 0.5 \approx 4.6$ V.

If the input voltage v_{in} were slowly raised from zero volts, $Q1$ would be OFF and $Q2$ ON until v_{in} reached the threshold at 4.6 V. At this v_{in} the regenerative switching action previously described would take place, and the circuit would switch rapidly to a new state. To find the values for this new state consider $Q1$ ON, $Q2$ OFF, and $v_{\text{in}} = 4.6$ V. Note that this assumption seems marginal with reference to the previous calculation, but we shall see what happens. For these assumptions

$$V_{EN1} = V_{EN2} = 4.6 - 0.7 = 3.9\text{ V} \tag{5-9}$$

$$I_{C1} = -I_{E1} = \frac{3.9}{2\text{ k}\Omega} = 1.95\text{ mA} \tag{5-10}$$

To find V_{CN1} the collector circuit of $Q1$ may be reduced as shown in Fig. 5-15. If we know I_{C1}, the collector voltage of $Q1$ may be found

$$V_{\text{eq}} = \frac{10 \cdot 20\text{ k}\Omega}{1\text{ k}\Omega + 20\text{ k}\Omega} = 9.53\text{ V}$$

$$V_{CN1} = 9.53 - 1.95 \cdot 0.952\text{ k}\Omega = 7.66\text{ V}$$

$$V_{BN2} = \frac{V_{CN1}}{2} = 3.83\text{ V}$$

The original assumption that $Q2$ is OFF is thus found to be true since $V_{BE2} = 3.83 - 3.9\text{ V} = -0.07$ V. The rather peculiar situation which is characteristic of bistable circuits arises that *either* assumption—$Q1$, ON; $Q2$, OFF or $Q1$, OFF; $Q2$, ON—is valid at $v_{\text{in}} = 4.6$ V. To find the mini-

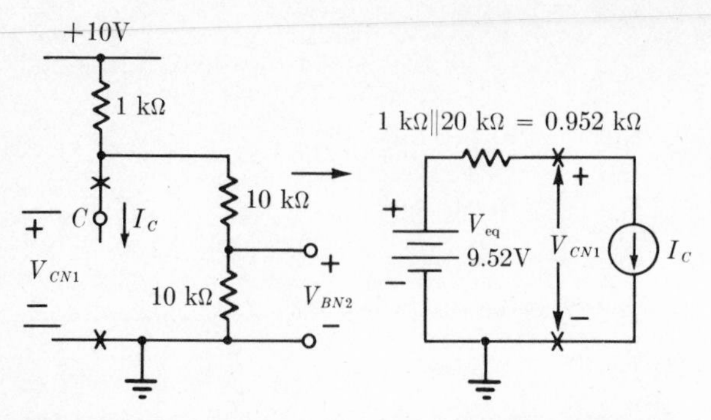

Fig. 5-15 The collector circuit of $Q1$ from Fig. 5-14, and its Thevenin equivalent. Note $Q2$ is assumed OFF and $I_{B2} = 0$.

mum value of v_{in} that will keep $Q1$ ON and $Q2$ OFF, we can write an expression for V_{BE2} and set it equal to the value that just begins to turn $Q2$ ON. The general expression for I_{C1} is

$$I_{C1} = \frac{-h_{fb}(v_{\text{in}} - V_{BE1})}{R_E} \approx \frac{v_{\text{in}} - 0.7}{2\text{ k}\Omega} \tag{5-11}$$

The collector voltage is again found using the Thevenin equivalent of Fig. 5-15.

$$V_{CN1} = 9.52 - 0.95\text{ k}\Omega \cdot I_{C1} = 9.52 - \frac{0.952\text{ k}\Omega(v_{\text{in}} - 0.7)}{2\text{ k}\Omega} \tag{5-12}$$

and

$$V_{BN2} = \frac{V_{CN1}}{2} = 4.76 - \frac{0.952\text{ k}\Omega(v_{\text{in}} - 0.7)}{2 \cdot 2\text{ k}\Omega} \tag{5-13}$$

$$V_{EN} = v_{\text{in}} - 0.7$$

$$V_{BE2} = V_{BN2} - V_{EN} = V_{BN2} - (v_{\text{in}} - 0.7) \tag{5-14}$$

$$= 4.76 - \frac{0.952\text{ k}\Omega(v_{\text{in}} - 0.7)}{4\text{ k}\Omega} - (v_{\text{in}} - 0.7) \tag{5-15}$$

If we assume $V_{BE} \geq 0.5$ V will begin significant transistor current, then we may set V_{BE2} to this value to find the transition value for v_{in}

$$V_{BE2} = 0.5 = 4.76 - \frac{4.95\text{ k}\Omega(v_{\text{in}} - 0.7)}{4\text{ k}\Omega} \tag{5-16}$$

$$v_{\text{in}} = 4.14\text{ V} \tag{5-17}$$

If $v_{\text{in}} \leq 4.14$ V, then the circuit will be in the state $Q2$ ON, $Q1$ OFF. For $4.14 \leq v_{\text{in}} \leq 4.6$ V the circuit can be in either state. The situation is shown graphically in Fig. 5-16 where the state of the bistable multivibrator is indicated by the collector voltage of $Q2$, the normal output terminal. If the input v_{in} starts low, say, 0 V, the output is in the low state and switches to the high state as v_{in} increases past 4.6 V. The output stays in the high

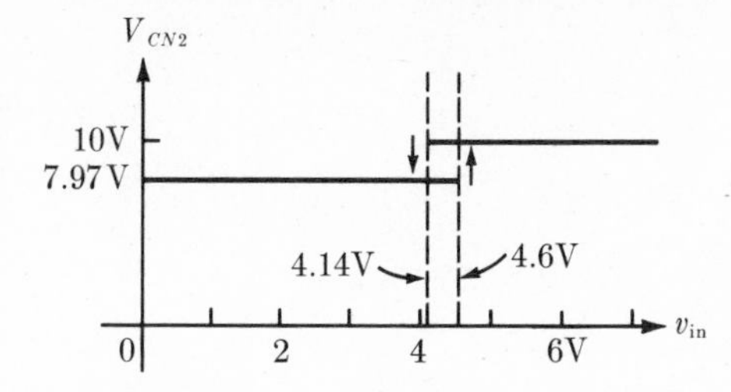

Fig. 5-16 Transfer characteristic for an emitter-coupled binary.

state if v_{in} is subsequently decreased, until v_{in} is decreased below 4.14 V. The difference between the regenerative switching points for inputs going in opposite directions is sometimes called *hysteresis.*

The preceding behavior is well illustrated if v_{in} is made a sine wave with peak amplitude greater than 4.6 V, as is shown in Fig. 5-17. The output of the circuit is then a square wave with abrupt transitions at t_1 and t_2. The speed of these transitions is almost entirely independent of the amplitude and frequency of the input sine wave; therefore this is a good circuit (compared to the clipper circuits described previously) to make a square wave from a low-frequency sine wave. To make the output square wave symmetrical (equal times for the two parts of the cycle) the sine wave must be biased so that its zero axis is in the center of the hysteresis zone. In this case 4.37 V connected in series with the sine wave source would give a symmetrical output.

The amount of hysteresis in the emitter-coupled binary can be varied by adjusting the amount of gain in the two stages, as shown in Fig. 5-18. A way of looking at this figure is to consider the MV as a two stage amplifier with an amount of dc feedback controlled by the position of the arm on the potentiometer R_E. If the arm is placed at the ground end of R_E ($a = 0$), there is no feedback, and $Q1$-$Q2$ act like an ordinary two stage dc amplifier. The transfer characteristic for this condition is shown in Fig. 5-19*a*, where the 8 and 10 V levels represent the output of the amplifier in a limiting condition. The sloping line between the two levels

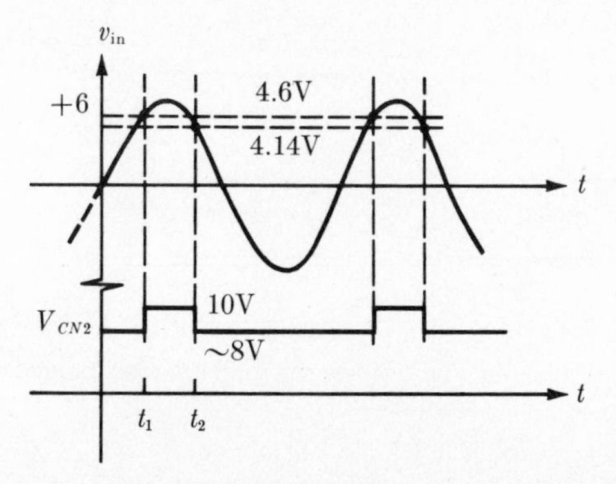

Fig. 5-17 Output from the emitter-coupled bistable MV when the input is a sine wave.

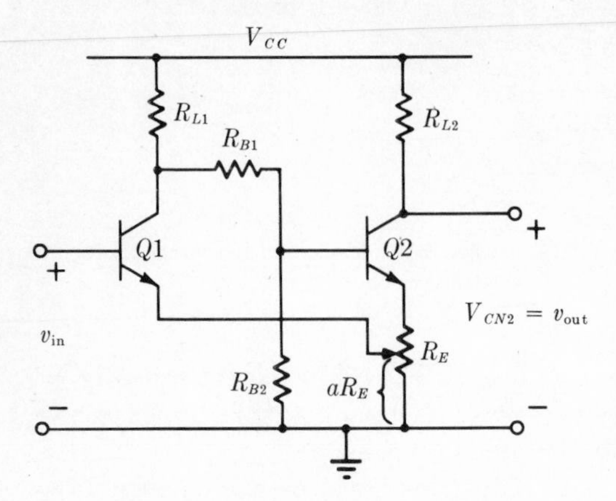

Fig. 5-18 The emitter-coupled bistable MV with an adjustable gain control added. The gain is increased as a is increased $(0 < a < 1)$.

is a region where both stages are operating linearly, and the slope is equal to the gain of the two stages. As the arm of the potentiometer is raised, thereby increasing a, the overall gain of the two stages is increased because of the addition of the positive feedback. At some point ($a \approx 0.9$) enough positive feedback is applied to make the gain of the two stages infinite. At this point the slope of the transfer characteristic also becomes infinite as shown in Fig. 5-19*b*. This may seem like a good way to achieve very high gain, but the situation is very unstable, and temperature or aging effects may cause the gain to change by very large amounts. In the last situation the a has been increased still further, and hysteresis results as shown in Fig. 5-19*c*. The amount of the hysteresis varies smoothly from the amount shown in 5-19*c* to zero as a is decreased from unity to 0.9.

5-7 THE MONOSTABLE MULTIVIBRATOR

The second type of multivibrator we shall investigate is the monostable version, which is permanently stable in one state but only temporarily stable in the other. In such a multivibrator one of the two interstage couplings is usually an ac coupling (capacitor-resistor) and the second is a

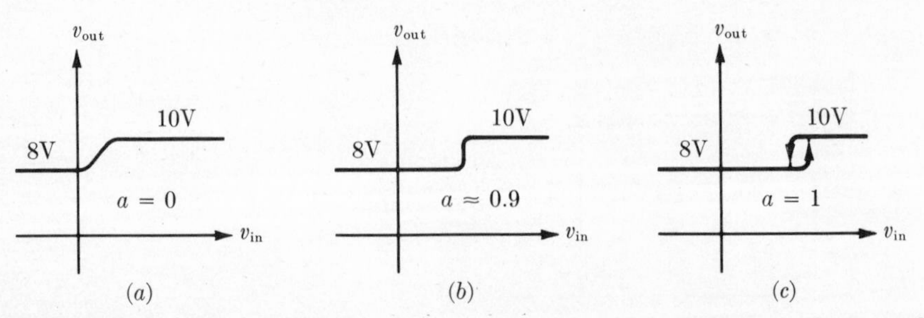

Fig. 5-19 Transfer characteristic of the circuit in Fig. 5-18 as the loop gain is changed with the emitter potentiometer.

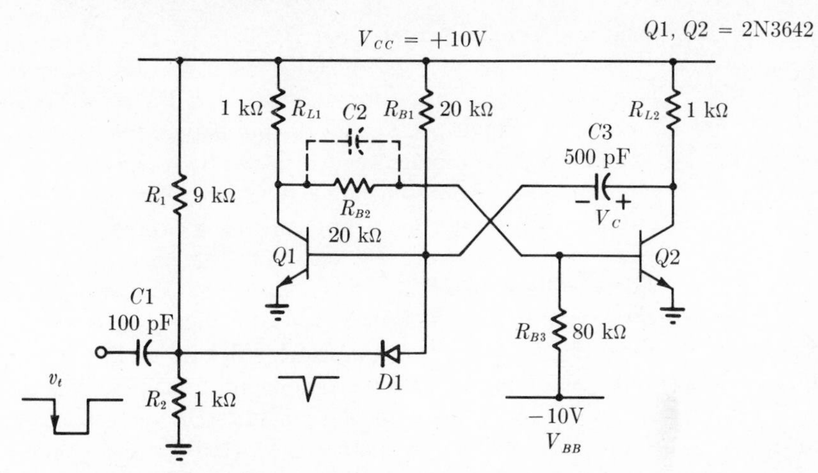

Fig. 5-20 A collector-coupled monostable multivibrator. The initiating trigger is v_t, a negative pulse.

dc coupling; however, such is not always the case. The monostable circuit of Fig. 5-20 has one such dc coupling (R_{B2} and R_{B3}) and one ac coupling (R_{B1} and $C3$). The circuit is very similar to the basic bistable circuit of Fig. 5-9*c* with one of the couplings changed to an *RC* coupling. The permanently stable state of the circuit could conceivably be either with $Q1$ OFF and $Q2$ ON, or with $Q1$ ON and $Q2$ OFF. The first condition, however, is impossible because the base of $Q1$ would have to be kept negative in order to keep $Q1$ OFF, and there is no way to keep the base permanently negative. In fact, the resistor R_{B1} is connected in such a way as to make the base positive.

Let us assume that $Q1$ is ON and $Q2$ is OFF. If this assumption is true, the following voltages appear in the circuit, and can be calculated in the order given.

$$V_{CE1} \approx 0 \text{ V} \tag{5-18}$$

(The ratio of R_{L1} to R_{B1} is such that $Q1$ will be saturated if its h_{FE} is greater than 20, that is, $h_{FE} > R_B/R_L$ insures saturation.)

$$V_{BE2} = \frac{-10 \cdot 20 \text{ k}\Omega}{20 \text{ k}\Omega + 80 \text{ k}\Omega} = -2 \text{ V} \tag{5-19}$$

$$V_{CE2} = +10 \text{ V} \tag{5-20}$$

$$V_{BE1} \approx +0.7 \text{ V (from 2N3642 curves)} \tag{5-21}$$

$$I_{B1} = \frac{10 - 0.7}{20 \text{ k}\Omega} = 0.465 \text{ mA} \tag{5-22}$$

$$I_{C1} \approx \frac{10}{1 \text{ k}\Omega} = 10 \text{ mA} \qquad \text{(neglecting current through } R_{B2}) \tag{5-23}$$

$$V_C = 10 - 0.7 = 9.3 \text{ V} \tag{5-24}$$

Transistor $Q1$ *is* saturated as may be seen from Fig. 3-14 for the values of I_{C1} and I_{B1} given above. Transistor $Q2$ is OFF as supposed because $V_{BE2} = -2$ V. These voltages and currents establish the initial conditions in the circuit. To make the circuit go into its temporarily stable (quasi-stable) state, the current in $Q1$ is momentarily reduced by a trigger pulse v_t provided at $t = 0$. The pulse is differentiated by the combination of C_1 and $R_1 \| R_2$, and the negative-going edge is applied to the cathode of $D1$. If the pulse is large enough to make $D1$ conduct, the base of $Q1$ is also driven negatively. The current in $Q1$ is thus reduced causing its collector voltage to rise toward V_{CC}. The rising collector voltage is coupled by the dc coupling R_{B2}-R_{B3} to the base of $Q2$, which will turn on if its base voltage rises to about +0.5 V, at which time its collector voltage begins to drop. This drop is coupled directly through the capacitor $C3$ to the base of $Q1$ and completes the turn-off process that was started by the trigger signal. At the moment the triggering action is completed, $Q1$ is being held off by the charge on $C3$. However, this charge will not stay constant because of the current flowing down through R_{B1} and into the left side of $C3$. The base voltage of $Q1$ begins to rise toward +10 V and sooner or later $Q1$ will begin to conduct again. This will happen when V_{BE1} rises to about +0.5 V. At this moment current will begin to flow in $Q1$ causing its collector voltage to drop and beginning to turn off $Q2$. This action starts another regenerative cycle culminating in $Q1$ saturated and $Q2$ OFF.

The details of the operation after the trigger pulse is applied at $t = 0$ may be studied in detail with the aid of the equivalent circuit of Fig. 5-21a. Two equivalent circuits are required to give all the voltages and currents in the circuit during the quasi-stable period. In the left-hand circuit we have

$$\left.\begin{aligned} I_1 &= \frac{10 - 0.7}{1\text{ k}\Omega + 20\text{ k}\Omega} = 0.443\text{ mA} && \text{(5-25)} \\ I_2 &= \frac{10 + 0.7}{80\text{ k}\Omega} = 0.134\text{ mA} && \text{(5-26)} \\ I_{B2} &= I_1 - I_2 = 0.309\text{ mA} && \text{(5-27)} \\ V_{CE1} &= 10 - 1\text{ k}\Omega(0.443) \approx 9.56\text{ V} && \text{(5-28)} \end{aligned}\right\} \quad t = 0^+$$

In the right-hand circuit the transistor $Q2$ is assumed to be saturated, an assumption that we shall have to prove later. The currents and voltages in this circuit at $t = 0^+$ are

$$I_3 = \frac{10 + 9.3}{20\text{ k}\Omega} = 0.97\text{ mA} \tag{5-29}$$

$$I_{C2} = I_3 + \frac{10}{1\text{ k}\Omega} = 0.97 + 10 = 10.97\text{ mA} \tag{5-30}$$

For $Q2$ to be saturated requires an $h_{FE} > 10.97/0.309 \approx 36$. Reference to Fig. 3-14 shows that the 2N3642 is indeed saturated. The equivalent cir-

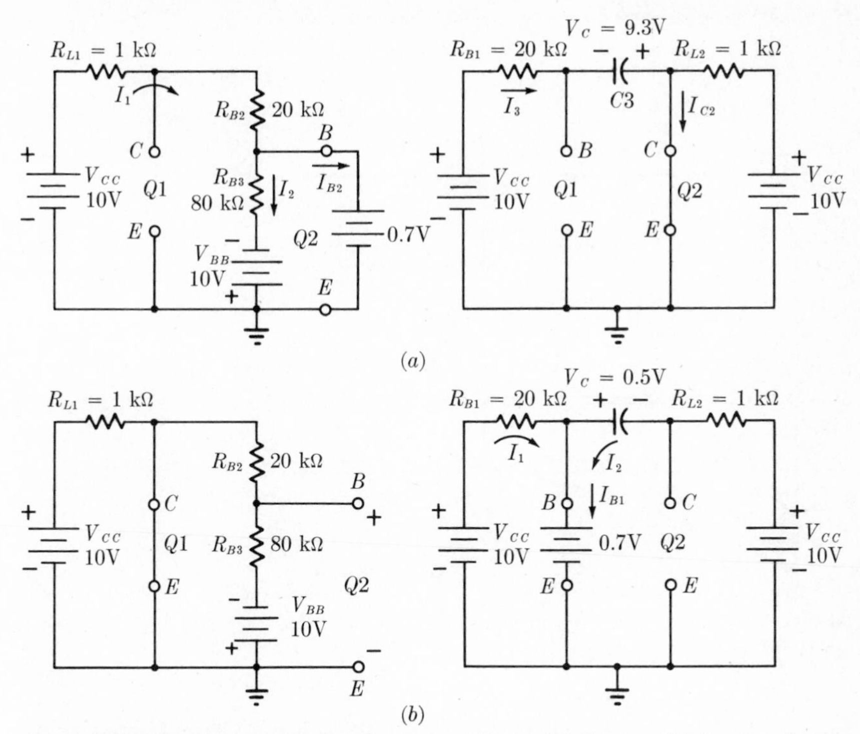

Fig. 5-21 Equivalent circuits for the monostable multivibrator. (a) for $0 < t < t_1$ (values for V_C given at $t = 0^+$). (b) For $t > t_1$.

cuit shows that the voltage on the base of $Q1$, V_{BE1} is going to be an exponential with an initial value of -9.3 V and a final value of $+10$ V. The time constant for the exponential is $\tau_1 = 20 \text{ k}\Omega \cdot 500 \text{ pF} = 10\ \mu\text{s}$. The actual equation for the base voltage, which is valid as long as $Q1$ does not conduct, is

$$V_{BE1}(t) = 10 - (9.3 + 10)\epsilon^{-t/\tau_1} \tag{5-31}$$

The end of the quasi-stable state occurs when $Q1$ begins again to conduct at a $V_{BE1} \geqq 0.5$ V. The time t_1 at which this occurs may be found by equating $V_{BE1}(t)$ to 0.5 V.

$$0.5 = 10 - (9.3 + 10)\epsilon^{-t_1/\tau_1} \tag{5-32}$$

$$t_1 = \tau_1 \ln\left(\frac{10 + 9.3}{10 - 0.5}\right) = 0.71\,\tau_1$$

$$\cdots \qquad \cdots = 7.1\ \mu\text{s} \tag{5-33}$$

The waveforms which result in this interval are shown in Fig. 5-22. The operation of the circuit is not completely finished at $t = t_1$ even though

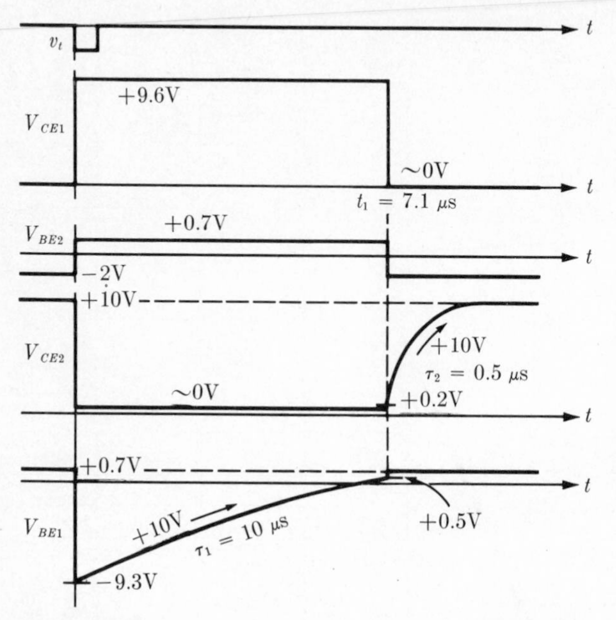

Fig. 5-22 Waveforms for the collector-coupled monostable multivibrator.

the original condition of $Q1$ ON and $Q2$ OFF is reestablished, because the capacitor has not yet regained its original charge. For $t < 0$ the capacitor was charged to 9.3 V, but the capacitor charge at t_1 is only -0.5 V. Because of the recharging of $C3$ necessary after t_1, the time interval following t_1 is known as the recovery time of the multivibrator and is important because the circuit cannot be retriggered for another cycle of operation until the end of the recovery or recharging time. To compute the behavior of the circuit for $t > t_1$ it is necessary to have two new equivalent circuits as shown in Fig. 5-21*b*. (These equivalent circuits would also have been valid for $t < 0$ with the appropriate value for charge on $C3$.) Again simple equivalent circuits have been chosen for the saturated transistor $Q1$ and the cutoff transistor $Q2$. The left-hand circuit gives no new information—$V_{CE1} = 0$ V, and $V_{BE2} = -2$ V—but the right circuit determines the recovery time. For $t = t_1^+$ a large base current flows in $Q1$:

$$I_{B1} = I_1 + I_2 = \frac{10 - 0.7}{20 \text{ k}\Omega} + \frac{10 + 0.5 - 0.7}{1 \text{ k}\Omega} = 10.3 \text{ mA} \qquad (5\text{-}34)$$

and $Q1$ is in a highly saturated condition. The time constant for recharging $C3$, the timing capacitor, is $R_{L2}C3 = 10^3(5 \times 10^{-10}) = 0.5\ \mu\text{s} = \tau_2$. The value of V_{CE2} immediately after t_1 is $0.7 - 0.5 = 0.2$ V, and is charging toward +10 V. With the model shown V_{BE1} stays constant at +0.7 V, but with a more sophisticated model we would see a slightly greater value at t_1^+ decaying to a final value of 0.7 V. The time constant, however, would be almost exactly as previously calculated, because r_{bIII} is small compared to R_{L2}. The values computed for the recovery time interval are also shown in the waveforms of Fig. 5-22.

It is important to recognize the effect of triggering the circuit before it is fully recovered, as it will be for $t \geq 5t_2 + t_1 = 2.5 + 7.1 = 9.6$ μs. If the circuit were to receive a sufficiently large trigger pulse before this time, triggering would result, but the capacitor $C3$ would not have had a chance to completely recharge. The effect of the incomplete recharging can be deduced from the waveform for V_{BE1}, the timing waveform, because the initial value after triggering would be less negative than the −9.3 V shown; therefore, the period would be reduced.

Several components in Fig. 5-20, which are not essential to the operation of the MV, can now be explained. One of these is the capacitor $C2$ across R_{B2}, which serves the dual function of speeding up the transitions between states by turning on and off $Q2$ more rapidly and of making the circuit somewhat more readily triggered by coupling the high-frequency components of the trigger from the collector of $Q1$ to the base of $Q2$. Oddly enough, however, ease of triggering is undesirable in many applications because the circuit may become inadvertantly triggered by small stray pulses. This problem is the reason for adding R_1, which biases the cathode of $D1$ slightly positive. The input trigger must then be at least big enough to forward bias $D1$ before it will appear on the base of $Q1$ and trigger the rest of the circuit. Another reason for biasing the diode is to make sure that none of the dc base current supplied to $Q1$ through R_{B1} is lost in $D1$.

The monostable circuit may be triggered in many ways other than the one shown thus far. Figure 5-23 shows some of the possible ways and the polarity of trigger signal required. Point a is the poorest point for triggering because the trigger signal is fed into the very low impedance of the saturated transistor. Point b is a very good point because the circuit is immediately disconnected from the trigger source by reverse biasing of the trigger coupling diode. Therefore, the circuit is not likely to be disturbed by signals on the trigger line during the operating period of the MV. Point c has similar advantages but uses a transistor to couple and invert the trigger signal. Point d is satisfactory if the capacitor $C2$ is not so large as to effectively short the trigger to ground through the saturation resistance of $Q1$. The diodes shown may sometimes be omitted and the trigger signal fed directly into the indicated points via a small capacitor, but the isolation provided by the diode during the MV period is lost, of course, and the circuit may be falsely triggered during the operation of the MV.

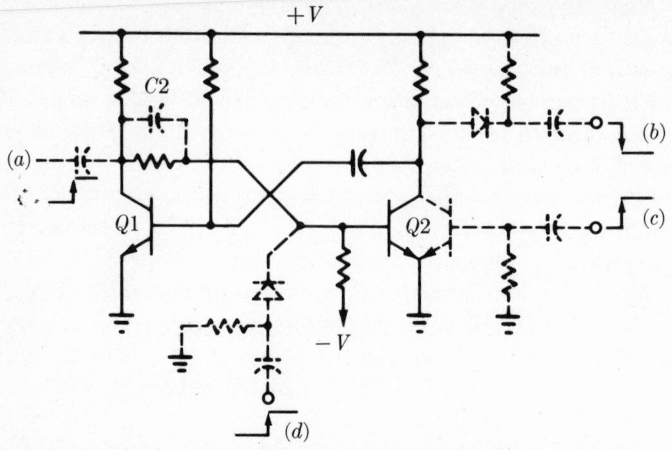

Fig. 5-23 Alternative methods of triggering the monostable multivibrator of Fig. 5-20.

5-8 AN EMITTER-COUPLED MONOSTABLE MULTIVIBRATOR

The preceding circuit, although satisfactory in many respects, does require both positive and negative power supplies and does not have an output terminal which can be loaded without significantly affecting the operation of the circuit. A circuit which does not have these drawbacks, and which has fewer components, is the emitter-coupled version of Fig. 5-24, which has a resemblance to the binary of Fig. 5-14 with R_{B1} replaced by C and V_{in} by a constant voltage. For practice and variety this time we

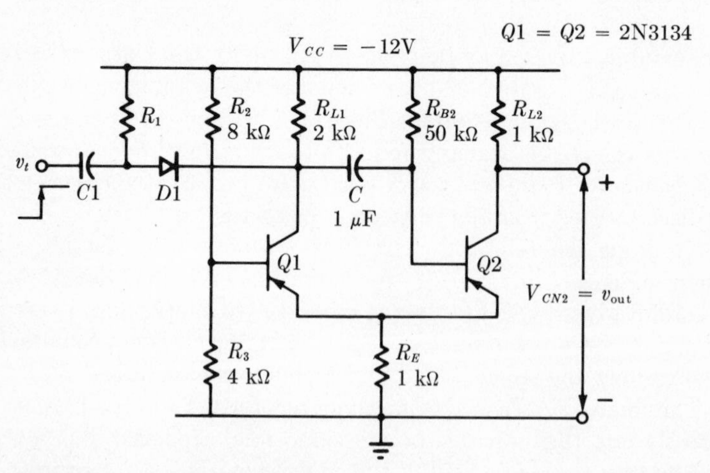

Fig. 5-24 An emitter-coupled monostable multivibrator using PNP transistors.

shall use PNP transistors instead of NPN; the supply voltage this time is negative. If the circuit is monostable as supposed, then in steady state $Q2$ must be ON since its base is returned through R_{B2} to -12 V. Therefore make the initial assumptions that $Q1$ is OFF and that $Q2$ is ON and saturated. If these assumptions are true, the circuit will be as in Fig. 5-25 where a simple equivalent circuit has been used for the saturated transistor. The voltages and currents in this circuit may be computed as follows assuming that the circuit is in steady state conditions (untriggered):

$$V_{CN1} = -12 \text{ V} \tag{5-35}$$

$$V_E = R_E(I_B + I_C) = R_E\left(\frac{-12 + 0.6 - V_E}{50 \text{ k}\Omega} + \frac{-12 - V_E}{1 \text{ k}\Omega}\right) \tag{5-36}$$

By solving for V_E we get

$$V_E = -6.05 \text{ V} \tag{5-37}$$

$$V_{BN2} = V_E + V_{BE} = -6.05 - 0.6 = -6.65 \text{ V} \tag{5-38}$$

$$V_{CN2} \approx V_E = -6.1 \text{ V} \tag{5-39}$$

$$I_{B2} = \frac{V_{CC} - V_{BN2}}{R_{B2}} = \frac{-12 + 6.7}{50 \text{ k}\Omega} = -0.106 \text{ mA} \tag{5-40}$$

$$I_{C2} = \frac{V_{CC} - V_{CN2}}{R_{L2}} = \frac{-12 + 6.1}{1 \text{ k}\Omega} = -5.9 \text{ mA} \tag{5-41}$$

The fact that $Q2$ is saturated may now be verified by referring to Fig. 3-32 and checking the above values of I_B and I_C. The one remaining important value from Fig. 5-25 is the charge on C.

$$V_X = V_{BN2} - V_{CN1} = -6.65 + 12 = 5.35 \text{ V} \tag{5-42}$$

The fact that $Q1$ is OFF is shown by computing its base voltage

$$V_{BN1} = \frac{(-12)(4 \text{ k}\Omega)}{4 \text{ k}\Omega + 8 \text{ k}\Omega} = -4 \text{ V} \tag{5-43}$$

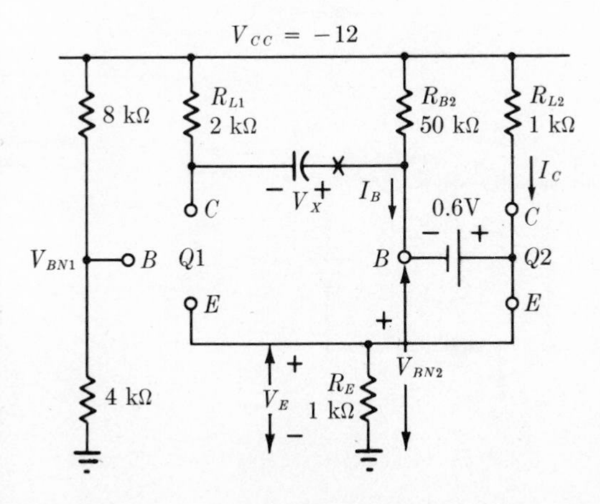

Fig. 5-25 An equivalent circuit for the emitter-coupled MV in the quiescent (untriggered) state.

and comparing with the emitter voltage, −6.05 V. The base is more positive than the emitter; therefore, the transistor is OFF.

The circuit can be triggered by applying a *positive* pulse to C_1 which turns $Q2$ OFF and $Q1$ ON. Conditions in the circuit are now as in Fig. 5-26, with $Q1$ ON, but probably not saturated because of the large emitter resistor. Let us assume that it is not saturated and calculate the collector current. The process used in Sec. 3-20 is useful here because $Q1$ operates very much like a common-base stage. Figure 5-26*a* shows the source for the base of $Q1$ reduced to its Thevenin equivalent of −4 V and 2.67 kΩ. If we neglect I_{B1} as a first approximation,

$$V_E \approx -4 - V_{BE1} = -4 + 0.6 = -3.4 \text{ V} \tag{5-44}$$

Then $I_E = -V_E/R_E \approx +3.4$ mA. For a first approximation

$$I_C = -I_E = -3.4 \text{ mA} \tag{5-45}$$

and $V_{CE1} \approx -12 - (-3.4)(3 \text{ k}\Omega) = -1.8$ V. Since this is a normal voltage for a PNP transistor in Region II, the transistor is *not* saturated. A more exact set of operating conditions may now be obtained by finding the actual base current from Fig. 3-32. From the figure $I_B = -26\ \mu$A, and

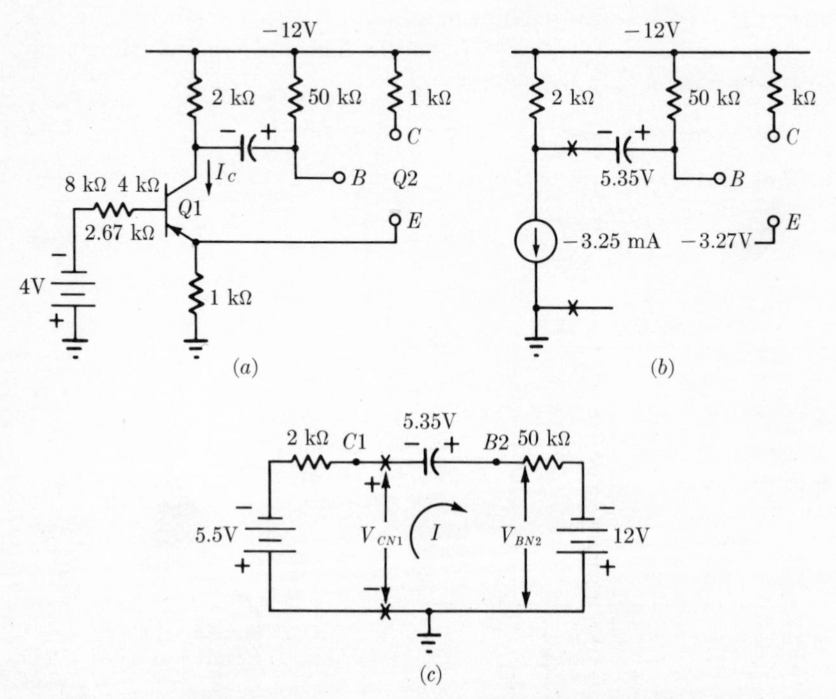

Fig. 5-26 Equivalent circuits for $0 < t < t_1$ for the emitter-coupled multivibrator.

a corrected value of V_{BN1} is $-4 - (-0.026)(2.67\ \text{k}\Omega) = -3.93$ V. Using the value of V_{BE} from the curves for $I_B = -26\ \mu\text{A}$, we obtain for

$$V_E = -3.93 - (-0.66) = -3.27 \text{ V} \tag{5-46}$$
$$I_E = +3.27 \text{ mA} \tag{5-47}$$
$$I_C = -3.27 + 0.026 = -3.25 \text{ mA} \tag{5-48}$$

These values are about as accurate as can be read from the curves. The emitter current of $Q1$ is very nearly completely independent of the collector voltage since the current is largely determined by the -4 volts and R_E. Therefore $Q1$ behaves like a transistor operating with common base and an emitter current of $+3.27$ mA. The collector current is very nearly represented by an ideal current generator of value -3.25 mA, as shown in the circuit of Fig. 5-26*b*. The current generator accurately represents the current removed from the collector node no matter where the other end of the current generator is connected. The most convenient place for the bottom end of the current generator is ground, since we already know what the value of V_E is: -3.27 V.

The equivalent circuit for finding the timing waveform is Fig. 5-26*c*, which is the same as Fig. 5-26*b* with the portion to the left of x-x replaced by its Thevenin equivalent. The current I for $t > 0$ is

$$I(t) = \frac{12 + 5.35 - 5.5}{2\ \text{k}\Omega + 50\ \text{k}\Omega}\,\epsilon^{-t/\tau_1}$$
$$= 0.228\epsilon^{-t/\tau_1} \text{ mA} \tag{5-49}$$
$$\tau_1 = (52\ \text{k}\Omega)(10^{-6}\ \text{F}) = 52 \text{ ms} \tag{5-50}$$

The collector voltage of $Q1$ and the base voltage of $Q2$ may be found knowing the value of $I(t)$;

$$V_{CN1} = -5.5 - (2\ \text{k}\Omega)(0.228\epsilon^{-t/\tau_1})$$
$$= -5.5 - 0.456\epsilon^{-t/\tau_1} \tag{5-51}$$
$$V_{BN2} = -12 + (50\ \text{k}\Omega)(0.228\epsilon^{-t/\tau_1})$$
$$= -12 + 11.4\epsilon^{-t/\tau_1} \tag{5-52}$$

These values may be used together with the initial values previously found to begin plotting the waveforms in the circuit as shown in Fig. 5-27. From the waveforms one can see that the positive step in V_{CN1} that occurred when the circuit was triggered is transmitted through the timing capacitor to the base of $Q2$ and considerably reverse biases it. However, the capacitor is discharging through R_{B2}, and the base voltage V_{BN2} is falling toward the emitter voltage V_E. The circuit will switch back to its original state when V_{BN2} has fallen sufficiently to permit $Q1$ to conduct again. From Fig. 3-32 this occurs at a base voltage of about -0.55 V. Therefore

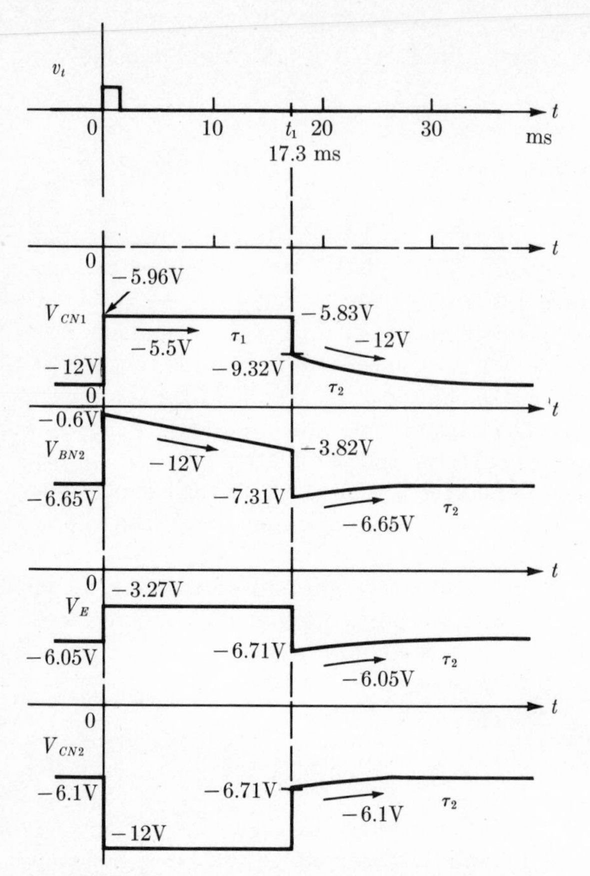

Fig. 5-27 Waveforms for the emitter-coupled PNP multivibrator: $\tau_1 = 52$ ms, $\tau_2 = 2.5$ ms.

the time t_1 that the circuit switches may be found by equating V_{BN2} to $V_E - 0.55$ V:

$$V_{BN2}(t_1) = -12 + 11.4\epsilon^{-t_1/\tau_1} = -3.27 - 0.55 = -3.82 \text{ V} \tag{5-53}$$

$$t_1 = \tau_1 \ln \frac{11.4}{8.18} = 17.3 \text{ ms} \tag{5-54}$$

This is the useful timing interval of the multivibrator. After t_1 the circuit is again in the state $Q1$ OFF, $Q2$ ON, but the original circuit conditions are not immediately reestablished because the timing capacitor takes a finite time to recharge. To find the circuit conditions after t_1 the circuit of Fig. 5-25 is valid, but the rather complicated resistive network to the right of X (next to the timing capacitor) needs to be replaced by a

Thevenin equivalent. This replacement has been made in the equivalent circuit of Fig. 5-28, which gives the details of the loop containing the timing capacitor for $t > t_1$. Note that the voltage in the Thevenin equivalent, -6.65 V, is the same as was obtained for V_{BN2} for $t < 0$. This must true for the initial and final resting conditions of the circuit to be the same. Again proceeding from an equation for the current $I(t)$ we can write the equations for the important voltages in the circuit:

$$I(t) = \left(\frac{12 - 6.65 - 2.01}{2\text{ k}\Omega + 0.495\text{ k}\Omega}\right)\epsilon^{-(t-t_1)/\tau_2}$$

$$= 1.34\epsilon^{-(t-t_1)/\tau_2}\text{ mA} \tag{5-55}$$

$$V_{CN1} = -12 + (2\text{ k}\Omega)(1.34\epsilon^{-(t-t_1)/\tau_2})$$

$$= -12 + 2.68\epsilon^{-(t-t_1)/\tau_2} \tag{5-56}$$

$$V_{BN2} = -6.65 - 0.495\text{ k}\Omega[I(t)]$$

$$= -6.65 - 0.66\epsilon^{-(t-t_1)/\tau_2} \tag{5-57}$$

$$\tau_2 = (2\text{ k}\Omega + 0.495\text{ k}\Omega)(10^{-6}\text{ F}) \approx 2.5\text{ ms} \tag{5-58}$$

These voltages are also plotted as the waveforms in Fig. 5-27 after t_1. The two remaining unknown voltages are the emitter voltage and the collector voltage of $Q2$.

$$V_{CN2} \approx V_E = V_{BN2} + 0.6\text{ V} = -6.05 - 0.66\epsilon^{-(t-t_1/\tau_2)} \tag{5-59}$$

This completes the solution of the multivibrator cycle, and shows that the circuit will recover and be ready for another trigger pulse in about $3\tau_2 \approx 8$ ms.

The recovery time can be shortened by using an emitter follower to help recharge the timing capacitor as shown in Fig. 5-29*a*. The effect of the follower during the timing interval is small since it merely repeats the collector waveform of $Q1$ at its emitter. The small exponential appearing on V_{CN1} is nearly removed, however, and τ_1 is reduced from 52 to about 50 ms. During the recovery period the circuit is greatly influenced because

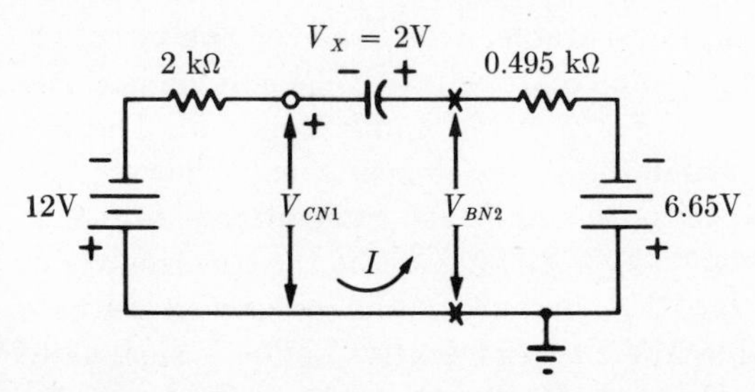

Fig. 5-28 The circuit for computing the waveforms in the recovery time interval $t > t_1$.

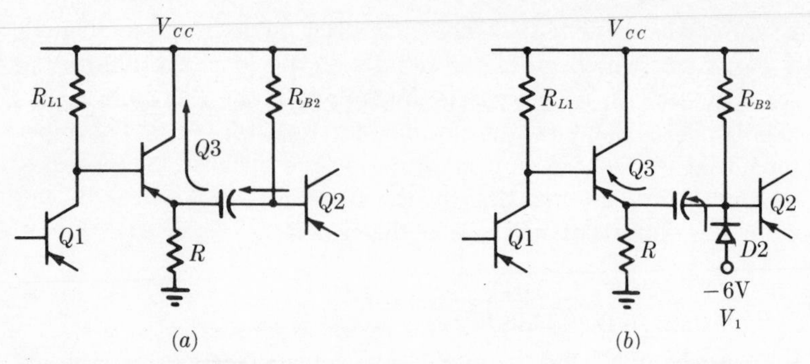

Fig. 5-29 Modifications to the basic emitter-coupled MV to speed up the recovery time. The arrows show the path of the recharging currents.

the effective impedance for recharging C becomes $R_{L1}/h_{fe3} + 0.495$ kΩ instead of $R_{L1} + 0.495$ kΩ, as before. Therefore the contribution of R_{L1} to the recovery time is greatly reduced, but the impedance looking from C into $Q2$ is unchanged. To reduce the latter impedance, a diode may be placed to conduct the recharging current to a constant-voltage source V_1 as shown in Fig. 5-29*b*. This diode has the effect of removing the exponential seen on V_E, V_{BN2}, and V_{CN2} after t_1 so that the greatest effect of the recovery-time constant will be seen at the collector of $Q1$. The diode $D2$ also can be useful by slightly reducing the magnitude of V_1. This will reduce the magnitude of V_{BN2} in the quiescent state of the MV so that $Q2$ will not be in the saturated state. Such an MV can be operated with very short transition times because both transistors operate in either Region I or II, never in Region III. Since the transistors are never saturated, the storage times associated with saturation do not occur.

5-9 THE TRANSISTOR ASTABLE MULTIVIBRATOR

At the opposite extreme from the bistable circuit is the astable version, which will not remain permanently in either of its stable states, i.e., the states are only temporary or quasi-stable. One form of the circuit is shown in Fig. 5-30 together with some representative element values. The contrast with the bistable circuit of Fig. 5-2 is readily apparent. There is no resistive or dc coupling between the active elements, only the coupling capacitors $C1$ and $C2$. The capacitive coupling will transmit instantaneous voltage transitions; but following any switching action the base voltage of either transistor will rise toward V_{CC} because of the resistors in the base circuits. In this type of astable MV the regenerative action is such as to drive one of the transistors into cutoff (Region I) and the other transistor into saturation (Region III). The half period of the MV will be determined by the length of time that the cutoff transistor remains cut off since it will sooner or later have to come into conduction.

Beginning the analysis of an astable circuit is always somewhat of a puzzle because there is no firm starting point as there is in the case of the monostable circuit. The best way is to assume a state, perform the analysis for one cycle, and check to see if the assumed initial conditions exist. If the assumed initial conditions agree with the calculated ones, then all is well, but if there is disagreement, the calculation has to be done over again with a new set of initial conditions. Usually the second set of initial conditions can be estimated very accurately from the results of the first calculation. For the circuit of Fig. 5-30 let us assume that $Q1$ is OFF and $Q2$ is ON for $t < 0$. The collector voltage of $Q1$ must then be $+6$ V, and the base voltage negative. If we assume further that $Q2$ is saturated when ON, then its collector voltage will be approximately 0 V and its base voltage (from the 2N3642 data) about $+0.7$ V. These values are indicated on the waveforms of Fig. 5-31 and for $t < 0$. As discussed previously the bases cannot be permanently negative to hold the transistors off because of the presence of R_{B1} and R_{B2}. Therefore, the base of $Q1$, which we assumed negative, must be rising toward $+6$ V, and sooner or later $Q1$ will turn ON. The turn-on point for a 2N3642 is about $+0.55$ V, so let us assume this voltage has been reached at $t = 0$. $Q1$ then begins to conduct and to turn off $Q2$ in the normal regenerative switching process. At the end of the switching process $Q1$ will be ON (and assumed saturated), and $Q2$ OFF. The collector voltage of $Q1$ therefore drops from $+6$ to 0 V, and the base of $Q2$ must change a similar amount—from $+0.7$ to -5.3 V. The half period of the MV is determined by the time it takes the base of $Q2$ to change from -5.3 V to the switching point, $+0.55$ V.

The details of the operation may be found by drawing an equivalent circuit for the two loops of the MV, which is valid for $0 < t < t_1$ as shown in Fig. 5-32. The essential timing circuit is Fig. 5-32a from which the

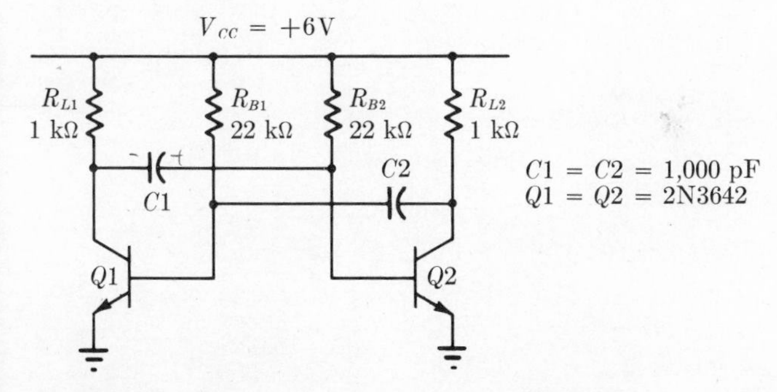

Fig. 5-30 A collector-coupled saturating astable MV.

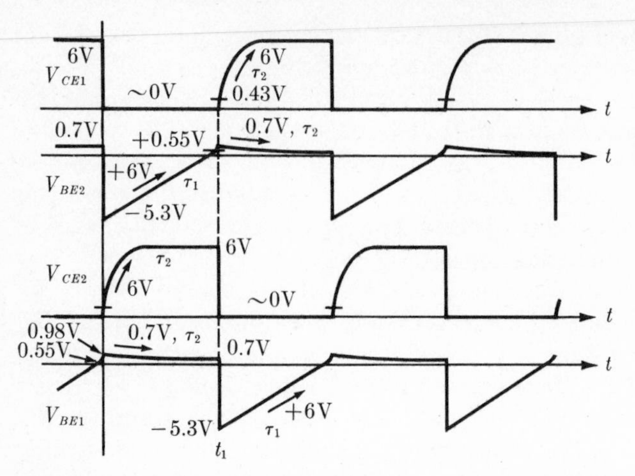

Fig. 5-31 Waveforms in the collector-coupled astable MV: $\tau_1 = 22\ \mu\text{s}$, $\tau_2 = 1.05\ \mu\text{s}$

equation for the timing waveform can be written by inspection, if we assume $R_{\text{sat}} \approx 0$.

$$V(t) = V_{\text{final}} + (V_{\text{initial}} - V_{\text{final}})\epsilon^{-t/\tau} \tag{5-60}$$

$$V_{BE2} = 6 + (-5.3 - 6)\epsilon^{-t/\tau_1}$$

$$= +0.55 \Big|_{t=t_1} \tag{5-60a}$$

$$\tau_1 = (22\ \text{k}\Omega)(10^{-9}) = 22\ \mu\text{s} \tag{5-61}$$

$$t_1 = \tau_1 \ln \frac{11.3}{5.45} = 16\ \mu\text{s} \tag{5-62}$$

This calculation is based upon the assumption that the timing capacitor $C1$ was charged fully at $t = 0$. Because of the symmetry of the circuit this assumption may be tested by finding the behavior of the second timing capacitor for the interval succeeding: $0 < t < t_1$. The circuit for this calculation is Fig. 5-32*b*, which can be simplified by replacing the portion to the left of *x-x* by the Thevenin equivalent of Fig. 5-32*c*. The time constant for this circuit is $\tau_2 = (1\ \text{k}\Omega + 0.05\ \text{k}\Omega)(10^{-9}) = 1.05\ \mu\text{s}$. Therefore, there is ample time in the 16 μs half period to recharge $C2$ (or $C1$ in the preceding half period) to the full supply voltage. To find the details of the recharging we must first find $V_{X2}(0)$:

$$V_{X2}(0) = V_{X1}(t_1) = V_{BE1}(0^-) - V_{CE2}(0^-) = +0.55 - 0$$

$$= 0.55\ \text{V} \tag{5-63}$$

$$I(0^+) = \frac{6 + 0.55 - 0.7}{1.05\ \text{k}\Omega} = \frac{5.85}{1.05\ \text{k}\Omega} = 5.57\ \text{mA} \tag{5-64}$$

$$V_{BE1}(0^+) = 0.7 + (0.05)(5.57) = 0.98\ \text{V} \tag{5-65}$$

$$V_{CE2}(0^+) = V_{BE1} - V_{X2} = 0.43\ \text{V} \tag{5-66}$$

Note also that the final value of V_{BE1} is the same as the initial value assumed for V_{BE2}, and that this value is actually attained since there are $16/1.05 = 15.2$ time constants for the capacitors to recharge. The assumption that the transistors are saturated when ON is easily verified by computing the actual base and collector currents. Just after the switching transient, the base and collector currents are very nearly equal so that the transistor is exceedingly well saturated. The minimum base current flows at the end of the half period; therefore the conditions for saturation should be checked at this time.

The preceding calculations for one half period are also valid for the succeeding half period if the subscripts 1 and 2 are interchanged. The total period of the MV is $2 \times 16 = 32\ \mu s$, or the frequency of operation is $1/32\ \mu s = 36$ kHz. The period is directly proportional to either the R_B's or C's if both resistors or both capacitors are changed together. The period may also be altered by changing the voltage to which the R_B's are returned. When changing either the return voltage or R_B's as a means of changing period, care must be taken to insure enough base current to keep the transistors saturated.

This type of astable MV has one problem which can lead to trouble: in addition to the two normal quasi-stable states, the circuit possesses a permanently stable state with *both* $Q1$ and $Q2$ ON and saturated. The gain of either $Q1$ or $Q2$ in this condition is near zero so that the circuit is per-

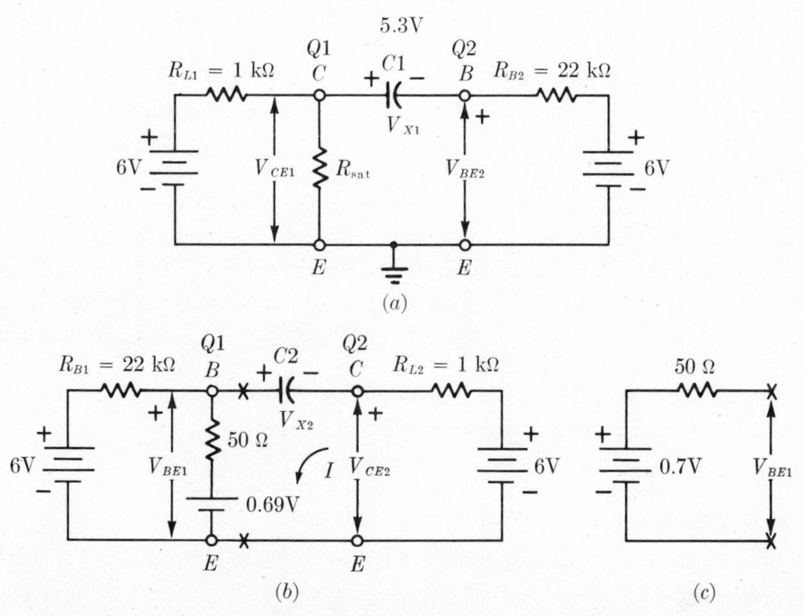

Fig. 5-32 Equivalent circuits valid for Fig. 5-30 in the time interval $0 < t < t_1$.

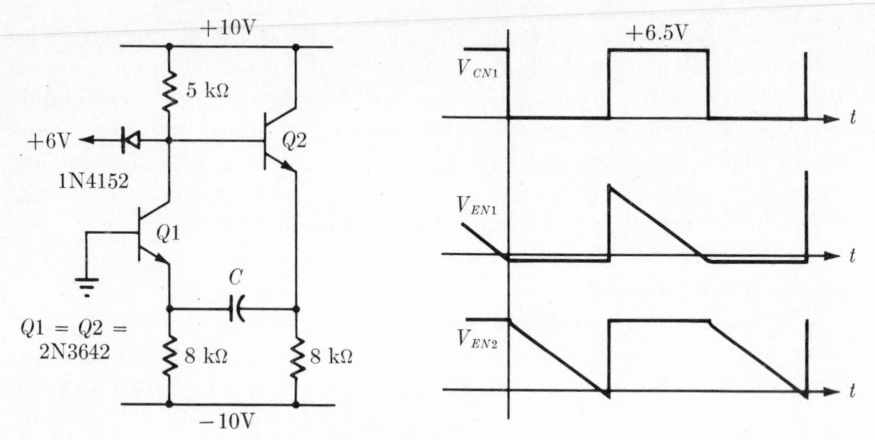

Fig. 5-33 An astable MV for generating a good square wave without slow charging exponentials.

fectly stable. Whether the circuit goes into this nonoscillating state or into the normal oscillating state when first turned on depends upon the symmetry of the circuit. If the circuit is being synchronized by an external signal, the synchronizing pulse usually will start the circuit; otherwise, if dependable starting is required, a nonsaturating form of astable MV like that in Prob. 5-6 is preferable.

Not all astable MV's require two timing capacitors; in fact, the monostable MV of Fig. 5-24 will become astable if the voltage at the base of $Q1$ is lowered to about -7 V. With this base voltage $Q1$ cannot be permanently held off, but will be OFF for a part of the normal recovery period. Therefore, the output of the MV will be a long pulse followed by a short pulse about as long as the normal recovery time. In Fig. 5-33 is shown a form of astable MV which appears to be very unsymmetrical but which nevertheless produces a very nearly perfect square wave with equal ON and OFF times. The waveforms for this circuit are shown in the figure, but the details of analysis are left for a problem. One point should be mentioned: the switching point for the half period when $Q1$ is conducting occurs when $Q1$ goes from Region III (saturated) to Region II *and* $Q2$ goes into conduction. The useful output from the circuit is V_{CN1}, which does not have the usual charging exponentials to give the waveform a long rise time as found in the collector waveforms of Fig. 5-31.

5-10 A FIELD-EFFECT TRANSISTOR MULTIVIBRATOR

A multivibrator may be constructed using almost any form of active device as the gain element, but there are particular advantages to particular devices. The junction transistor used in the previous examples is readily available and cheap, but suffers from the fact that considerable base current must be supplied to it in the ON condition. This limits the magnitude of resistor that may be placed in the base circuit to less than R_L/h_{FE} in practice. If very long periods are desired, a high value of base resistance is required to keep the size of the timing capacitor within reasonable bounds. If a field-effect transistor is used instead, the base resistor (in this case, the gate resistor) may be made much larger because only the leakage current to the gate need be supplied through the resistor. An example of such a circuit is shown in Fig. 5-34, which is an astable MV using FET's. Note that the ratio between the load R_L and timing resistors R_G is 10,000 kΩ/3 kΩ = 3,333. Even this large a ratio is not the limit, but

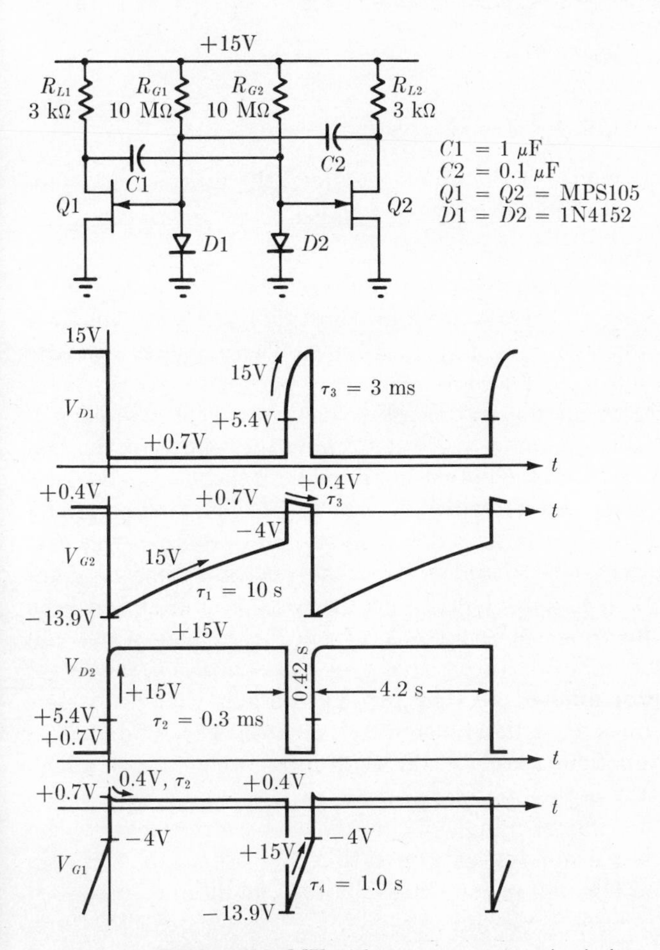

Fig. 5-34 An FET astable MV with 10:1 asymmetry in timing.

is about an order of magnitude greater than can be attained with junction transistors.

The waveforms for the circuit are also shown in Fig. 5-34 and can be calculated in a manner similar to those for the transistor astable MV. Two significant differences should be noted, however. First, the FET operated as shown is not quite as perfect a switch as the junction transistor, so that the drain waveform (corresponding to the collector waveform) does not go quite as close to ground. Second, the FET goes from Region I (cutoff) to Region II (linear) at a negative gate voltage; therefore the timing waveform does not terminate near zero volts as it did with the junction transistors. The FET curves in Fig. 3-39 show that cutoff occurs in the given FET at about -4 V on the gate; therefore the solution for the timing in the first half period is

$$V_{G2}(t) = 15 - 28.9\epsilon^{-t/\tau_1} = -4 \bigg|_{t=t_1} \tag{5-67}$$

$$t_1 = \tau_1 \ln\left(\frac{28.9}{19}\right) = 0.42\tau_1 \tag{5-68}$$

Because of the large values of the timing resistors, the time constants for the "half periods" are almost exactly $\tau_1 = R_{G2}C_1 = 10$ s, and

$$\tau_4 = R_{G1}C_2 = 1 \text{ s}$$

Because of the diodes, the recovery time constants are approximately $\tau_2 = R_{L2}C_2 = 0.3$ ms and $\tau_3 = R_{L1}C_1 = 3$ ms. These values are also shown on the waveforms in Fig. 5-34.

This MV is also an illustration of an asymmetrical astable MV because the two time constants which determine the length of the half periods are different. One of the difficulties in making such a circuit is also illustrated; that is, the capacitor which determines the long half period $C1$ must be recharged during the subsequent short half period. In this case the 0.42 second interval is adequate since the recovery time is about $5\tau_3 = 15$ ms $\ll 0.42$ s. If a 10:1 difference between τ_1 and τ_4 had been attempted in the junction transistor MV of Fig. 5-30, there would not have been sufficient time for the recovery of the charge on the larger capacitor, and a 10:1 difference in the two half periods would not have actually occurred for a 10:1 difference in $C1$ and $C2$. The fundamental problem with the junction transistor MV that does not allow such a high degree of assymmetry is the limitation on the ratio of R_B to R_L.

The FET MV as shown in Fig. 5-34 may also have a starting problem because the device is in a very low-gain operating point when ON, and both FET's could remain ON and in the nonoscillating condition as described for the saturated junction transistor astable MV. This may easily be prevented by tying both FET source leads together and connecting a small resistor between the sources and ground. A value for this resistor of a few hundred ohms is sufficient to place both FET's in the active region should both attempt to be ON at the same time. With both FET's in the active region the circuit is very unstable and will quickly regenerate to the condition of one FET ON and one FET OFF.

5-11 OTHER FORMS OF MULTIVIBRATORS

Many hybrid forms of MV may be created by mixing device types; for example, junction and FET transistors. However, there are other types which appear rather fundamentally different. One such is the binary shown in Fig. 5-35*a* which uses both a PNP and NPN transistor. This type of binary, or its monostable and astable versions, has the property that both transistors are OFF together or ON together. This is easily seen in Fig. 5-35*a* if $Q1$ is assumed OFF. Then its collector current is zero as is the base current of $Q2$ which is therefore OFF also. Since $Q2$ is OFF, its collector current is also zero, and $Q1$ is OFF as supposed because it has no base current. The opposite situation, that is, both transistors ON and saturated can be shown possible if $Q1$ is assumed ON. If this is true, a large base current flows into $Q2$, which will also turn it ON thereby providing $Q1$ with a large base current also. Therefore $Q1$ is ON as hypothesized, and in fact both transistors will be saturated.

Essentially the same circuit with the addition of the 7-V zener diode $D1$ is used in Fig. 5-35*b* to make an astable oscillator. If the supply voltage is assumed to be turned ON at $t = 0$, this circuit begins with both $Q1$ and

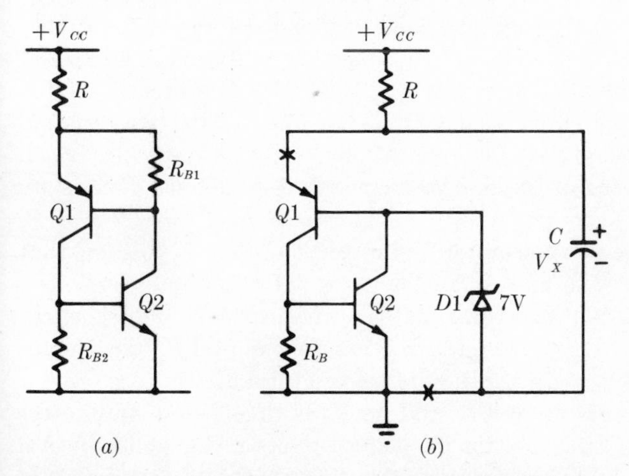

Fig. 5-35 (*a*) A form of complementary bistable MV where both transistors go ON and OFF together. (*b*) A complementary astable multivibrator.

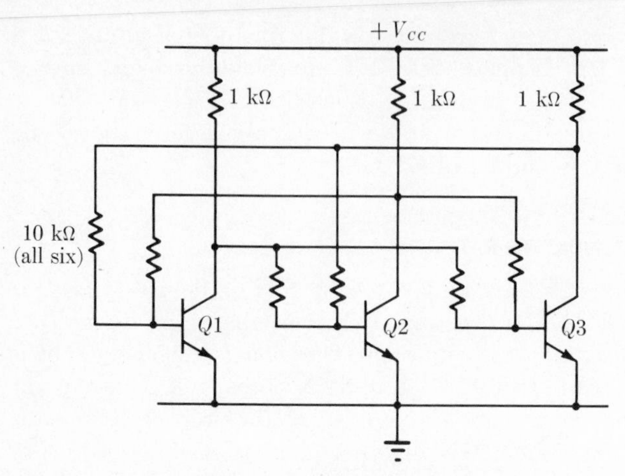

Fig. 5-36 A "tristable" multivibrator; $Q1$, $Q2$, or $Q3$ may be OFF with the other two transistors ON.

$Q2$ nonconducting and C discharged. $Q1$ remains OFF as the voltage across C rises toward V_{CC}. When the capacitor voltage V_X rises about 0.5 V more than the zener breakdown voltage, current begins to flow in the emitter of $Q1$. This also begins to turn ON $Q2$, and a regenerative switching process begins to produce large currents in $Q1$ and $Q2$ to discharge C. When C is nearly completely discharged, and $Q1$ and $Q2$ are saturated, a reverse switching process takes place because not enough current goes through R on a steady-state basis to keep $Q1$ and $Q2$ in saturation. (The large currents flowing previously were provided by the discharge of the capacitor.) At the end of the second switching process $Q1$ and $Q2$ are OFF, and the cycle begins again with C recharging toward V_{CC}. This kind of astable circuit is perhaps more easily analyzed on the basis of its two-terminal V-I characteristics, which in this case display negative resistance. Such an analysis is conducted in Chap. 8 where negative-resistance devices are discussed.

The second type of unusual multivibrator is presented to show that circuits may be built with more than the usual pair of stable states. The tristable MV of Fig. 5-36 has three stable states in which one transistor is OFF and two are ON. For example, if $Q1$ is assumed OFF, its collector voltage will be nearly V_{CC}. This voltage is applied through one base resistor to $Q2$ and another resistor to $Q3$, and keeps both of these transistors saturated. (The other base resistor in each case goes to the collector of a saturated transistor, or approximately zero volts.) The base resistors of $Q1$ both go to collectors of saturated transistors, or about zero volts, so that $Q1$ is kept OFF as supposed. A similar line of reasoning will show that $Q2$ may be kept OFF if $Q1$ and $Q3$ are ON. Similarly, $Q3$ may be the only transistor OFF. The practical use of the circuit requires additional resistors to the bases of the transistors to turn OFF the desired transistors by means of an external signal.

5-12 FREQUENCY DIVISION AND SYNCHRONIZATION IN MULTIVIBRATORS

One important function of an MV is that of frequency division, that is, operating with an output frequency which is some submultiple of the input or synchronizing frequency. The use of a chain of binaries to divide an input frequency has already been discussed. This mode of frequency division has the advantage of great stability, i.e., the constant of division will not change with either time or input frequency, but has the disadvantage that each circuit only divides by two. If a large frequency division is required, many circuits must be employed. Astable or monostable circuits, on the other hand, may be made to divide by integers greater than two, thereby reducing the number of circuits required for a given division ratio.

An example of the use of a monostable MV to perform frequency division is illustrated in Fig. 5-37 where the timing waveform V_{BN2} of the

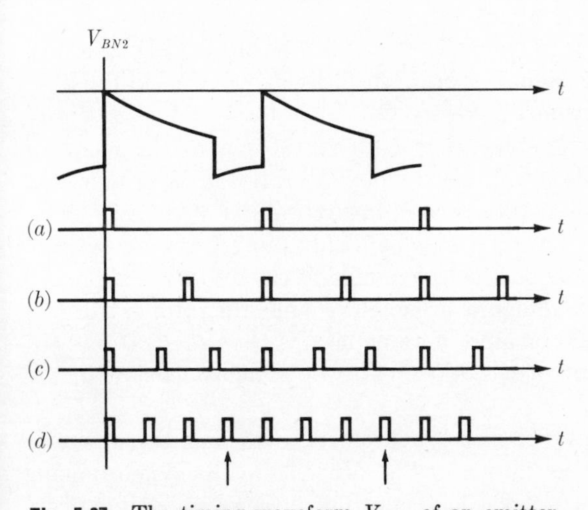

Fig. 5-37 The timing waveform V_{BN2} of an emitter-coupled MV (see Figs. 5-24 and 5-27) and various triggering waveforms to illustrate frequency division. (*a*) The MV runs at the same frequency as the trigger. (*b*) The MV is dividing the frequency by two. (*c*) The MV is dividing the frequency by three. (*d*) The indicated pulse may or may not trigger the circuit, thus leading to erratic operation.

emitter-coupled MV of Fig. 5-24 is shown for reference. In Fig. 5-37*a* a trigger waveform v_t is shown which triggers the circuits once each cycle and allows minimal time for the circuit to recover between pulses. This is the normal, nonfrequency dividing mode for the circuit. The trigger waveform of Fig. 5-37*b* is twice the frequency of the preceding trigger, and the second trigger pulse during the cycle occurs when the trigger coupling diode $D1$ is reverse biased. Therefore the second trigger pulse is not coupled into the MV and does not affect it. The third trigger pulse does trigger the MV so that it runs at the same frequency as it did before, but with twice the input frequency. In Fig. 5-37*c* the trigger pulses are occurring at triple the original frequency, and two pulses occur when the trigger coupling diode is reverse biased; therefore the circuit triggers on the first, fourth, and seventh pulses, and is operating at one third the trigger frequency. The situation in Fig. 5-37*d* is not so conclusive, however, because one of the trigger pulses which should not trigger the circuit (the fourth) occurs at the end of the MV period when it is recovering. If the trigger is big enough, the circuit may then be triggered even though it has not completely recharged the timing capacitor. If this occurs, the next period will be shorter than the one shown, and the triggering of the next cycle will be even more in doubt. Under these conditions the MV will operate erratically, and the frequency division obtained will be very sensitive to small changes in any part of the circuit.

From the preceding it may be seen that the monostable MV operates well as a divider as long as no pulses occur in the recovery time of the MV. Therefore an MV with a short recovery time is desirable if high ratios of division are required in one stage. The multivibrator with the recovery aids shown in Fig. 5-29*b* should stably divide with ratios of 10 or more. At such high ratios the input frequency and natural period of the MV must be very stable, or the division ratio may suddenly jump from one ratio to the same ratio plus or minus one. (For example, if the divider is supposed to divide by 10, a change in temperature might change the ratio to 9 or 11.) Very high ratios such as 20 or more may be used if the exact ratio is not important, but the output of the MV has to be synchronized with the high trigger frequency.

An astable circuit may also be synchronized with a pulse voltage or even a sinusoidal voltage as shown in Fig. 5-38. Here the synchronizing voltage is inserted in series with the base resistors so that some of it appears on the timing waveforms at the bases. The effect of the added sinusoid on the timing is shown in Fig. 5-38*b* where the period is shown to be slightly shortened so that two periods of the sine wave correspond to half the period of the MV. The MV is therefore dividing by four. Depending upon the phasing of the sine wave, the period of the MV can also be lengthened by the addition of the sine wave. If, in practice, the natural

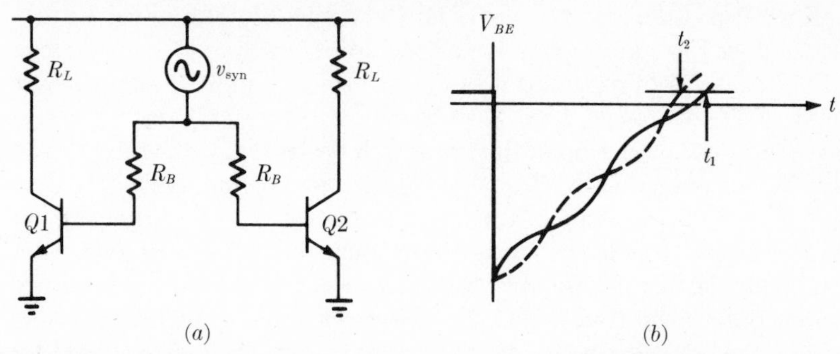

Fig. 5-38 (a) An astable MV with a sinusoidal synchronizing voltage v_{syn} added. (b) The timing waveform (at either base) as modified by the synchronizing voltage.

unsynchronized period of the MV were 1 kHz, then synchronization might be effected with an input frequency of 3.9 to 4.1 kHz, and, of course, at other multiples of 1 kHz. The synchronizing ability of an MV also can be a nuisance in that an MV may synchronize with a noise voltage which is close by the desired frequency or a harmonic. Synchronization becomes more and more probable as the desired and noise frequencies come closer together.

5-13 STABILITY OF THE MULTIVIBRATOR PERIOD

Multivibrators are usually regarded as rather unstable oscillators in terms of their frequency stability, and if they are constructed from run-of-the-mill components and with no regard to stability, this sentiment is probably appropriate. With careful attention to the components and knowledge of what in the circuit can produce drift, a very stable oscillator can be built (not a competitor to the quartz crystal, however). A generalized equation for the partial period t_1 of an MV is

$$t_1 = \tau_1 \ln \left(\frac{V_i - V_f}{V_t - V_f} \right) \tag{5-69}$$

where τ_1 is the time constant, V_i is the initial value of the timing waveform, V_f is the final value of the timing waveform, and V_t is the level at which the circuit triggers.

Since the time t_1 is directly proportional to the time constant, the values of R and C making up this time constant should have long-term stability and either zero temperature coefficients or should be chosen so that their temperature coefficients are equal and opposite. As an example, metal film resistors and metallized mica capacitors can be found with temperature coefficients of less than 0.005%/°C.

In a saturating MV the value of V_i is usually determined by the supply voltage V_{CC} and R_{sat}. If V_t and R_{sat} were zero, and $V_f = V_{CC}$, or if all the voltages were proportional to V_{CC}, the time t_1 would be independent of supply voltage. Since these conditions are not usually met exactly, there is some dependence upon the supply; therefore, for high stability the supply should be regulated. R_{sat} is usually so small compared to R_L that its variation has little effect on the time. The point at which regeneration takes place V_t is stable over long periods of time, but changes about -2.5 mV/°C. This change is probably the major contributor to temperature instability in a silicon transistor MV. The effect of the decreasing V_t with increasing temperature is to decrease the period. In a nonsaturating multivibrator V_i and V_t are in part determined by the resistors in the circuit. Therefore these must also be stable if the period is to be stable. The temperature dependence of V_t may be reduced considerably by canceling V_{BE} against the drop in an ordinary diode as shown in Fig. 5-39. The current through R_E must be larger than the transistor emitter current to keep the diode turned ON. To the extent that $V_{BE} = V_D$, the voltage $V_{BN} = 0$ and will stay zero over wide ranges of temperature.

The last thing to affect the time t_1 is variation of V_f because of current flowing in the base of the cutoff transistor. This current is the leakage current which varies exponentially with temperature. The effect of the current is to *increase* the value of V_f as temperature increases; therefore, the period shortens with increasing temperature. The effect is minimized by keeping R_B relatively small (which is bad for recovery time) and using silicon devices with low leakage currents. Compensation for this effect is usually not feasible because leakage currents are not stable over long periods of time. Therefore the best solution is to design the circuit so that the effect is tolerable at the maximum temperature likely to be encountered. With silicon devices and the usual range of R_B, there is not much trouble from this source until temperatures of 80 to 100°C are reached.

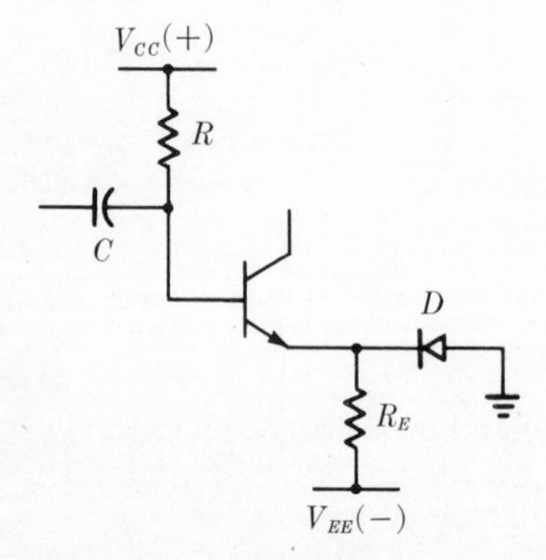

Fig. 5-39 A way to compensate for changes in V_{BE} with temperature in an MV. The emitters of both transistors may be connected to the same diode.

PROBLEMS

5-1. One of the problems in the design of bistable multivibrators is the provision for sufficient margin in the values of the transistors.

(*a*) As an example assume the transistors in Fig. 5-9c have a base leakage current of -20 nA at $T = 25°C$. Find the reduction in the turn-off signal by computing the V_{BE} of the OFF transistor at $T = 125°C$. (Assume I_{BO} doubles for each ten degree rise in temperature).

(*b*) With the circuit operating normally at 25°C, what is the minimum allowable trigger *current* required through $D1$ to initiate the regenerative cycle and initiate triggering? Note that for the regeneration to begin *both* $Q1$ and $Q2$ must be in Region II.

5-2. (*a*) The circuit shown is a simple monostable multivibrator. Assume the circuit is triggered at $t = 0$; calculate and carefully sketch the waveforms at each transistor terminal.

(*b*) Write an equation in terms of the R's, C, V_{CC}, and transistor potentials which gives the delay period of the MV.

(*c*) If the delay were to be adjustable, discuss the relative merits of changing C, R_{B2}, and V_{CC}. A good way would be to make a table of advantages and disadvantages. How large and how small may R_{B2} and C be practically made? (Note that an exact answer to the last question may not be possible.)

(*d*) What is the minimum peak trigger voltage that will successfully trigger the circuit?

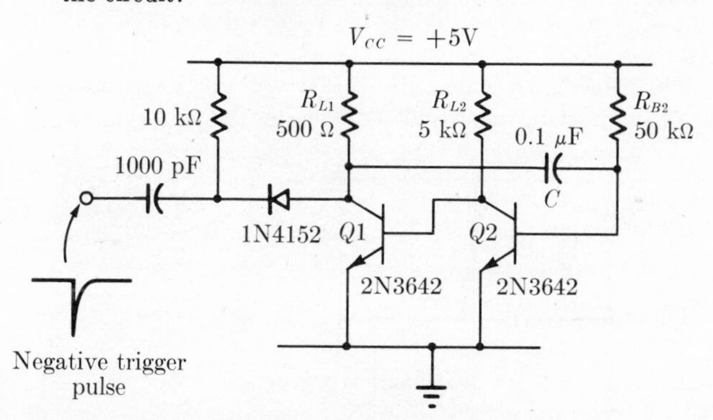

5-3. The circuit is a monostable multivibrator with values as shown.

(*a*) Compute the electrode potentials for the circuit at rest (untriggered).

(*b*) Assuming the trigger source has zero impedance and neglecting diode drop, what is the minimum magnitude and polarity of trigger pulse? Which way should the diode D be connected?

(*c*) Compute the waveform appearing at each electrode. Sketch and label the same.

(*d*) If the trigger pulse is just larger than that needed to trigger the circuit, what range of input frequencies will be divided by two by the circuit? Assume that for the best operation of the circuit, the natural pulse length of the MV should not be altered by the trigger.

(*e*) Prove that $Q2$ is saturated immediately after the end of the pulse generated by the circuit (i.e., before C reacquired its normal charge).

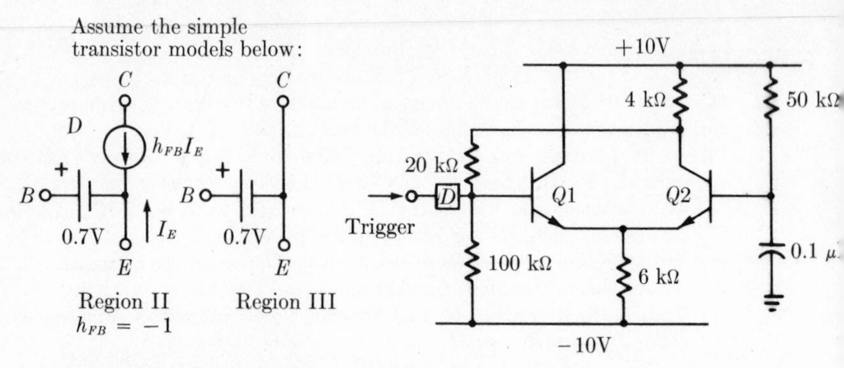

5-4. Calculate, sketch, and carefully label the waveforms occurring in the astable circuit of Fig. 5-33. Find C to give a frequency of 1 kHz.

5-5. A Schmitt trigger (Fig. 5-14) is desired in which the point at which $Q1$ switches ON is +5 V.

(a) What resistors may be changed in the circuit to provide this switching point? (Change only one resistor at a time.)

(b) Supposing R_{B1} were the resistor changed, what would its new value be and at what input voltage would $Q1$ switch OFF?

5-6. The astable MV shown is a variety which will always begin to oscillate.

(a) Show why this MV does not have a non-oscillatory state like the MV in Fig. 5-30.

(b) Calculate and sketch the waveforms in the circuit.

(c) What is the maximum value of $C2$ that will not appreciably alter the portion of the period primarily determined by $C1$?

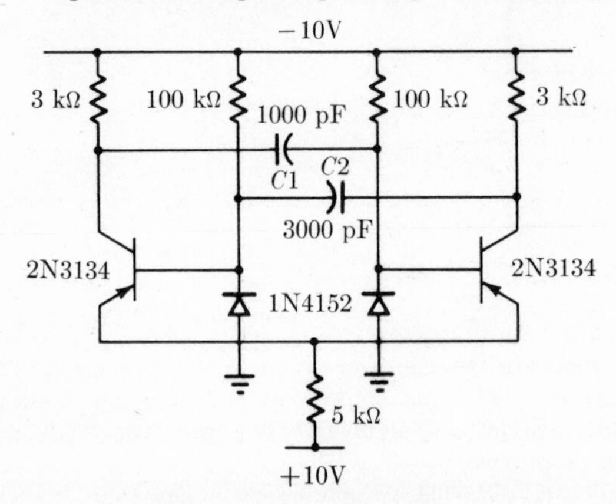

5-7. The circuit shown is a FET astable multivibrator in which the transistors are not saturated and the circuit will reliably start.

(a) Begin by finding the static operating point of the ON FET. Assume the other FET is OFF, of course.

(b) At what gate-to-ground voltage (V_{GN}) will the OFF FET just begin to conduct?

(c) Calculate and sketch the waveforms occurring in the circuit.

(*d*) The circuit is to be synchronized by a short negative pulse. Disconnect the cathode of both diodes from ground and connect to the synchronizing pulse generator. The pulse will then appear on the gate of the ON FET and therefore at its collector as a positive pulse. This latter pulse appears on the timing waveform where it can affect synchronization. What are the minimum and maximum values of the pulse if the pulse frequency is 30 Hz, and the synchronized multivibrator is to operate at 15 Hz?

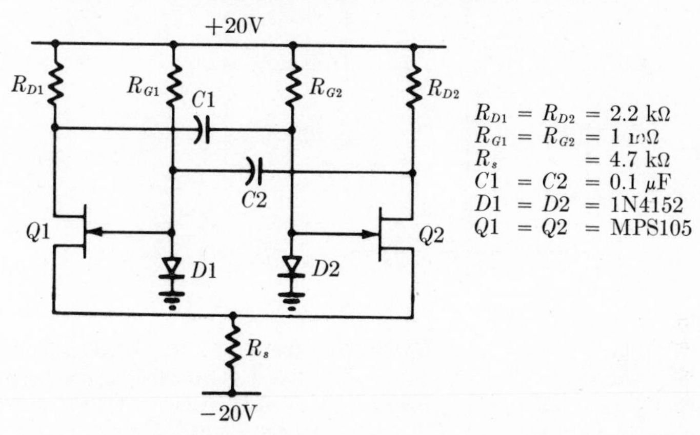

5-8. (*a*) The circuit is a *non*-saturating bistable multivibrator which is capable of high-speed operation. Assuming it to be in the state *Q*1 ON and *Q*2 OFF, calculate all of the node voltages in the circuit and verify that this state is possible.

(*b*) What is the minimum transistor dc gain (I_C/I_B) that will keep the OFF transistor reverse biased ($V_{BE} < 0$)?

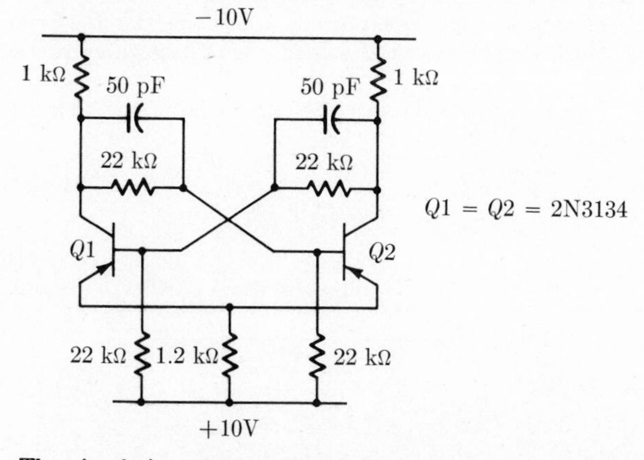

5-9. The circuit is a monostable multivibrator in which neither transistor is ever saturated. The period is controlled by *R* and *C*. The resistor R is to be of such a value as to permit a diode current of at least 0.1 mA when *Q*2 is on. The circuit is triggered by the negative step v_1.

(*a*) What is the maximum value R may have?

(*b*) Calculate and plot the waveforms at each electrode of $Q1$ and $Q2$. Plot with respect to normalized time t/τ where $\tau = RC$ (assume $R \gg 1$ kΩ).

(*c*) What is the minimum value of v_1 to trigger the circuit?

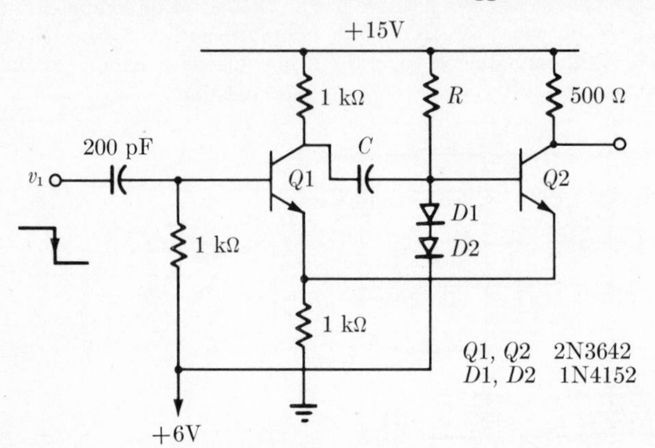

5-10. The bistable circuit of Fig. 5-2 is to drive a grounded 2 kΩ load at each collector. The minimum output voltage required is +3 V. Give the element values for the circuit (R_L and R_B) to minimize the current drain on the power supply assuming that the minimum h_{FE} for $Q1$ and $Q2$ is 100.

5-11. To the multivibrator circuit shown in Fig. 5-30 add a diode in series with each emitter, cathode end connected to ground. The purpose of these diodes is to prevent the reverse breakdown of the emitter-base junction which might occur when the supply voltage is raised.

(*a*) Give the values for the resistors and capacitors for the modified circuit to operate from $V_{CC} = +20$ V, an average current drain of about 5 mA, a recovery time constant as short as possible, and a frequency of 200 Hz. Assume that the minimum h_{FE} of any 2N3642 is 30.

(*b*) Show a method using a single potentiometer which will vary the frequency but not change the symmetry of the waveforms. (You may need to reduce the R_B's.)

(*c*) Would the frequency of the MV increase or decrease if the added diodes were removed (shorted)? Why?

6

Circuits for Generating Linear Voltage Slopes

Techniques for generating a voltage that rises (or falls) linearly with time comprise only a few basic categories, even though the specific forms of circuits derived from these basic techniques are too numerous even to tabulate. Two circuits will be examined in detail, namely, the Miller integrator and the bootstrap.

The reasons for examining these circuits are several. First, as with circuits previously examined, they furnish good examples whereby the techniques of analysis advanced in the earlier chapters are effectively employed in providing quantitative details of the circuit operation. Second, these circuits are exceedingly efficient in accomplishing their intended purpose and are therefore useful in themselves. Third, they represent a rather high degree of circuit refinement, and by retracing their apparent course of development from elemental beginnings, one gets a glimpse of how a "trick" circuit can evolve through gradual stages without requiring superhuman inventive genius.

6-1 SIMPLE *RC* INTEGRATOR

The most elementary way of generating a linearly rising voltage is to charge a capacitor from a constant-voltage source in series with a resistor, as in the integrator circuit of Fig. 2-7. A transistor can be used as a switch to start and finish the operation (Fig. 3-22). The rising curve of voltage versus time is, of course, an exponential, but the initial portion is approximately linear. Hence, if the time constant of the charging circuit is made very large compared with the desired interval wherein the linear rise should take place, the approximation can be very accurate indeed.

To see how linear a portion of an exponential is for use as a linear sweep consider the series representation for the exponential:

$$V(t) = 1 - \epsilon^{-t/\tau} \tag{6-1}$$

$$V(t) = 1 - 1 + \frac{t}{\tau} - \frac{1}{2!}\left(\frac{t}{\tau}\right)^2 + \frac{1}{3!}\left(\frac{t}{\tau}\right)^3 \cdots \tag{6-2}$$

$$V(t) \approx \frac{t}{\tau}\left(1 - \frac{t}{2\tau}\right) \qquad \text{for } \frac{t}{\tau} \ll 1 \tag{6-3}$$

The latter equations are good approximations if the sweep is going to be anywhere nearly linear. If we consider using the generated voltage for a time t_1, the important parameter is the normalized time t_1/τ. The actual sweep voltage at $t = t_1$ is

$$V(t) \approx \frac{t_1}{\tau}\left(1 - \frac{1}{2}\frac{t_1}{\tau}\right) \tag{6-4}$$

The error in the sweep voltage compared to a straight line passing through zero and the final value is maximum at the time $t_1/2$, as is shown in Fig. 6-1. The equation for the straight line shown is

$$V_{\text{lin}}(t) = \frac{t}{\tau}\left(1 - \frac{1}{2}\frac{t_1}{\tau}\right) \tag{6-5}$$

and the error in the sweep voltage at the midpoint of the sweep $t = t_1/2$ is

$$V\left(\frac{t_1}{2}\right) - V_{\text{lin}}\left(\frac{t_1}{2}\right) \approx \frac{1}{8}\left(\frac{t_1}{\tau}\right)^2 \tag{6-6}$$

Therefore the percent nonlinearity of the sweep related to its final value is

$$\text{Percent error} \approx \frac{1}{8}\left(\frac{t_1}{\tau}\right)^2 \cdot \frac{\tau}{t_1} \times 100\% = \frac{1}{8}\frac{t_1}{\tau} \times 100\% \tag{6-7}$$

This is the maximum *displacement* error assuming that the initial and final values of the sweep voltage are correct. This definition of nonlinearity would be useful in defining the performance of a sweep generator used in an oscilloscope because the normal procedure would be to make the beginning and end points of the sweep in the correct places on the screen of the

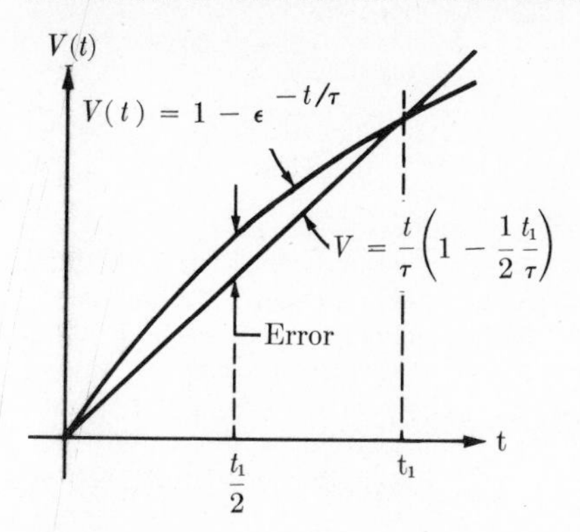

Fig. 6-1 An exponential sweep voltage $V(t)$ compared to an ideal linear sweep (ramp) having identical start and end points.

oscilloscope. The displacement error then gives the maximum error of the beam position anywhere on the face of the oscilloscope.

In some other applications different definitions of linearity are appropriate. For example, the ideal sweep might be defined as a straight line having the same slope as the initial slope of the actual sweep. The departure of the sweep voltage from this line is more than in the previous case, and the error is

$$\text{Percent error} = \frac{1}{2}\frac{t_1}{\tau} \times 100\% \tag{6-8}$$

Whatever the definition of error, it will be increased proportionally to the fraction of the effective time constant that is used (assuming the two-term representation of the exponential is adequate). If a simple RC circuit is used to generate the sweep, the amplitude of the output will be inversely proportional to the linearity obtained. Of course an amplifier may be used to increase the amplitude of the output, but there is the disadvantage that the slope of the output waveform from the amplifier is directly proportional to the gain of the amplifier as well as the slope of the input waveform to the amplifier. This means that the amplifier must be a relatively sophisticated, gain-stabilized form if the sweep is to be of constant magnitude and slope over long periods of time. This method of generating a large-amplitude, accurate sweep is therefore not often used unless the amplifier must be employed for some other reason. For example, an oscilloscope usually needs a horizontal deflection amplifier for signals other than the sweep; therefore the same amplifier can be used to amplify either a low-level sweep signal or any other horizontal input signal.

6-2 LINEAR SWEEP GENERATORS USING CURRENT SOURCES

An obvious way to increase the linearity of the simple RC sweep generator is to replace the resistor with a constant-current source delivering the

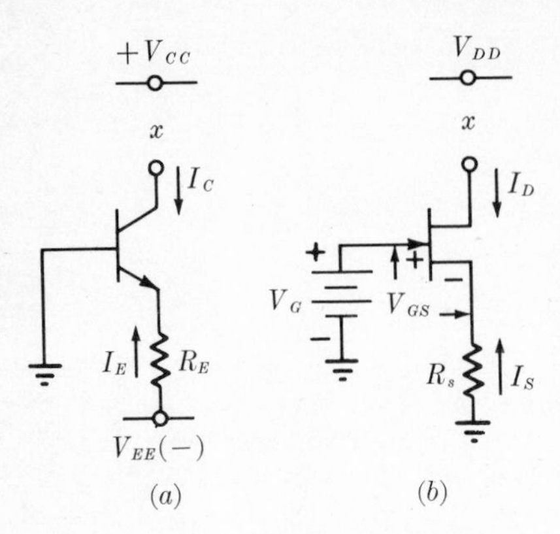

Fig. 6-2 Two circuits which simulate constant current generators at the points x. (*a*) The transistor is operated with constant emitter current to give a nearly constant collector current $I_C \approx h_{FB} I_E$. (*b*) The drain current is $(V_G - V_{GS})/R_s$ and is approximately constant if $|V_G| > |V_{GS}|$.

same current to the capacitor at $t = 0$ that the resistor and voltage source did. The theory is good, but in practice there are problems. The first problem arising is how to take an output from the circuit, since any resistive element placed across the capacitor will limit the linearity of the sweep because the current source will no longer behave as a pure current source. Therefore the output must be taken from the circuit via an amplifier with high input impedance and constant gain. This amplifier is conveniently a gain of one in most cases, and is therefore not difficult to design, but some of the previous amplifier problems return. The second difficulty is obtaining a current source which is stable and not too dependent upon device parameters.

Two circuits are shown in Fig. 6-2 which make good current sources. In Fig. 6-2*a* a transistor is used with a constant-emitter current

$$I_E = \frac{V_{EE} - V_{EB}}{R_E}$$

The collector current is $I_C \approx h_{FB} I_E \approx -I_E$, since $h_{FB} \approx -1$. This type of current generator used in a sweep circuit was previously discussed in Sec. 3-12 and was illustrated in Fig. 3-37. The deviation from a perfectly linear sweep in this case would be caused by the nonzero output conductance of the common-base transistor h_{ob}, and the loading caused by whatever means are employed to take the signal off the terminals of the capacitor.

The second constant-current generator shown (Fig. 6-2*b*) uses an FET to produce the nearly constant current. The resistor in the source, R_s, and the gate voltage V_G are used to make the current more independent of the device parameters. The drain current, assuming that the drain is a few volts positive with respect to the source (i.e., negligible gate current flows) is

$$I_D = \frac{V_G - V_{GS}}{R_s} \tag{6-9}$$

To the extent that $V_{GS} \ll V_G$ the drain current is determined mainly by the passive elements in the circuit. The actual current for a given operating condition can be found from the bias-line construction of Eq. (3-31) and Sec. 3-13. In some cases V_G and R_s may be omitted if a high degree of precision in I_D is not required. To see how well the FET serves as an ideal current source, we must find the impedance which shunts the current found from the bias line. A suitable equivalent circuit for the FET in this case is Fig. 6-3*a*, which is to be reduced to the Norton equivalent Fig. 6-3*b*. If the terminals *x*-*x* are open-circuited, the resulting open-circuit voltage is

$$V_{oc} = -\frac{I_0 + g_{fs}V_G}{g_0} \qquad \text{(Note } V_{GS} = V_G\text{)} \tag{6-10}$$

If the terminals *x*-*x* are short-circuited, the current which flows is

$$\begin{aligned} I_{sc} &= -\frac{(I_0 + g_{fs}V_{GS})(1/g_0)}{R_s + 1/g_0} \\ &= -\frac{I_0 + g_{fs}V_{GS}}{1 + g_0R_s} \end{aligned} \tag{6-11}$$

The unknown gate voltage V_{GS} can be expressed in terms of the short-circuit current:

$$V_{GS} = V_G - I_DR_s = V_G + I_{sc}R_s \tag{6-12}$$

Using the value of V_{GS} from Eq. (6-12) in Eq. (6-11), the short-circuit current is found to be

$$I_{sc} = -\frac{I_0 - g_{fs}V_G}{g_0R_s + 1 + g_{fs}R_s} \tag{6-13}$$

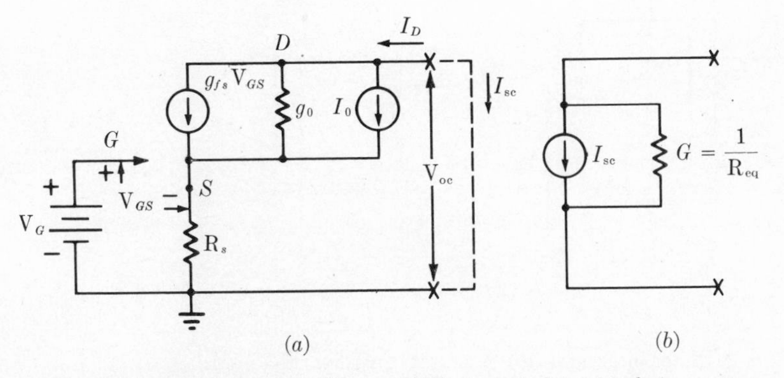

Fig. 6-3 An equivalent circuit for an FET connected to simulate a constant-current source as in Fig. 6-2*b*.

The quotient of the open-circuit voltage and the short-circuit current is the output impedance of the circuit R_{eq}.

$$R_{eq} = \frac{V_{oc}}{I_{sc}} = \frac{1 + R_s(g_0 + g_{fs})}{g_0} \tag{6-14}$$

For $g_0 = 1/50\ \text{k}\Omega$, $R_s = 1\ \text{k}\Omega$, and $g_{fs} = 1/0.2\ \text{k}\Omega$, the equivalent output resistance R_{eq} is 300 kΩ. The FET connected in this manner therefore makes a reasonably good current source—much better than would be expected from the values of g_0 alone.

6-3 THE MILLER INTEGRATOR

Many of the problems with the preceding circuits, such as linearity and loading, are overcome in the so-called "Miller integrator"[1] sweep circuit, which has become the standard method of generating high-precision sweep or ramp-voltage waveforms. The basic circuit is as shown in Fig. 6-4*a*, where an ideal amplifier has feedback provided by the capacitor C, and an input current provided by V_1 and R. If the capacitor is initially discharged, then $v_1 = v_2 = 0$. If the switch is closed at $t = 0$, a current V_1/R flows in the input resistor, but because of the high input impedance of the amplifier, no current flows into it, and all the current flows into C beginning to charge it. The capacitor voltage begins to change at the rate $dv_2/dt = i/C = V_1/RC$. If the gain of the amplifier is large, v_1 remains

[1] This class of circuits derives its name from the so-called "Miller effect," observed by J. M. Miller in 1919, in which the input impedance of an amplifier with a feedback capacitance has an apparent capacitance approximately equal to the voltage gain of the amplifier times the feedback capacitance.

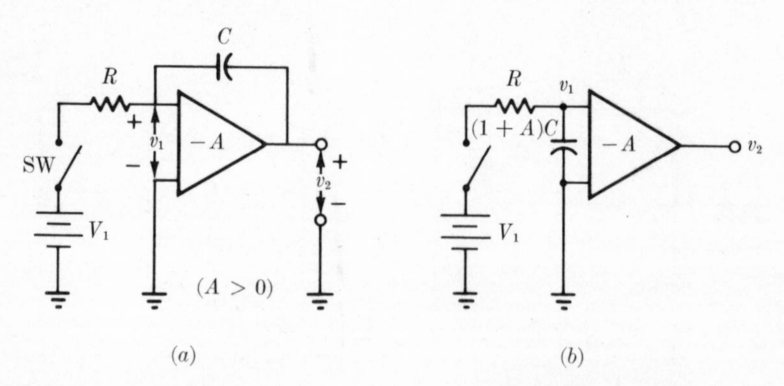

Fig. 6-4 (*a*) A Miller integrator. (*b*) A circuit equivalent to the Miller integrator. The amplifier is ideal in that $Z_{in} = \infty$, $Z_{out} = 0$, and $v_2 = -AV_1$ $(A > 0)$.

small regardless of the output voltage. Therefore the current in the capacitor remains nearly V_1/R, and the output voltage changes linearly with time.

The action of the circuit may be seen more clearly in the equivalent circuit Fig. 6-4*b* where the feedback capacitor has been replaced by a fictitious capacitance $(1 + A)C$ at the input of an amplifier without external feedback. The waveform appearing at the input terminals of the amplifier is

$$v_1 = V_1(1 - \epsilon^{-t/(1+A)RC}) \tag{6-15}$$

and the output waveform is

$$v_2 = -AV_1(1 - \epsilon^{-t/(1+A)RC}) \tag{6-16}$$

From Eq. (6-16) one would get the impression that the output from the circuit varies directly with the gain A, a situation which would be very undesirable. However, differentiating Eq. (6-16) to get the initial slope of the output waveform shows that the waveform is quite independent of gain as long as only a small fraction of the effective time constant $RC(1 + A)$ is used.

$$\left.\frac{dv_2}{dt}\right|_{t=0^+} = -\frac{AV_1}{(1+A)RC} \approx \frac{V_1}{RC} \qquad (\text{if } A \gg 1) \tag{6-17}$$

Equation (6-17) shows that the gain should be large if the initial slope is to be independent of the gain. Increase of the gain also has the desirable effect of giving a larger effective time constant $[= (1 + A)RC]$ so that the linearity of the output waveform is improved.

In some applications the amplifier is of rather elaborate design and is well represented by the concept of an ideal amplifier. An integrated-circuit operational amplifier is one such realization which could easily give

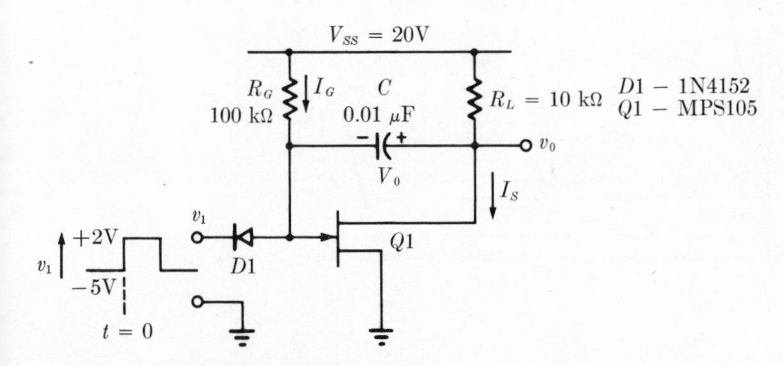

Fig. 6-5 A simple FET Miller integrator used to develop a linear-sweep output voltage.

an open-loop gain $A = 10^4$ or greater. Many simpler circuits, however, can be used when the requirements are not so stringent. One such simple circuit, which is both useful and illustrative, is illustrated in Fig. 6-5 where a single FET is used to provide the amplifying function. In addition to the basic sweep generator a diode gate has been added to enable a gating waveform v_1 to begin the sweep at the desired moment. The voltage v_1 in Fig. 6-4 is provided by the supply V_{SS}. Initially the circuit is at rest with the FET cutoff by the negative voltage on the gate provided by v_1 through diode $D1$. The current I_G flows through $D1$ to the gate generator v_1. The current through the capacitor is zero so that the FET drain voltage is equal to the supply voltage. At $t = 0$ the diode is turned off so that the gate voltage is free to rise and turn on the FET. The current I_G now flows into C so that the current through R_L is the sum of I_G plus I_S. As the FET turns ON, the rise in gate voltage is opposed by the fall in drain voltage, which is coupled by the capacitor back to the gate. In order to find the initial operating point at $t = 0^+$, an equivalent circuit (Fig. 6-6) valid only for this moment in time may be used. In this circuit the capacitor is

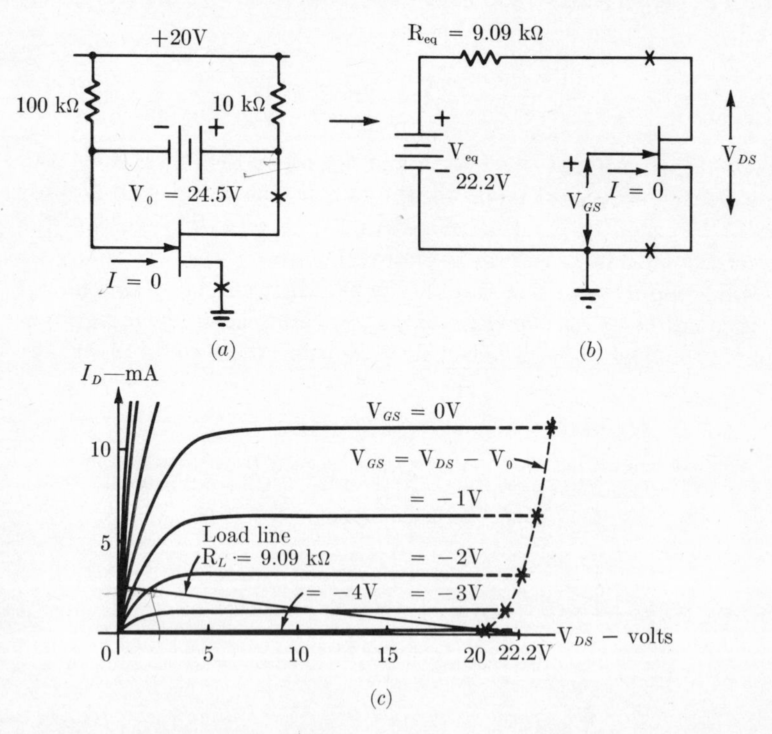

Fig. 6-6 (*a*) The circuit conditions in Fig. 6-5 at $t = 0^+$. (*b*) A circuit equivalent to Fig. 6-6*a*. (*c*) The FET drain characteristics with the two graphical constructions required to find $V_{DS}(0^+)$ and $V_{GS}(0^+)$.

regarded as a battery with a value equal to the initial charge on the capacitor V_0.

$$V_0 = V_{SS} - v_1 - V_{D1} = 20 + 5 - 0.5 = 24.5 \text{ V}$$

The circuit as seen by the FET drain terminal is reduced to its Thevenin equivalent, and is used to find an operating load line valid at $t = 0^+$. The effective supply voltage 22.2 V is greater than the actual supply because of the charge on C. Since the gate voltage is also unknown, an additional construction line is needed to give a solution. The gate voltage may be expressed as

$$V_{GS}(0^+) = V_{DS}(0^+) - V_0 \tag{6-18}$$

This equation may also be plotted on the FET characteristics as shown in Fig. 6-6*c*. The intersection of this bias line and the load line gives the initial operating point. As shown, the FET is initially very nearly, but not quite, cutoff.

As time increases the capacitor charges, V_{DS} decreases, and V_{GS} increases. The drain voltage decreases to nearly zero, and the maximum drain current $I_D \approx 20/10\text{ k}\Omega = 2$ mA. Therefore the region of operation during the generation of the sweep is: cutoff $< V_{GS} < -2.5$ V and $1 < V_{DS} < 20$ V. An equivalent circuit for this region is shown in Fig. 6-7*a*, where the g_{fs} is taken to be the average value for V_{GS} between -2 and -4 V, and g_0 is assumed negligible compared to R_L. The first step is to redraw the circuit as a series circuit as shown in Fig. 6-7*b*. (The original circuit—Fig. 6-7*a*—is purposely drawn in an extended form to facilitate the drawing of the simple series circuit.) The quantity $g_{fs}R_L \triangleq A = 16.5$ is the voltage gain of the FET stage.

The unknown gate voltage V_{GS} may be expressed in terms of I, that is, $V_{GS} = V_{SS} - R_G I$, and this value may then be placed in the equivalent circuit as in Fig. 6-7*c* and the following complete circuit, Fig. 6-7*d*. From Fig. 6-7*d* all the information concerning the linear sweep may be obtained.

$$I(0^+) = \frac{AV_{SS} + I_0R_L + V_0}{(1 + A)R_G + R_L} \approx \left(\frac{V_{SS} + I_0R_L/(1 + A)}{R_G}\right) \tag{6-19}$$

$$I(0^+) = \frac{330 + 67 + 24.5}{17.5 \cdot 100\text{ k}\Omega + 10\text{ k}\Omega} = 0.24 \text{ mA} \tag{6-20}$$

$$V_{GS}(0^+) = 20 - 100\text{ k}\Omega \cdot 0.24 = -4.0 \text{ V} \tag{6-21}$$

$$V_{DS}(0^+) = V_{GS} + V_0 = 20.5 \text{ V} \tag{6-22}$$

The values for $V_{DS}(0^+)$ and $V_{GS}(0^+)$ are in good agreement with the values obtained graphically. The effective time constant for the linear sweep is

$$\begin{aligned}\tau_1 &= [(1 + A)R_G + R_L]C \approx (1 + A)R_GC = (10^{-8})(17.5)(10^5)\\ &= 17.5 \text{ ms}\end{aligned} \tag{6-23}$$

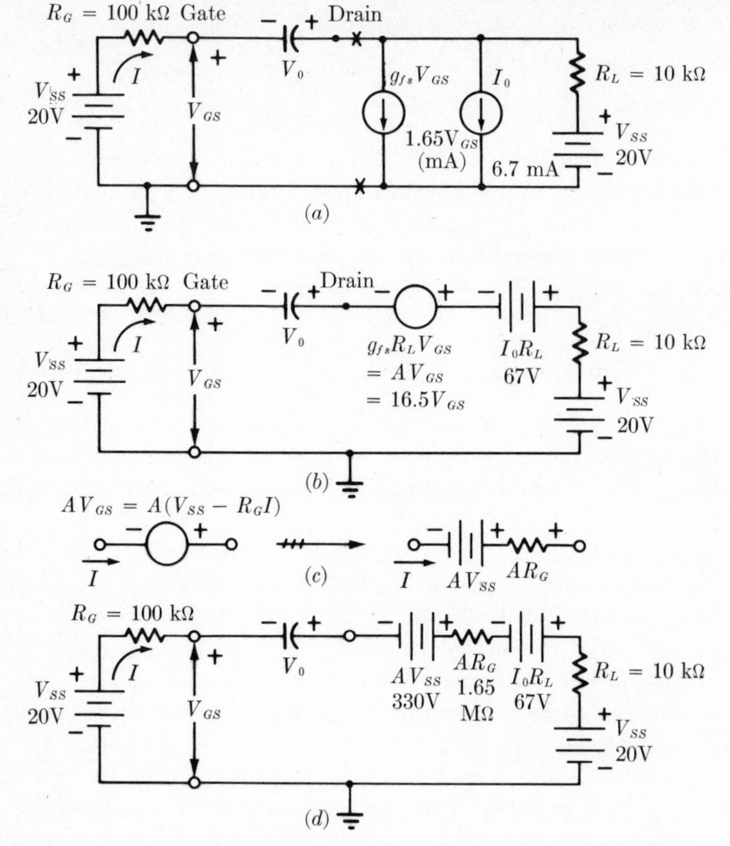

Fig. 6-7 The development of an equivalent circuit to give the details of the sweep generation in a Miller integrator (Fig. 6-5).

and the final value for the linear portion is

$$
\begin{aligned}
V_{DS}(\infty) &= (1 - A)V_{SS} - I_0R_L \\
&= -377 \text{ V}
\end{aligned}
\tag{6-24}
$$

These values are shown on the waveforms of Fig. 6-8.

Since the gain A appears in the time constant and the final value, the circuit may appear very dependent upon the device parameters. We can see that this is not true by computing the slope of the output waveform.

$$
\begin{aligned}
\frac{dV_{DS}(0^+)}{dt} &= \frac{V_{DS}(\infty) - V_{DS}(0^+)}{\tau_1} \\
&= \frac{[(1 - A)V_{SS} - I_0R_L] - [V_{SS} - I(0^+)R_G + V_0]}{[(1 + A)R_G + R_L]C} \\
&\approx \frac{-AV_{SS} - [I_0R_L + I(0^+)R_G - V_0]}{(1 + A)R_GC}
\end{aligned}
\tag{6-25}
$$

To the extent that A is large, the first term in the numerator will predominate. For the circuit in question the values are

$$\frac{dV_{DS}(0^+)}{dt} \approx \frac{-330 - (67.5 \text{ V})}{(17.5)(10^{-3})} = -22.6 \text{ V/ms} \tag{6-26}$$

$$\frac{dV_{DS}(0^+)}{dt} \approx \frac{-AV_{SS}}{(1 + A)R_G C} \approx \frac{-V_{SS}}{R_G C} \tag{6-27}$$

The last approximation [Eq. (6-27)] becomes better as A is made large. The waveform at the drain would go to zero volts at $t = 20.5/22.6 = 0.91$ ms if the sweep had constant slope and the device were linear all the way to $V_{DS} = 0$.

The slope of the gate waveform may be found similarly by calculating

$$\frac{dV_{GS}(0^+)}{dt} = \frac{V_{GS}(\infty) - V_{GS}(0^+)}{\tau_1} \tag{6-28}$$

$$\frac{dV_{GS}(0^+)}{dt} = \frac{20 - (-4)}{17.5} = 1.37 \text{ V/ms} \tag{6-29}$$

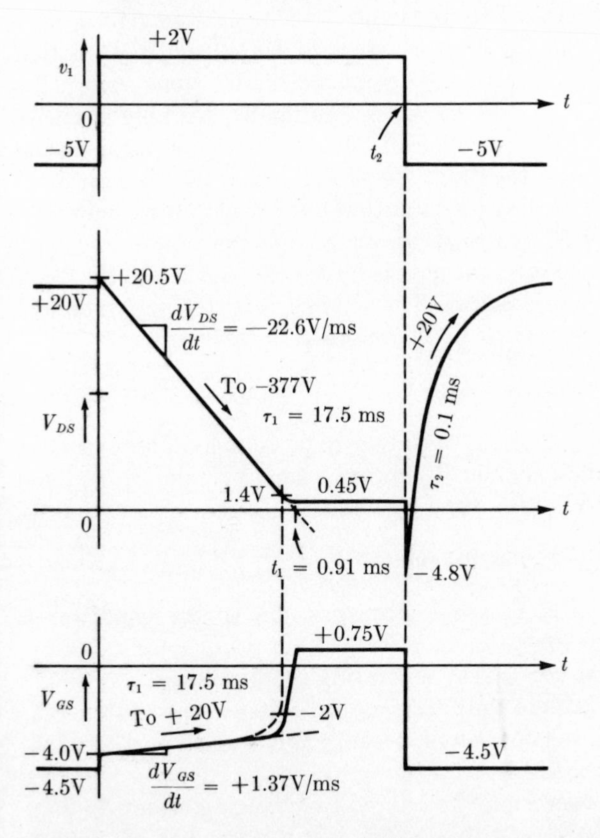

Fig. 6-8 The waveforms in the Miller integrator of Fig. 6-5.

At $t_1 = 0.91$ ms the gate voltage would be

$$V_{GS}(t_1) = -4 + 1.37 \cdot 0.91 = -2.75 \text{ V} \tag{6-30}$$

The initial values and the values at t_1 may be used to plot the gate and drain waveforms as shown in Fig. 6-8. Note that the voltage at the drain will be a nearly linear ramp since less than $0.91/17.5 = 0.052$ or 5.2 percent of the time constant is used. The ramp will not continue all the way to zero with the linearity given because the device becomes nonlinear as V_{DS} becomes less than about two volts. However, the rate of change of V_{cap} does not change significantly until the FET saturates because I_G (or the current discharging the capacitor) is very nearly V_{SS}/R_G as long as $V_{GS} < 0$ (the gate is nonconducting), but the gate voltage begins to change more rapidly as $V_{DS} \to 0$. The actual values may be read off the load line of Fig. 6-6c for known values of V_{DS}, that is, for $V_{DS} = 1.4$ V, $V_{GS} = -2$; for $V_{DS} = 0.8$ V, $V_{GS} = -1$; and for $V_{DS} = 0.5$ V, $V_{GS} = 0$. The minimum value V_{DS} reaches is about 0.45 V. (This value may be found more accurately from Fig. 3-41.) The gate waveform may be found graphically from the known values of $V_{DS}(t)$ and the relation between V_{DS} and V_{GS} given above. The construction for one such point ($V_{DS} = 1.4$ V, $V_{GS} = -2$ V) is shown in the figure. Actually the gate waveform is not as linear as shown because the device is not linear during the entire sweep. However, the drain waveform is as linear as shown.

The circuit maintains nearly zero output until the input gating waveform v_1 returns to -5 V at $t = t_2$. If we assume the diode and gate generator have zero impedance, the gate is forced to -4.5 V and the drain is forced to change a like amount. The change in gate (and therefore drain) voltage is $\Delta V_{GS} = -5.25$ V. Therefore

$$V_{DS}(t_2^+) = 0.45 - 5.25 = -4.8 \text{ V}$$

The FET is OFF because of the negative gate voltage. The capacitor charges through the source, v_1 and R_L toward a final value of 24.5 V, and V_{DS} rises toward $+20$ V with a time constant

$$\tau_2 = R_L C = (10^4)(10^{-8}) = 0.1 \text{ ms}$$

The circuit is ready for a new operation as soon as the capacitor is recharged or about 5 to $10\tau_2$ after t_2.

The circuit is quite efficient in generating the linear sweep in that a sweep voltage almost equal to the entire supply voltage is generated. The linearity is good, as pointed out before, because only a small portion of the effective time constant is used.

The slope of the sweep may be altered by changing either R_G or C. Usually C is changed in steps with a switch if it is altered at all. Smaller

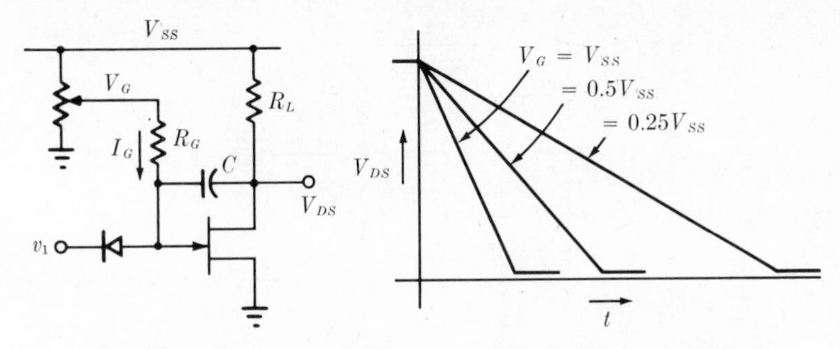

Fig. 6-9 A modified Miller integrator to provide variable output slopes.

changes in τ_1 may be made by varying R_G but a fixed minimum value should always be provided to prevent destroying the FET by excessive gate current. Both of these methods of varying the slope suffer from the control element being in the signal circuits where the capacitance to ground of the control may adversely affect the circuit. Another control method shown in Fig. 6-9 overcomes this objection and makes use of a lower resistance potentiometer. (Wire-wound potentiometers are the most stable and are typically available in the range of 10^1 to about 10^4 ohms.) The controlled voltage is V_G, which controls the current going into the capacitor during the linear rundown. The capacitor current is then equal to I_G since the gate current is substantially zero.

$$I_G = \frac{V_G - V_{GS}}{R_G} \tag{6-31}$$

This current is almost directly proportional to V_G since $V_{GS} \ll V_G$; hence the output voltage slope is also almost directly proportional to V_G. The circuit should not be operated with V_G too near zero to give very small output slopes because I_G will become very dependent upon the exact value of V_{GS}, and therefore will become dependent upon the device parameters.

6-4 IMPROVED MILLER INTEGRATOR

The circuit of Fig. 6-5 is useful, but there are a number of practical problems that limit its usefulness. Some of these are the requirement for a separate driving generator, difficulty in generating sweep voltages of very short time intervals (or equivalently, the generating of very short time delays), and the rather long recovery time compared to the sweep time. If we begin by thinking of a circuit which generates its own gating waveform, we might first notice that the useful time in the circuit is only that of the linear rundown occurring in the drain circuit. The gate waveform is seen to

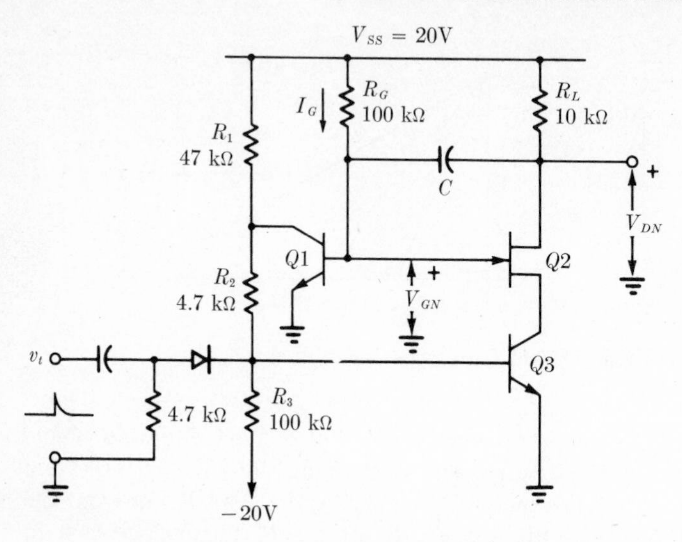

Fig. 6-10 The Miller integrator with transistors added to provide a regenerative self-gating circuit.

terminate abruptly with the end of the useful time, and therefore might be used to make a driving generator v_1 if the voltage were properly shaped.

A circuit which accomplishes this function appears in Fig. 6-10 in which $Q2$, R_L, R_G, and C comprise the basic Miller integrator as before, and transistors $Q1$ and $Q3$ produce the necessary recycling action. The function of $Q1$ is to amplify the waveform appearing at the gate of the FET and apply the amplified waveform to the base of the switch transistor $Q3$. The conditions in the circuit while at rest may be seen with reference to the waveforms plotted in Fig. 6-11. The transistor $Q1$ is normally ON and saturated by the current I_G flowing in its base. The collector voltage of $Q1$ is nearly zero, thereby making the base of $Q3$ slightly negative. $Q3$ is OFF and keeping the source current to the FET $Q2$ at zero. Since both the source and drain of the FET are positive with respect to the gate, there is no current flowing in the FET, and the drain voltage is +20 V.

At $t = 0$ a trigger pulse v_t is applied, which momentarily turns ON $Q3$ and $Q2$. Because of the drain current the collector voltage of $Q2$ drops[1] and causes V_{GN} to become negative, thus turning OFF $Q1$. The turnoff of $Q1$ causes a rise in its collector voltage and maintains the ON state of $Q3$ initiated by the trigger pulse. After $t = 0$ the circuit operates exactly as

[1] The initial value of V_{DN} may be found as before. In this case the initial value of V_{DN} is less than V_{SS} because the initial charge on C is less than that shown in the circuit of Fig. 6-5.

before because $Q1$ and $Q3$ are, respectively, open and short circuits. At $t = t_1$, however (which is slightly less than before because the starting point of V_{DN} is lower than before), the rising gate voltage V_{GN} causes $Q1$ to again conduct and to turn OFF $Q3$. The operation of the circuit is abruptly terminated at t_1 by the transistors returning to their initial states, thereby also turning the FET OFF. As before, the circuit is completely recovered when C has recharged through R_L and the base-to-emitter junction of $Q1$. The circuit now generates both the linear sweep and a rectangular waveform with a length equal to the run-down time of the sweep.

The linearity of the circuit may be improved by increasing the value of R_L, which will in turn increase the value of the gain A. However, increasing the value of R_L will also increase the recovery time proportionately. A very useful method of decreasing the time for recharging a timing capacitor is incorporated in the circuit of Fig. 6-12 where the current required for recharging C is obtained from the emitter follower $Q2$. During the timing portion of the circuit operation the emitter follower operates as a linear amplifier with a voltage gain of approximately one; therefore this portion of the circuit operation is essentially as before (except that the current I_G does not flow in R_L but instead is part of the emitter current of the emitter follower).

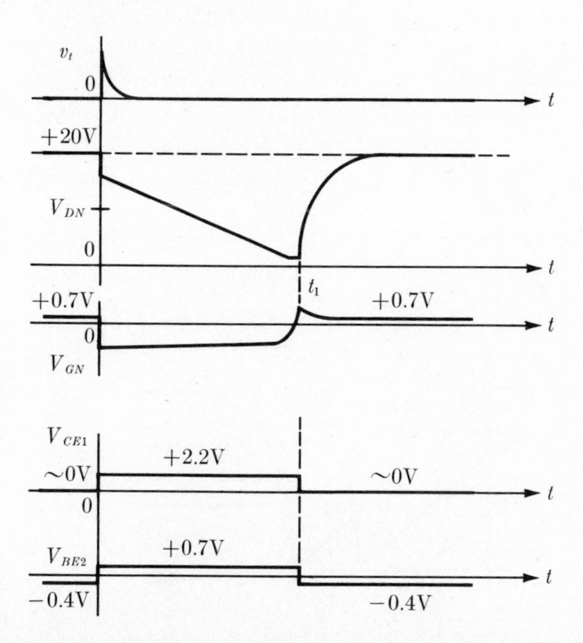

Fig. 6-11 Waveforms in the circuit of Fig. 6-10 assuming the circuit is triggered by v_t at $t = 0$.

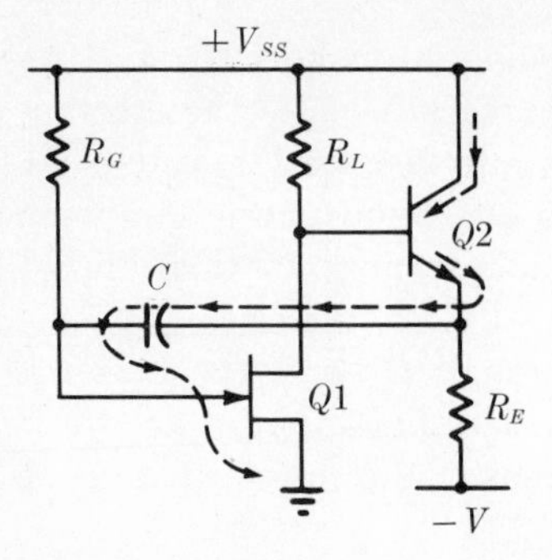

Fig. 6-12 The addition of an emitter follower to the basic Miller integrator to decrease the recovery time. The dotted line shows the path of the current to recharge C during the recovery interval. The resistance of this path is much less than R_L so that the recovery time is decreased by a factor of roughly $h_{FE}(Q2)$.

Another method of speeding recovery, but which more importantly controls the delay obtained, is shown in Fig. 6-13. Here the diode and voltage source V_1 set the initial value of the ramp occurring at the drain terminal. The slope of the ramp is independent of the setting of V_1; therefore the time required for the circuit to reach its final voltage is directly proportional to the voltage V_1. This is a valuable property of the circuit because a delay or time interval can be generated by the circuit which is precisely proportional to a shaft rotation. (In actual practice trimming resistors would have to be added to each end of P to compensate for the diode drop and to calibrate the delay versus potentiometer shaft position.) The diode also reduces the recovery time because the number of recovery time constants required for the drain voltage to reach its initial value is reduced as the voltage V_1 is reduced. As an example, the diode reduces the time required for the circuit to reach 99 percent of the equilibrium voltage from 4.6 time constants without the diode to 2.3 time constants with the diode and $V_1 = 0.9V_{SS}$. Proportionately greater decreases in recovery time are obtained as V_1 is reduced.

Figure 6-14 shows a complete circuit with the addition of all the improvements that have been discussed. In this circuit there is one addi-

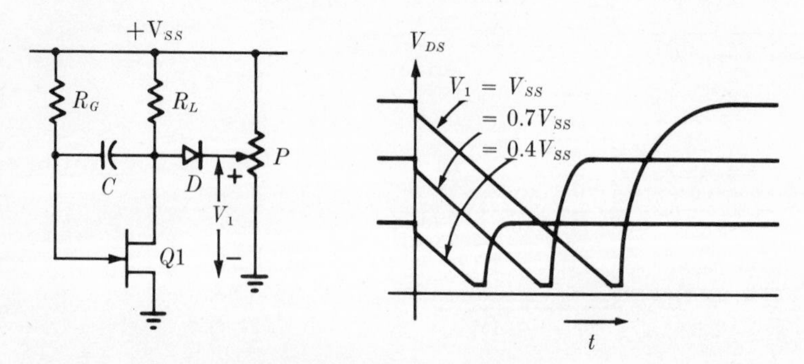

Fig. 6-13 A circuit added to the Miller integrator to make the sweep time proportional to a control voltage V_1.

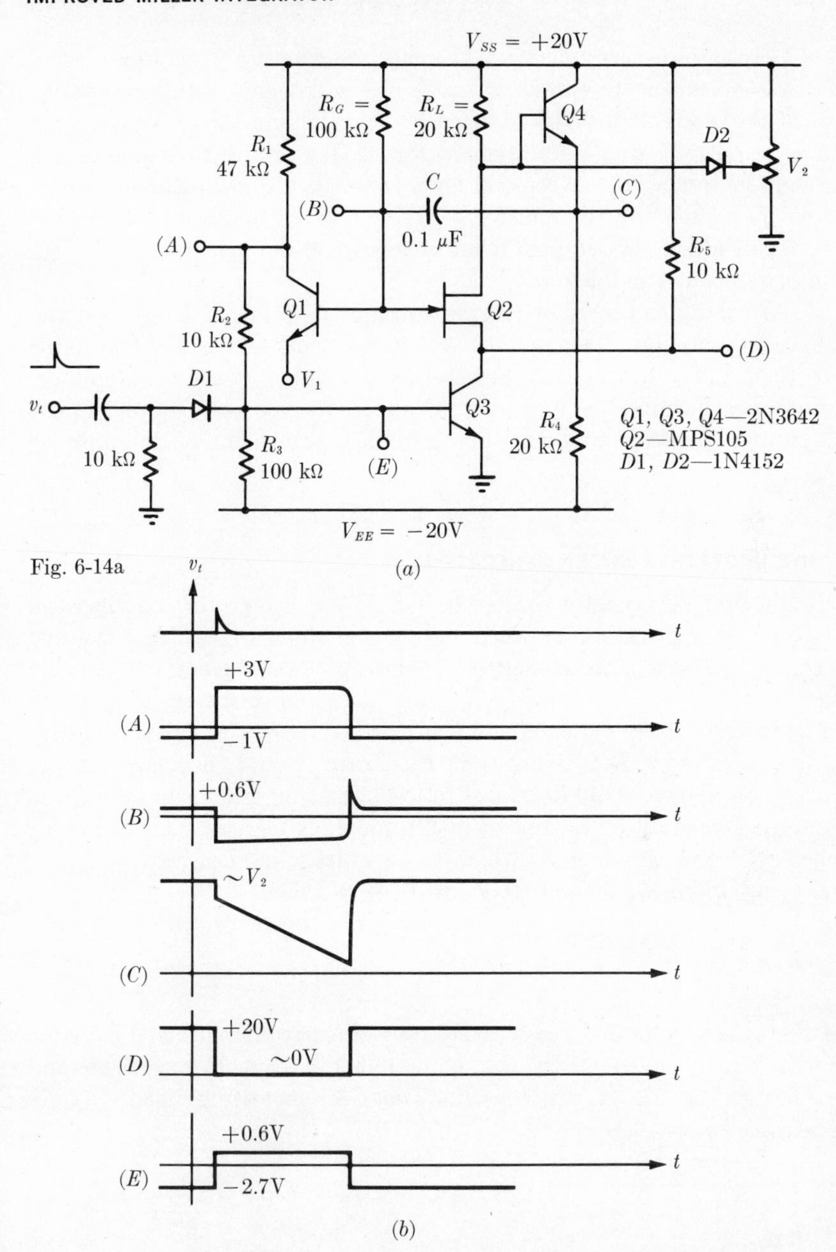

Fig. 6-14 (*a*) A complete regenerative Miller integrator with the circuits to decrease recovery time and control the delay time added. (*b*) The waveforms generated by the circuit.

tional feature that serves to reduce the magnitude of the initial jump in the drain waveform and to eliminate much of the flat end of the waveform. Both of these effects may be reduced by maintaining the quiescent base voltage of $Q1$ and the gate voltage of the FET closer to the value of the gate voltage during the sweep-rundown time. In Fig. 6-14 this is accomplished by connecting the emitter of $Q1$ to a small negative voltage V_1. The magnitude of this voltage must not be great enough to turn ON $Q1$ during the linear rundown.

The examples of Miller integrators shown so far all have used the FET as the amplifier. This is not necessary, of course, and the transistor or another active device may be used as the simple one-stage amplifier, or more direct-coupled stages may be added to give the required open-loop gain. The design problem then becomes largely one of dc amplifier design.

6-5 THE BOOTSTRAP SWEEP GENERATOR

A circuit that is basically similar to the Miller integrator, but has the advantage of generating a sweep voltage commencing at zero voltage referred to ground, is the so-called "bootstrap" sweep generator. A basic diagram for such a circuit is shown in Fig. 6-15. The circuit begins operation when the switch is opened and C begins to charge through R at a rate $dV_1/dt = I/C = V/RC$. Normally this rate would decrease as V_1 increases because I would decrease. In the bootstrap circuit, however, I is maintained constant by raising V_2 just as much as V_1 rises. This is accomplished with the amplifier A, which has a voltage gain of exactly unity, ideally (assuming $R_{\text{in}} = \infty$). If $V_2 = V_1 + V$, then

$$I = \frac{V_2 - V_1}{R} = \frac{V}{R} = \text{a constant}$$

and the sweep voltage V_1 rises absolutely linearly with time. The name for the circuit comes from the fact that the capacitor voltage and the voltage driving R rise together, thus V_1 is raising itself "by its bootstraps."

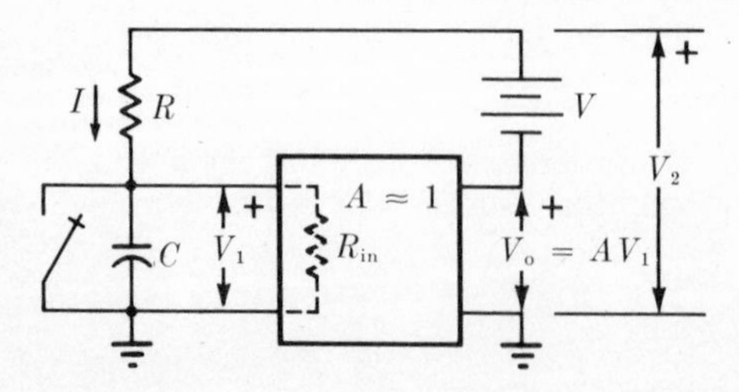

Fig. 6-15 The basic bootstrap sweep generator.

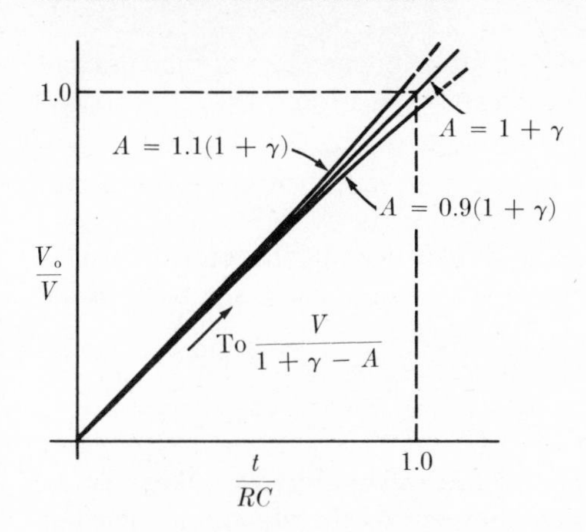

Fig. 6-16 The effect on the output of the bootstrap of having different amplifier gains: $A = 1 + \gamma$ gives the exactly linear sweep shown.

Since the amplifier will not really have exactly unity gain, and its input impedance will not be infinite (causing some of I to be shunted away from C), the sweep will not be exactly linear. The basic equation for the circuit may be written by summing currents at the input to the amplifier

$$I = \frac{V_2 - V_1}{R} = \frac{C\,dV_1}{dt} + \frac{V_1}{R_{\text{in}}}$$
$$= \frac{(V_o + V) - V_1}{R} = \frac{(A - 1)V_1 + V}{R} \tag{6-32}$$

or

$$\frac{C\,dV_1}{dt} = \frac{(A - 1)V_1 + V}{R} - \frac{V_1}{R_{\text{in}}} = \frac{V_1(A - 1 - \gamma) + V}{R} \tag{6-33}$$

where $\gamma \triangleq R/R_{\text{in}}$. The solution of this differential equation assuming the switch is opened at $t = 0$ is

$$V_1(t) = \frac{V}{1 + \gamma - A}\left[1 - \epsilon^{-t(1+\gamma-A)/RC}\right] \qquad (A \neq 1 + \gamma) \tag{6-34a}$$

and

$$V_1(t) = t\frac{V}{RC} \qquad (A = 1 + \gamma) \tag{6-34b}$$

Equation (6-34*b*) is the equation for a truly linear sweep, and is obtained only for the singular case $A = 1 + \gamma$. If the gain is less ($A < 1 + \gamma$), the sweep is a normal exponential with $\tau = RC/(1 + \gamma - A)$ and final value $V/(1 + \gamma - A)$ as shown in Fig. 6-16. The sweep can be made as close as necessary to linear by making A nearly equal to $1 + \gamma$. There is, of course, the possibility of making $A > 1 + \gamma$, and the waveform then becomes an exponential with a *positive* exponent. This is perfectly reasonable and can be utilized for some applications. The sweep voltage does not become infinite, of course, but is limited by the linear range of the amplifier.

The initial slope of the sweep is entirely independent of the amplifier (unless it draws a dc input current when $V_1 = 0$) and is

$$\frac{dV_1(0^+)}{dt} = \frac{V}{RC}$$

This constancy of initial slope and the fact that the output can be obtained from a reasonably low-impedance point make the bootstrap a very useful circuit.

6-6 A PRACTICAL BOOTSTRAP CIRCUIT

To realize the circuit practically we require an amplifier with large R_{in} (to make γ small) and nearly unity gain. The simplest realization of such an amplifier uses an emitter follower, as in Fig. 6-17.

In the circuit of Fig. 6-17*a* the emitter follower provides the gain-of-one element just as shown in the block diagram of Fig. 6-15, but the function of the battery V, which would require an ungrounded or floating power supply, is provided instead by the charge upon the capacitor C_2. This capacitor should be sufficiently large to keep a nearly constant voltage during the useful sweep time. When the circuit is at rest C is discharged through the switch (which would usually be provided by a saturated transistor) and the output V_o is about -0.6 V with respect to ground. The capacitor C_2 is thus charged to almost exactly V_{CC} since the drop in the diode D is about equal to the emitter voltage of $Q1$. When the switch opens, the sweep begins and the output voltage V_o begins to rise. Diode D immediately opens, and the circuit operates as the one in Fig. 6-15 did. The sweep voltage will continue to rise until it exceeds the value of the supply voltage, at which time the base-collector junction in $Q1$ will become forward biased and essentially short circuit C to the positive supply. A reasonably linear sweep of magnitude equal to the entire supply voltage is thus generated. The output of the circuit may be taken from the emitter terminal of the emitter follower without seriously loading the circuit; however, the load impedance should be as large as possible to keep the gain of the emitter follower as close to unity as possible.

After the sweep ends, the capacitor C must be discharged through the switch and the lost charge replaced on C_2. The first capacitor can be discharged very rapidly if a transistor is used for the switch because of the high-peak current capability of even small transistors, but C_2 in general will take a good deal longer to recharge. Since the current I is almost constant, the charge that is lost by C_2 is equal to IT_1 where T_1 is the length of time that the switch is left open. The current available to recharge C_2 is only the current through R_E because the current in $Q1$ cannot flow in the direction to recharge C_2 (also the transistor is usually held OFF immedi-

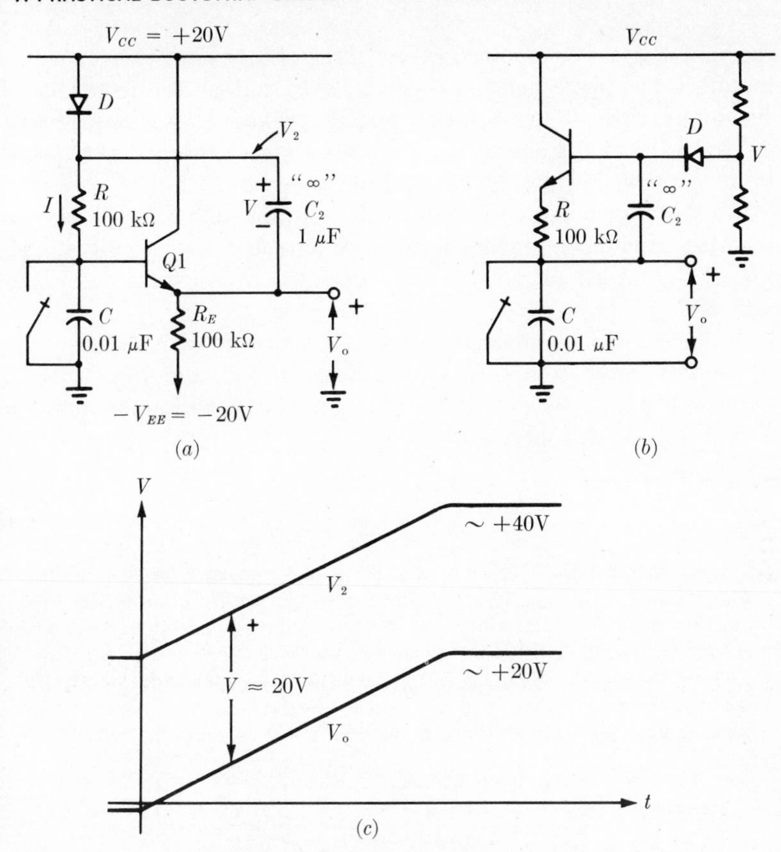

Fig. 6-17 (*a*), (*b*) Two practical forms of the bootstrap circuit. (*c*) Waveforms in the circuit of (*a*).

ately after the switch is reclosed by the slightly discharged capacitor). Therefore the time to recharge C_2 may be found by equating the charge changes during the discharge and recharging intervals.

$$\frac{T_2 V_{EE}}{R_E} = IT_1 = \frac{T_1 V_{CC}}{R} \tag{6-35}$$

where T_2 is the time for the circuit to fully recover. For the circuit values given the recovery time will be equal to the gate time (the time the switch is left open). This recovery time is surprisingly long and is not easily seen from the waveforms of the circuit. If, however, insufficient time is allowed for the circuit to recover, the sweep will not be as linear as predicted. This effect is caused by the amplifier operating nonlinearly during the initial part of the sweep. The recovery time can be reduced by lowering R_E, but this also reduces the linearity.

The bootstrap circuit shown in Fig. 6-17*b* does not have this recovery problem, but it does not generate as large a sweep voltage nor does it have the convenient emitter-follower output terminal that the preceding circuit had. The recovery of the circuit is facilitated by the facts that C_2 has a much lower current flowing out of it during the sweep (thus it loses less charge) and that C_2 can recharge though the low impedance of the switch and the voltage divider producing V. One advantage of this circuit is that the slope of the sweep may be easily controlled by varying V with a potentiometer.

Other forms of the bootstrap circuit may be made with other active elements or with more complicated amplifiers. If a dc amplifier with a stabilized gain of greater than unity is used, waveforms with an increasing slope $(1 - \epsilon^{+t/\tau})$ can be generated.

PROBLEMS

6-1. (a) The circuit shown is a Miller integrator using a transistor for the amplifier. The input gating waveform v_1 allows the transistor, Q, to conduct and produce the output ramp during the 10 ms interval. Sketch and label the output waveform v_2 including the recovery period.

(b) Redraw the circuit including another 2N3134 transistor as an emitter follower to decrease the recovery time of the circuit. With the emitter follower what is your estimate of the time at which a sweep may be re-initiated?

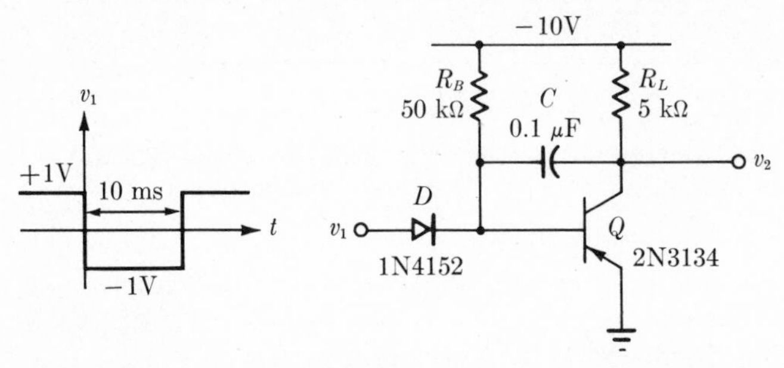

6-2. Using a current source and a capacitor, design and give all element values for a circuit giving a sweep voltage going from −10 V to 0 V in 10 μs. Use a transistor for the current source. Two power supply voltages, +10 and −10 V, are available. Assume a switch across the capacitor opens at $t = 0$ to initiate the sweep.

This is a typical engineering problem to which there are an infinite number of correct answers. In a *brief* paragraph justify your choice of capacitor size.

6-3. Sketch and label the waveforms in the modified Miller sweep circuit of Fig. 6-12. Assume $R_G = 100$ kΩ, $R_L = R_E = 10$ kΩ, $V_{SS} = +20$ V and $-V = -5$ V. Compare your answer to the waveforms in Fig. 6-8.

6-4. Calculate the waveforms appearing in the bootstrap circuit of Fig. 6-17a. Assume $Q1$ is a 2N3642 and that C_2 acts like a battery during the ramp. Be sure to calculate the effective time constant and final voltage for the sweep.

6-5. The circuit shown is a form of bootstrap which will generate a sweep with positive curvature if $R_1 > 0$. The triangle is an operational amplifier (an integrated version might be the 709) which has high open-loop gain and very high input impedance in the configuration shown. Assuming the open loop gain is much larger than the closed-loop gain, the closed loop (connected as shown) gain is closely $A = v_{in}/v_{out} = R_2/(R_1 + R_2)$ (for the 709 the open loop gain is about 10^4).

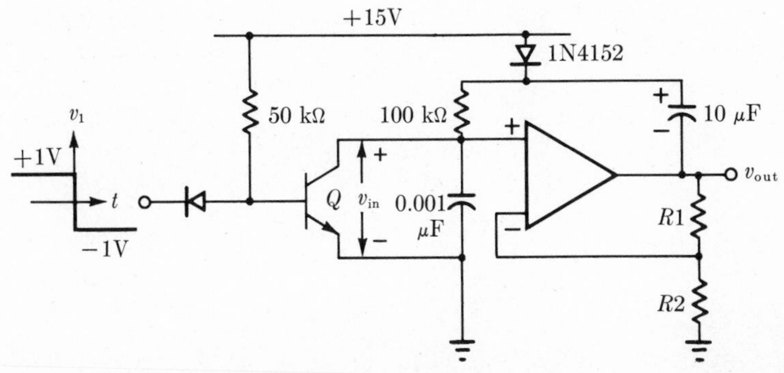

Assume $R_1 = R_2$ and that the amplifier input resistance is infinite. Sketch v_{out} for $t > 0$ assuming the amplifier is linear for $-12 < v_{out} < +12$ V, and limits at $v_{out} = \pm 12$ V. Calculate the time for the sweep to reach +10 V, and calculate the deviation of the sweep from a straight line passing through the beginning and 10 V points.

7 Circuits Containing Inductors or Transformers

We have come about as far as is desirable with circuits containing only R and C. Actually it is rather remarkable how much can be accomplished with simple RC combinations, employing one or more conventional active devices and usually only a single capacitor. Nonetheless, a whole new field of possibilities opens up when L is included in the circuit components.

7-1 THE RL CIRCUIT

The logical first step in introducing L is the consideration of the series RL circuit of Fig. 7-1, even though such a circuit is seldom realized among practical electronic circuits, primarily because the presence of stray capacitances in the circuit and the inductor windings results inevitably in RLC combinations. Such strays influence the operation more than in the circuits previously considered, and can indeed be utilized to accomplish some desired ends. The idealized RL circuit does, however, provide some insight into the behavior of inductances under transient conditions.

We note that the circuit acts like the RC differentiating or "peaking"

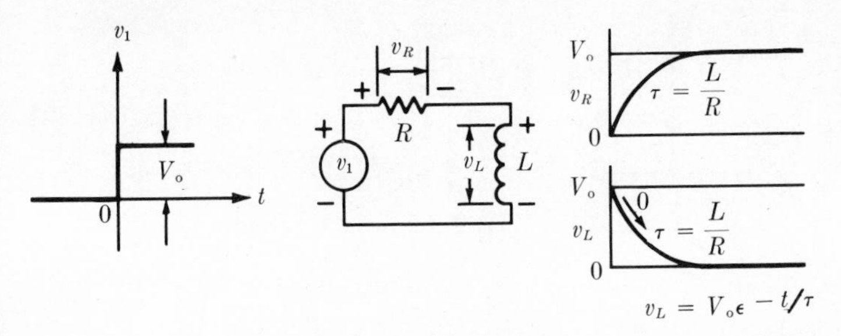

Fig. 7-1 Step-function response of *RL* circuit.

circuit of Fig. 2-6 and can be used to provide short pulses if L/R is small. We must reserve judgment on this as a practical possibility until we evaluate the influence of stray C across the inductance.

We do note, however, that at the initial instant the inductance acts like an open circuit, and the jump in v_1 is transmitted directly to v_2. This is to say that the initial *current* in L is zero (corresponding to an open circuit), and it cannot change instantaneously. This is the dual of the previous rule that a capacitor *voltage* cannot change instantaneously. Where there is a series resistance through which the current must flow, the inductor current or the capacitor voltage changes exponentially, with a time constant L/R in one case and RC in the other. If in Fig. 7-1 there had been a current flowing in L prior to time $t = 0$, that current would have continued at $t = 0^+$. It is thus helpful to keep in mind the following:

1. The current in an inductor cannot change instantaneously when the voltages are finite.
2. Sudden changes in the driving voltage (or current) in a circuit containing one L and one R result in exponential transient voltages (and currents) with time constant L/R.
3. The inductor cannot sustain a steady value of voltage without an unlimited source of current. Therefore the final value of voltage across the inductor will be zero, i.e., the inductor behaves like a short circuit for $t \to \infty$.
4. The usual relation $v_L = L\,di/dt$ of course applies.

A simple circuit which can be analyzed as an *RL* circuit is shown in Fig. 7-2*a* in which a relay coil having an inductance of 0.1 H and a resistance of 1,000 ohms is to be controlled by a transistor operating as a saturated switch. The maximum possible collector current is $+12/1\ \text{k}\Omega = 12$ mA, so the value of R_B should be chosen to insure saturation of the transistor with this level of collector current. In this case the base current is about 0.5 mA and the transistor is very saturated, as may be seen from

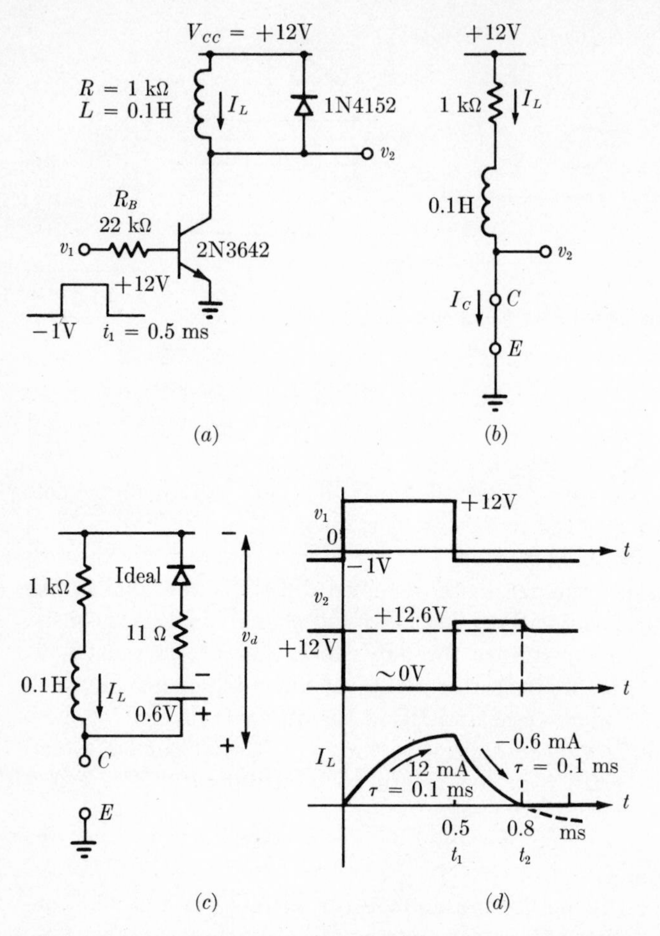

Fig. 7-2 (*a*) An inductive relay coil driven by a transistor switch. The input pulse v_1 saturates the transistor for 0.5 ms. (*b*) An equivalent circuit valid for $0 < t < 0.5$ ms. (*c*) An equivalent circuit valid for $t > 0.5$ ms. (*d*) The waveforms appearing in the circuit.

the transistor collector characteristics (or by specifying that a minimum $h_{FE} = 12/0.5 = 24$ is required in the transistor).

At $t = 0$ the transistor is switched from the OFF state to the ON state; since the inductor current was initially zero, it must be zero at $t = 0^+$ also. Therefore the initial value of the collector current is also zero, and the transistor is saturated at $t = 0^+$ no matter how large a base current is supplied to it. The circuit may be represented as shown in Fig. 7-2*b*, where the transistor is represented as a short circuit, and the diode—because it is reverse-biased—is shown as an open circuit. The current

$I_L = I_C$ and rises toward its final value of 12 mA with a time constant $L/R = 0.1$ ms. Since the transistor is turned on for a total time of 0.5 ms, the current rises to substantially its final value before the transistor is switched off at t_1. Notice that the slowly rising current in the relay coil will delay the operation of the relay since the magnetizing force in the relay has the same waveshape as the coil current. Whether the relay would actually operate in this case would depend upon the mechanical inertia of the relay.

At $t = t_1$ the transistor base is reverse biased to turn it OFF; however, if the diode were not provided, it would not be possible to turn the transistor off because of the current flowing in the inductor at this time. The 12-mA current would continue to flow in the collector even though the base were reverse biased. This could only happen at collector voltages high enough to cause collector breakdown. The energy stored in the inductor ($LI_L^2/2 = 14.4\ \mu$J) would then be largely dissipated in the transistor and would in all probability damage it. The diode is added to provide a return path for the current in the inductor and to prevent damage to the transistor. Figure 7-2c is an equivalent circuit valid for $t > t_1$ if $I_L(t_1^-) = I_L(t_1^+) = 12$ mA. The diode is shown as a combination of an ideal diode to emphasize that the current in the diode branch can only flow upward, a rather negligible resistor, and a 0.6 V battery.

The easiest way to analyze the circuit is to observe the behavior of the current, which is initially in the direction shown and is 12 mA. The final value of the current is found by considering the inductor to be a short circuit, that is, $I_L(\infty) = -0.6/1.011\ \text{k}\Omega \approx -0.6$ mA. The equation for the current is then

$$I_L = -0.6 + 12.6\epsilon^{-(t-t_1)/\tau} \qquad (\text{for } t > t_1) \tag{7-1}$$

where τ is $L/R \approx 0.1$ ms, as before. The current cannot ever actually attain the final value of -0.6 mA because to do so would require the diode to conduct backward; therefore the current decays toward -0.6 mA as shown in Fig. 7-2d, but when it reaches zero (and the stored energy in the inductor is also zero), the diode becomes an open circuit and the current does not reverse. To find the time for this to occur, equate I_L to zero in Eq. (7-1) and solve for $t_2 - t_1$.

$$\begin{aligned} t_2 - t_1 &= \tau \ln\left(\frac{12.6}{0.6}\right) = 3.04\tau \\ &\approx 0.3 \text{ ms} \end{aligned} \tag{7-2}$$

At this time the stored energy in the circuit is zero, the currents are zero, and the drop across the diode becomes zero. The diode voltage after t_1 is

$$\begin{aligned} v_d &= 0.6 + I_L(0.011\ \text{k}\Omega) \approx 0.6 \text{ V} && (0.5 < t < 0.8 \text{ ms}) \\ &= 0 \text{ V} && (t > 0.8 \text{ ms}) \end{aligned} \tag{7-3}$$

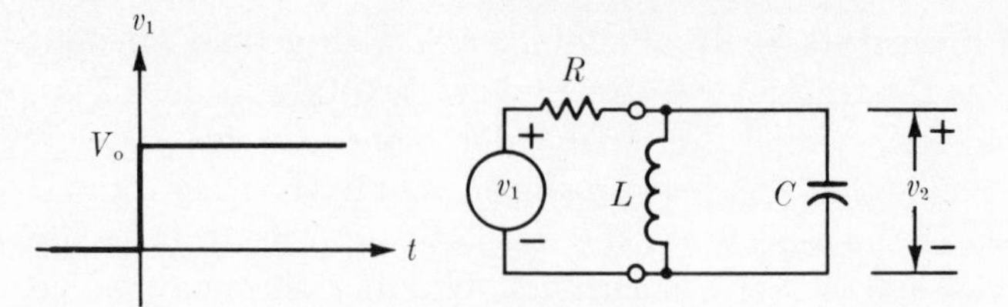

Fig. 7-3 Step of voltage applied to *RLC* circuit.

and the voltage $v_2 = +12 + v_d$, and is shown in Fig. 7-2*d*. The abrupt drop in the diode voltage at t_2 may be surprising, but is a consequence of the loss in stored energy in the circuit. The circuit is not recovered and ready for another operation until this energy is lost, and again the operation of the relay is delayed by the fact that the coil current does not immediately go to zero. The recovery of the circuit may be speeded by adding a resistor in series with the diode to decrease the time constant; for example, adding 1 kΩ in series with the diode will halve the time constant and the recovery time at the cost of about doubling the peak voltage across the switch transistor. In general the speed with which currents can be changed in inductive circuits is limited by the voltage rating of the device used to control the currents. The fastest possible recovery time is obtained by holding the voltage across the switch constant during the recovery period at the maximum voltage permissible for the switch. Such a circuit is shown in Prob. 7-1.

7-2 THE *RLC* CIRCUIT

Now let us add a small capacitance in parallel with L giving the circuit of Fig. 7-3 or its alternative form shown in Fig. 7-4.

In general, the behavior of v_2 with time will depend greatly upon the circuit proportions, there being three principal cases:

1. overdamped
2. critically damped
3. underdamped

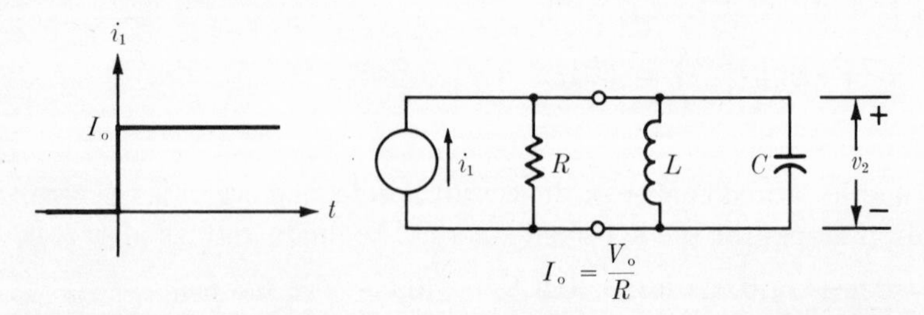

Fig. 7-4 Parallel *RLC* circuit derived from that of Fig. 7-3 by converting the voltage source to a current source.

It happens that critical damping corresponds to a Q of $\frac{1}{2}$, whereas overdamping calls for lower Q and underdamping for higher Q. The Q, of course, is inversely proportional to damping and can be defined as $R/\sqrt{L/C}$. (It is the familiar Q of resonant circuit theory.) If we start with C very small, the Q is very small, and the circuit is overdamped. The exact form of the solution is

$$v_2 = \frac{I_o}{C}\frac{\epsilon^{-\alpha t} - \epsilon^{-\gamma t}}{\gamma - \alpha} \tag{7-4}$$

where

$$\alpha = \frac{1}{2RC} - \sqrt{\left(\frac{1}{2RC}\right)^2 - \left(\frac{1}{\sqrt{LC}}\right)^2} \tag{7-5}$$

$$\gamma = \frac{1}{2RC} + \sqrt{\left(\frac{1}{2RC}\right)^2 - \left(\frac{1}{\sqrt{LC}}\right)^2} \tag{7-6}$$

Case I. Overdamping For extreme overdamping, that is, C extremely small

$$\frac{1}{2RC} \gg \frac{1}{\sqrt{LC}}$$

$$\sqrt{\left(\frac{1}{2RC}\right)^2 - \left(\frac{1}{\sqrt{LC}}\right)^2} = \frac{1}{2RC}\sqrt{1 - \left(\frac{1/\sqrt{LC}}{1/2RC}\right)^2}$$

$$= \frac{1}{2RC}\sqrt{1 - (2Q)^2} \tag{7-7}$$

$$\approx \frac{1}{2RC}(1 - 2Q^2) \qquad (Q \ll 1) \tag{7-8}$$

Then

$$\alpha \approx \frac{1}{2RC} - \frac{1}{2RC}(1 - 2Q^2)$$

$$\approx \frac{2Q^2}{2RC} = \frac{R}{L} \tag{7-9}$$

$$\gamma \approx \frac{1}{2RC} + \frac{1}{2RC}(1 - 2Q^2)$$

$$\approx \frac{1}{RC} \tag{7-10}$$

$$\gamma - \alpha \approx \gamma \approx \frac{1}{RC} \tag{7-11}$$

Hence

$$v_2 \approx I_oR[\epsilon^{(-R/L)t} - \epsilon^{-t/RC}] \qquad \left(\frac{L}{R} \gg RC\right) \tag{7-12}$$

For small t

$$\epsilon^{(-R/L)t} \approx 1 \qquad (7\text{-}13)$$

$$v_2 \approx I_o R(1 - \epsilon^{-t/RC}) = V_o(1 - \epsilon^{-t/RC}) \qquad (7\text{-}14)$$

For large t

$$\epsilon^{-t/RC} \approx 0 \qquad (7\text{-}15)$$

$$v_2 \approx I_o R \epsilon^{(-R/L)t} = V_o \epsilon^{(-R/L)t} \qquad (7\text{-}16)$$

We can confirm these results from physical intuition. By referring to Fig. 7-3, we see that the inductor current is zero at the initial instant; hence, it acts initially like an open circuit. Moreover, this current can change only at a relatively slow rate. Thus, at first, we have essentially an RC circuit. The capacitor voltage can rise exponentially to V_o with a time constant RC, if RC is so small that V_o is reached before appreciable current begins to flow in L. To state this in another way, during this first interval *all* the current flows into the capacitor and *none* into the inductor. After the capacitor has charged, the current in C will be zero or negligibly small, whereas the inductor current will increase slowly. Only a negligible current in C is needed to follow the slow changes in v_2 as the current in L builds up; hence, the presence of the capacitor can be ignored, leaving only a series RL circuit with time constant L/R. The appearance of the waveform for v_2 is shown in Fig. 7-5.

Still another interpretation of this analysis of the overdamped case corresponds to the conventional analysis of audio and video amplifiers. The response to abrupt changes (near $t = 0$) is dependent upon the high-frequency response, for which an equivalent circuit would show only R and C, with $j\omega L$ being infinite. On the other hand, the slow variations depend upon a low-frequency equivalent circuit, which would contain only R and L, with C being an open circuit. We shall have an application of the overdamped RLC circuit in the pulse transformer.

Case II. Critical damping This represents a Q of $\frac{1}{2}$, or $R = (\frac{1}{2})\sqrt{L/C}$. We can approach it from the overdamped situation by increasing C or decreasing L.

Suppose we wish to generate a short pulse in response to a step function of current or voltage. By looking at Fig. 7-5 we note that decreasing L is a move in the right direction. It happens that the critical damping case is about the best compromise between shortness of the pulse duration and height of the peak. The solution for v_2 is shown in Fig. 7-6 and is

$$v_2 = \frac{I_o}{C} t \epsilon^{-\alpha t}$$

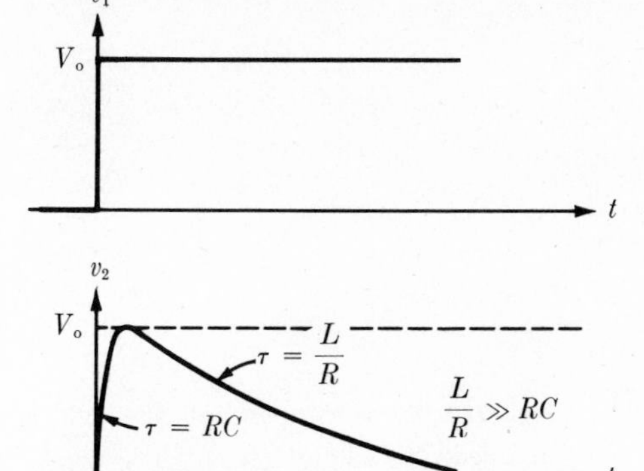

Fig. 7-5 Step response of the overdamped *RLC* circuit (Fig. 7-3), that is, when $L/R \gg RC$. The approximations of Eqs. (7-14) and (7-16) are incorporated.

where

$$\alpha = \frac{1}{2RC} = \frac{1}{\sqrt{LC}} \tag{7-17}$$

Maximum v_2 is attained when $t = \sqrt{LC}$.

$$v_2(\text{max}) = \frac{2I_oR}{\epsilon} = \frac{2}{\epsilon} V_o = \frac{I_o}{\epsilon} \sqrt{L/C} \tag{7-18}$$

Case III. Underdamping As the Q is increased above ½, the nature of the voltage waveform becomes oscillatory, as for example in Fig. 7-7.

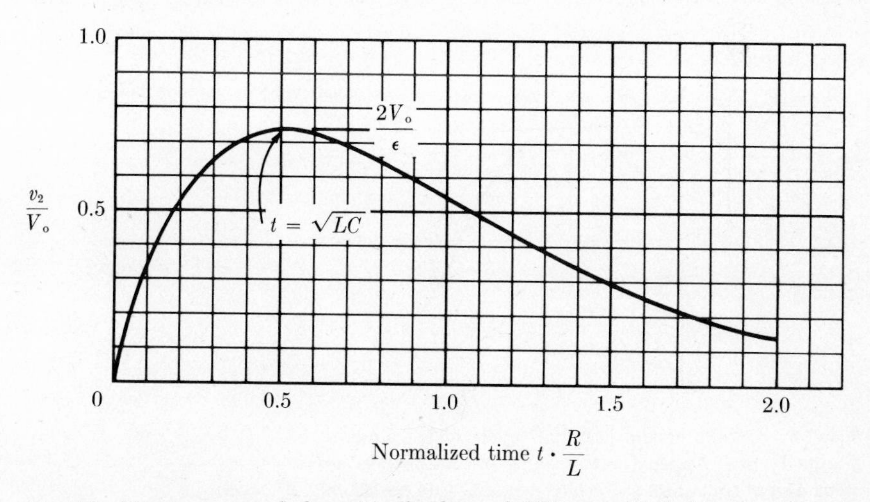

Fig. 7-6 Step response of the critically damped *RLC* circuit, that is, $Q = 0.5$, or $R = 0.5\sqrt{L/C}$.

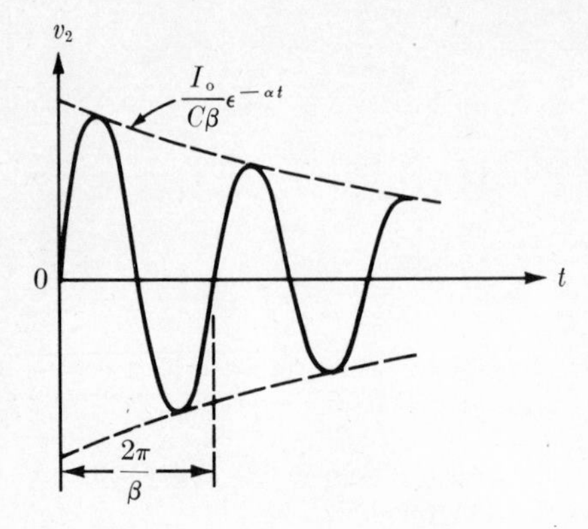

Fig. 7-7 Step response of the underdamped RLC circuit ($R \gg 0.5\sqrt{L/C}$).

The exact solution is as follows:

$$v_2 = \frac{I_o}{C}\left(\frac{1}{\beta}\epsilon^{-\alpha t}\sin\beta t\right) \tag{7-19}$$

where

$$\alpha = \frac{1}{2RC}$$

$$\beta = \sqrt{1/LC - (1/2RC)^2} = \omega_o\sqrt{1 - (1/2Q)^2} \qquad \left(\omega_0 = \frac{1}{\sqrt{LC}}\right)$$

When the Q is very high, the exponential decay is slight, and the values shown in Fig. 7-8 for the height of the first maximum and its time

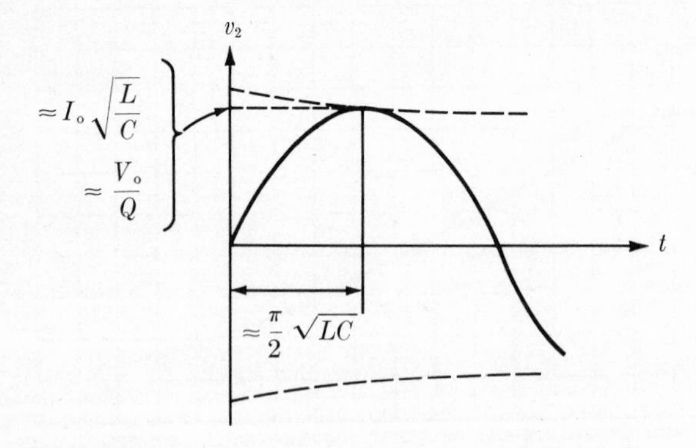

Fig. 7-8 Details of the first half cycle of the underdamped case. Approximate values for amplitude of peak and of time interval to this peak become exact for zero damping (infinite Q).

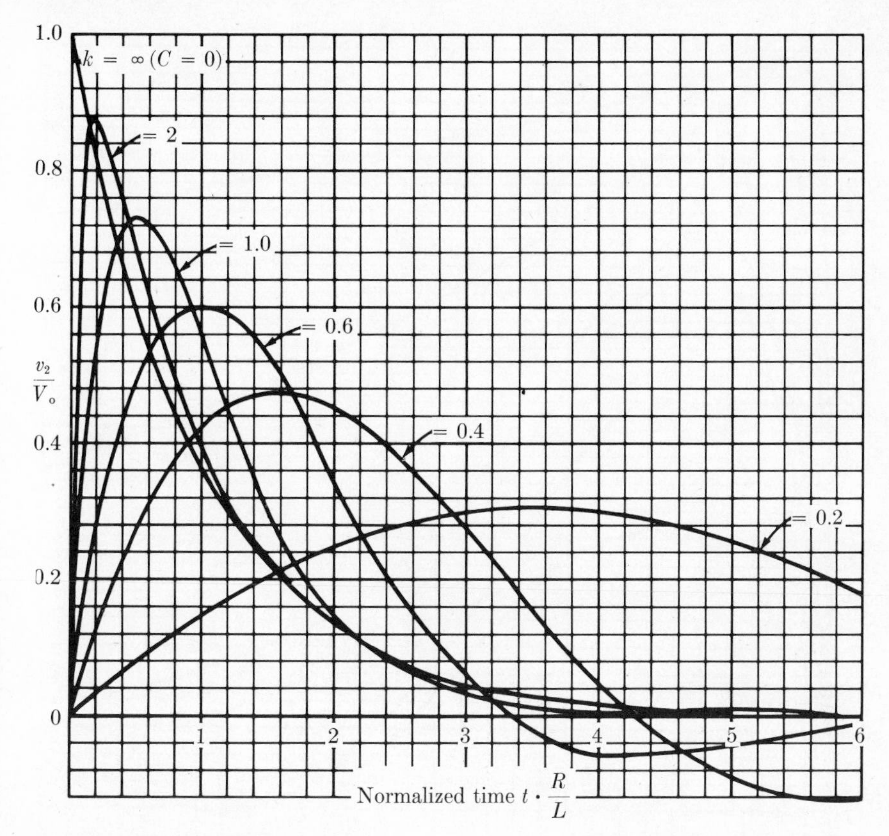

Fig. 7-9 Response of the circuit of Fig. 7-3 for varying values of C. The parameter k is related to the capacitance C by $k = \dfrac{\sqrt{L/C}}{2R} = \dfrac{1}{2Q}$. Critical damping occurs for $k = 1$.

of occurrence are reasonably exact. They are exact for the case of zero damping.

It is interesting to study how the response of the circuit varies while only one of the circuit parameters is varied. In Fig. 7-9 the inductance and resistance are kept constant while C is increased from zero. For $C = 0$ (corresponding to $Q = 0$, also) the response is a simple exponential with time constant R/L. For small values of C (or Q) the capacitance merely modifies the rise of the response while slightly reducing the peak amplitude. For sufficient C to give critical damping, the curve departs markedly from the simple exponential and still further increases in C give a broader peak with a damped oscillatory tail.

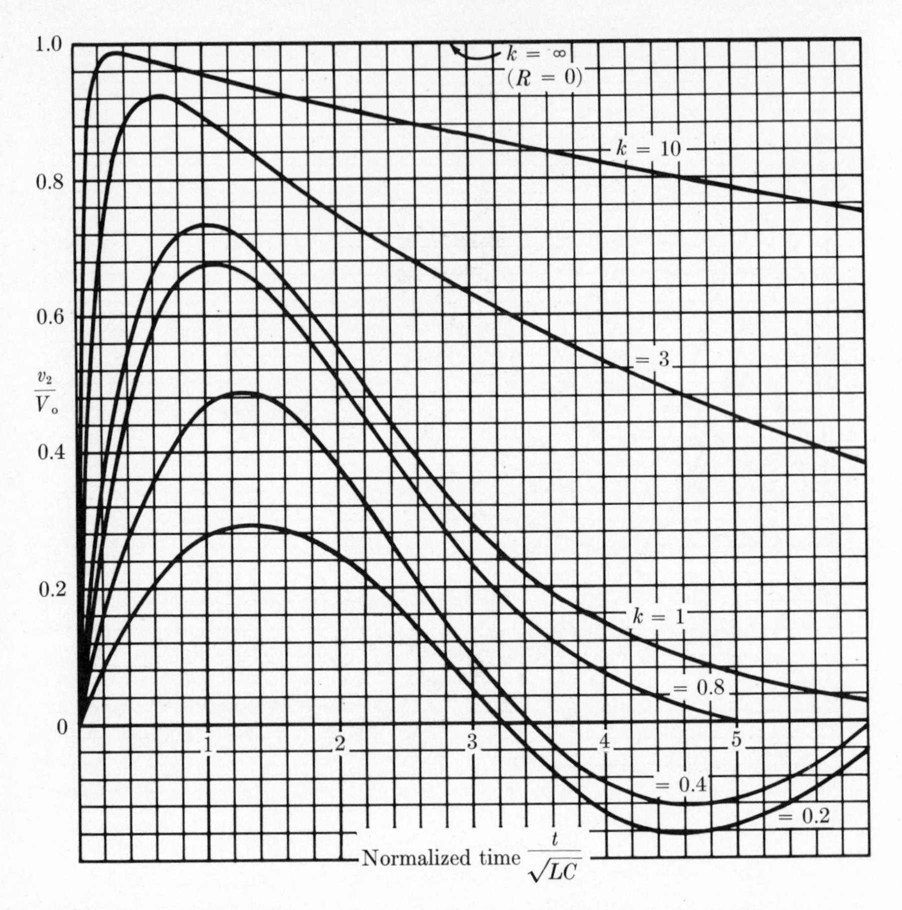

Fig. 7-10 Response of the circuit of Fig. 7-3 for varying values of R. The parameter k is related to the resistance R by $k = \frac{\sqrt{L/C}}{2R} = \frac{1}{2Q}$.

If, on the other hand, the values of L and C are maintained constant and R is varied from zero, the waveforms of Fig. 7-10 result. Here $R = 0$ corresponds to $Q = 0$, gives an output equal to the input unit step (an impossible situation because the initial current into the capacitor is infinite, and the inductor current rises linearly with time, eventually either saturating the inductor or overloading the source). As the resistance is increased the Q also increases. Note the change in the frequency as the Q increases past that required for critical damping. This is evidenced in the shift of the time of the zero crossings as the Q is increased. When $Q = 5$, the frequency is within about 0.5 percent of the ultimate value obtained with a Q of infinity.

7-3 THE RINGING OSCILLATOR

A high-Q circuit can be used as a timing standard for oscilloscopic measurements. The so-called "ringing circuit" is shown in Fig. 7-11. Instead of applying a positive current step from an external source as in Fig. 7-4, with no initial energy in the RLC circuit, the procedure is to establish an initial current I_o in the inductance prior to $t = 0$, at which time the external current is reduced to zero by the transistor acting as a switch. The RLC circuit then rings giving a waveform as in Fig. 7-7 but with reversed polarity. The period of ringing is independent of the transistor characteristics. It depends primarily on L and C, and only secondarily upon Q if the Q is at all high.

In the circuit shown, the transistor is originally in a saturated state so that the initial current in the inductor is determined by V_{CC} and R_C. At $t = 0$ the transistor is turned OFF by the input pulse, and the current which was flowing in the emitter now flows into the capacitor tending to charge it negatively; therefore, the output waveform starts with a negative half cycle. The input pulse must be sufficiently negative to keep the transistor turned off even though the emitter is made negative by the output

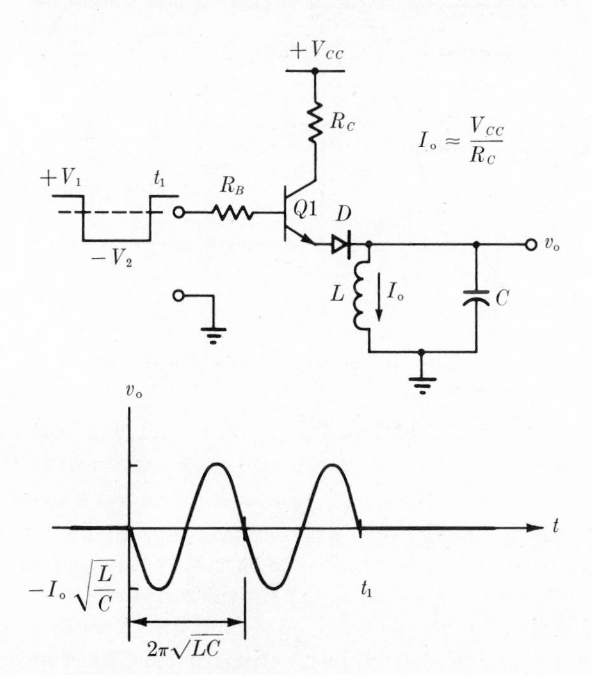

Fig. 7-11 A ringing circuit to generate a pulsed train of sine-wave oscillations starting exactly when the transistor $Q1$ is turned OFF.

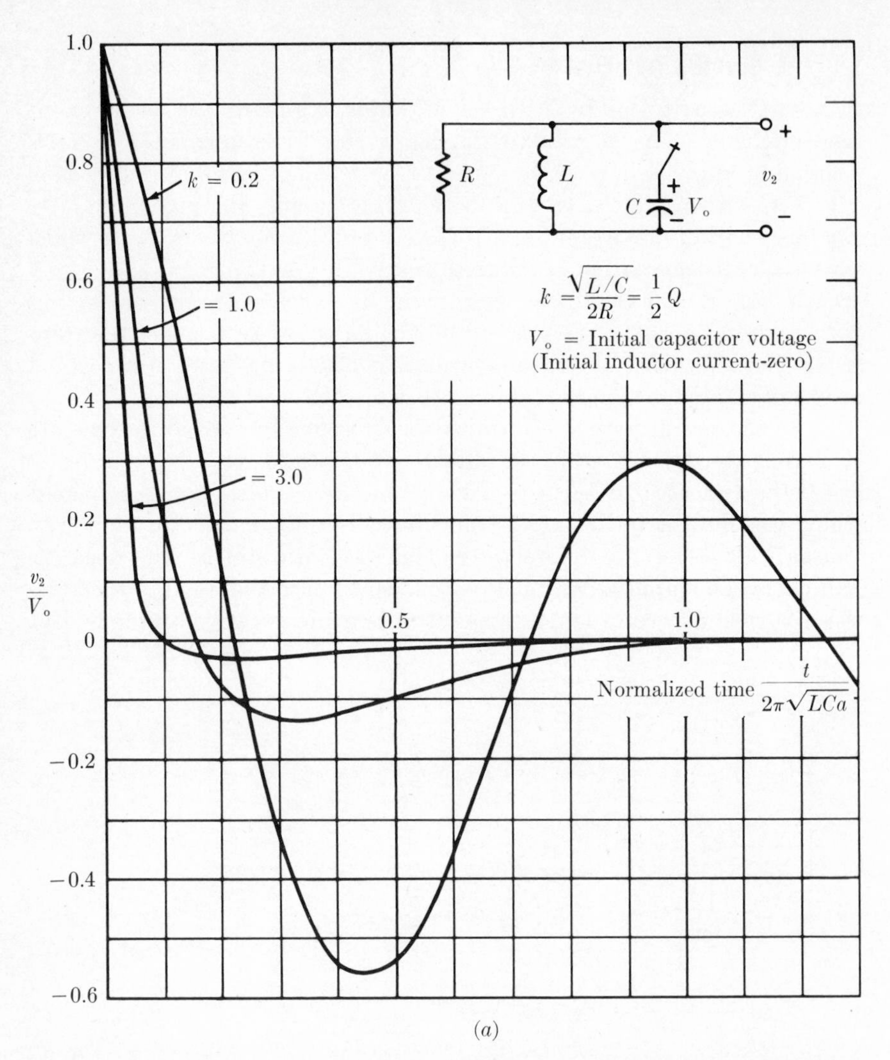

(*a*)

Fig. 7-12 (*a*) An *RLC* circuit with all the initial energy stored in the capacitor. Note the undershoot that occurs even with $k = 1$. ($k = 1$ corresponds to critical damping.)

sine wave. The purpose of the diode is to keep the emitter-base junction of the transistor from breaking down since it must otherwise withstand a voltage at least as great as the peak-to-peak voltage of the output. At the end of the pulse the transistor is turned back ON, and the circuit is damped by the resistors R_B and R_C effectively acting in parallel. The exact details of the output will then depend upon the current in L and the charge on C at t_1. For t_1 equal to two periods as shown, the charge on C is zero, and the current in the inductor at t_1 is exactly equal to the initial current I_o (assuming the Q is effectively infinite). With this set of conditions the circuit returns instantly to its initial conditions at $t = t_1$, and there is no transient following t_1.

7-4 OTHER *RLC* CIRCUIT CONFIGURATIONS

Having the solutions to several other types of *RLC* circuits will be useful in the analysis of the pulse transformer and other circuits. Figure 7-12*a* shows the same circuit as Fig. 7-4, but with different initial conditions. In

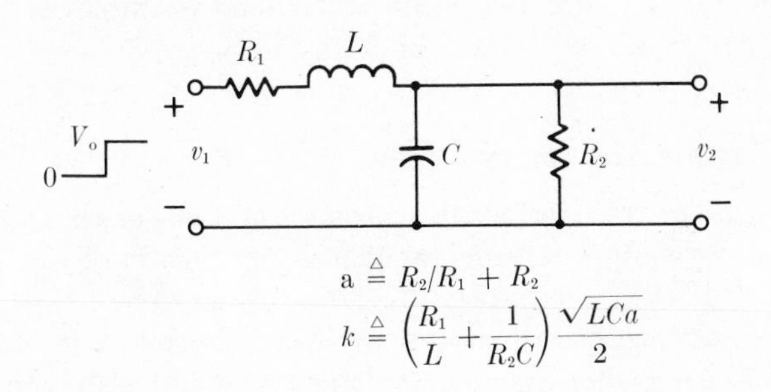

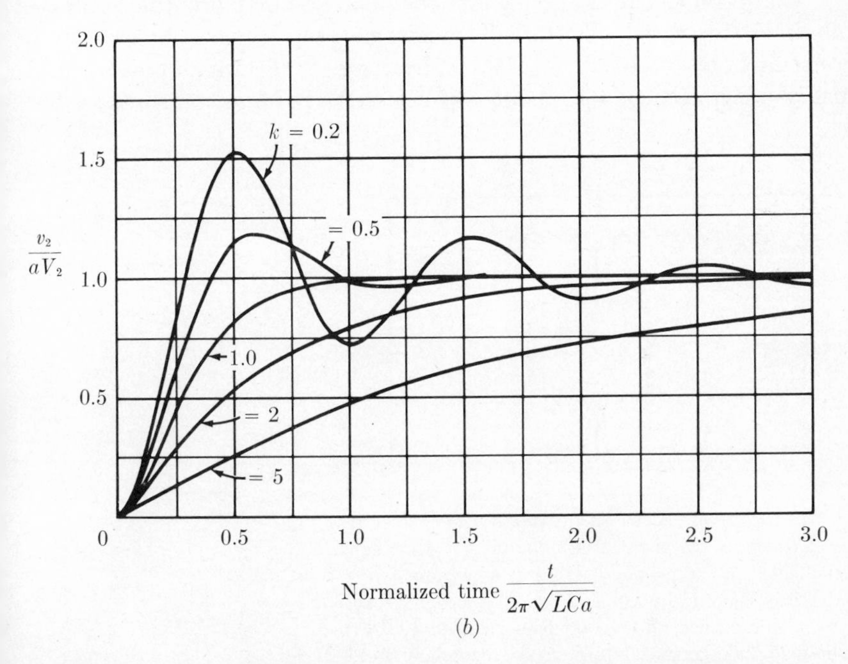

Fig. 7-12 (*b*) An *RLC* circuit with the input step applied through the inductor. The factor k is defined in the figure, and again $k = 1$ for critical damping.

this case the capacitor is initially charged and provides the only energy for the circuit. Again the dividing line between an underdamped or oscillatory response and an overdamped response is $Q = \frac{1}{2}$.

We have a somewhat more complicated case in Fig. 7-12*b* in that there are two resistors for damping, and the input step voltage is applied in series with the inductor. The output v_2 then has a final value equal to the input voltage time $aR_1/(R_1 + R_2)$, but has a rise time determined by the rate of charge of C through R_1 and L. The damping factor is somewhat more complicated in this case, but the familiar overdamped, critically damped, and underdamped cases still exist—this time for values of $k = 2$, 1, and 0.2, respectively. Explicit examples of the use of these particular RLC circuits will be found in Sec. 7-5.

7-5 THE PULSE TRANSFORMER

The pulse transformer is an iron-core, two- or three-winding transformer with its design parameters chosen for certain objectives. Otherwise, it is much like an audio or power transformer. The objectives in the choice of the pulse transformer are specified in terms of its desired transient performance rather than its frequency response, although the two are related.

Pulse transformers may be large or small and may provide a voltage step up or step down. With small transformers the primary objective may be accurate transmission of a waveform, e.g., a rectangular pulse, and little concern need be felt about power loss. With large transformers for

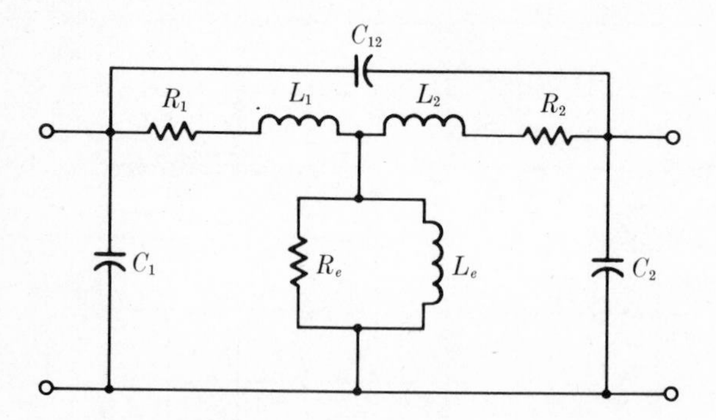

Fig. 7-13 Equivalent circuit for pulse transformer reduced to unity turns ratio. Elements: C_1 = primary capacitance, R_1 = primary winding resistance, L_1 = primary leakage inductance, L_e = magnetizing (excitation) inductance, R_e = core-loss resistance, L_2 = secondary leakage inductance, R_2 = secondary winding resistance, C_2 = secondary capacitance, C_{12} = interwinding capacitance.

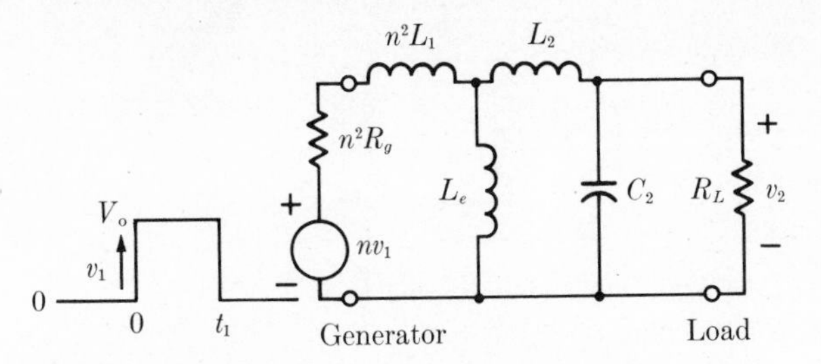

Fig. 7-14 Simplified equivalent circuit for a step-up transformer wherein C_2 is the influential capacitance and C_1 can be omitted.

high-power applications, the accuracy of the waveform transmission can be compromised in favor of smaller energy losses; we shall not examine this problem.

The behavior of the pulse transformer can be derived from the general equivalent circuit shown in Fig. 7-13. This circuit contains far too much detail, however, for in a well-designed transformer many of the elements included in Fig. 7-13 have negligible influence. The key to a successful analysis, i.e., one that is as simple as possible and which keeps the important transformer parameters sharply in focus, is *approximation.* In the approximation process we simplify Fig. 7-13 in two directions. (1) Step-up transformers have large secondary capacitance compared to primary C_1, hence we neglect the latter (vice versa in step-down). (2) We consider separate equivalent circuits for the rising and falling edges of the pulse and for the flat top of the pulse (this corresponds to high- and low-frequency equivalent circuits in the audio case). In all cases here we shall ignore the transformer resistances and the interwinding capacitance C_{12}.

7-6 THE STEP-UP TRANSFORMER

First let us consider a step-up transformer, thus neglecting the primary capacitance. The equivalent circuit referred to the secondary side of the transformer is given in Fig. 7-14.

When the pulse rises at time $t = 0$, the relatively high magnetizing

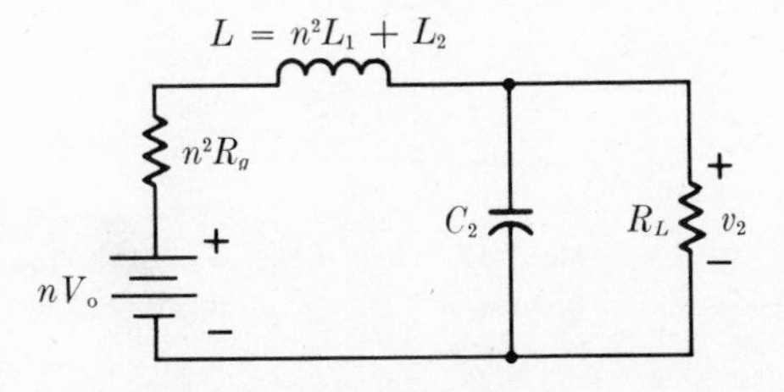

Fig. 7-15 Equivalent circuit to Fig. 7-14 reduced to elements governing response to leading edge of input pulse.

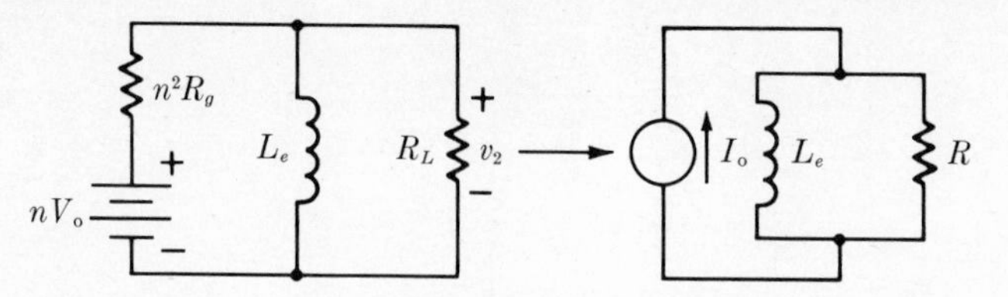

Fig. 7-16 Equivalent circuit to Fig. 7-14 for evaluating the response to the flat top of the input pulse. $R = n^2R_gR_L/(n^2R_g + R_L)$, $I_o = V_o/nR_g$.

inductance L_e remains essentially an open circuit during the short interval while the circuit is responding to the rising edge. The equivalent circuit for this interval is shown in Fig. 7-15, and is similar to the circuit of Fig. 7-12*b*. Therefore the response of the pulse transformer to the input step is that shown in Fig. 7-12*b* with the various degrees of damping produced by varying the resistances in the circuit while keeping L and C_2 constant.

7-7 DECAY OF THE PULSE TOP

After the top of the pulse is reached, current begins to flow in L_e, and the output voltage begins to decay slowly toward zero. The equivalent circuit of Fig. 7-16 covers this interval. It is a simple RL circuit, with a time constant L/R (L being the magnetizing inductance L_e, and R being the parallel combination of n^2R_g and R_L). The waveform of v_2 during the top of the pulse is shown in Fig. 7-17. Notice that the decay in the pulse top is caused by the magnetizing inductance, and can be minimized by using a transformer with greater inductance or reducing the resistance associated with the transformer. However, either method will also increase the rise time. This is because there is a constant relationship between the leakage inductance to the magnetizing inductance for a given core and physical coil size so that increasing the latter inductance will also, in general, increase the former inductance. Therefore for a given transformer design there is a compromise between having a fast rise time and a "flat top."

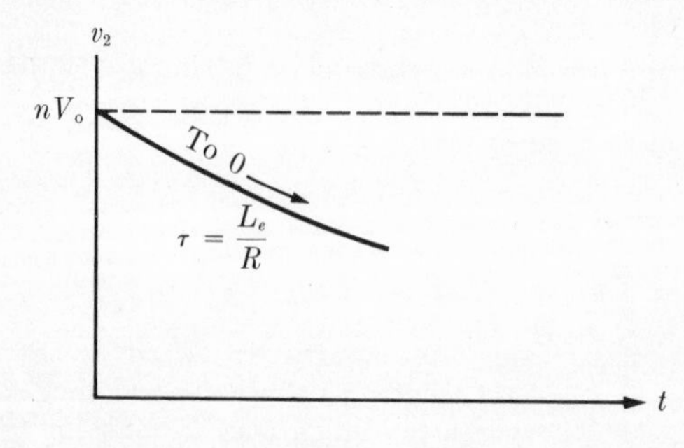

Fig. 7-17 Output waveform during flat top of pulse when input voltage is constant.

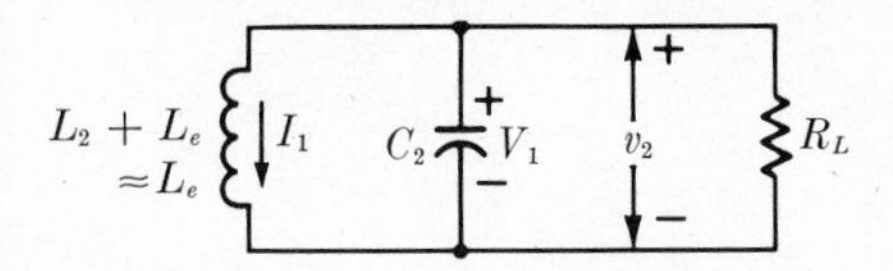

Fig. 7-18 Equivalent circuit governing the transformer response to the trailing edge of the pulse.

Note that C_2 was omitted because in a well-designed transformer the voltage changes only slowly during the flat top of the pulse, and negligible current flows in C_2.

7-8 PULSE TRAILING EDGE

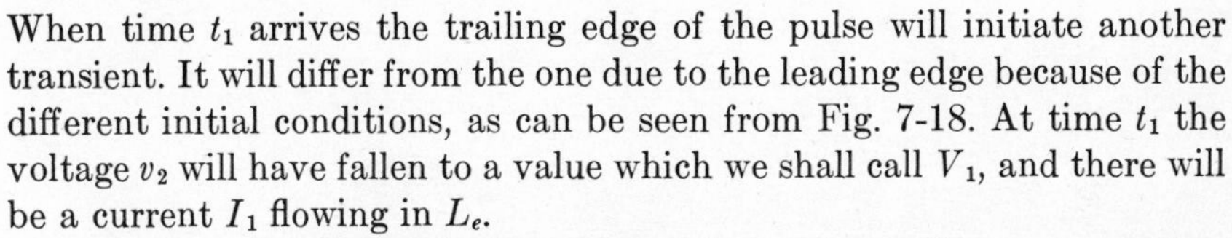

When time t_1 arrives the trailing edge of the pulse will initiate another transient. It will differ from the one due to the leading edge because of the different initial conditions, as can be seen from Fig. 7-18. At time t_1 the voltage v_2 will have fallen to a value which we shall call V_1, and there will be a current I_1 flowing in L_e.

There are two possible situations at time t_1, depending upon whether the generator resistance remains in the circuit when the driving voltage goes to zero. In many electronic circuits the trailing edge results from a transistor being cut off, leaving an open circuit at the "generator" terminals. This simple situation leads to the equivalent circuit of Fig. 7-18. The solution of this circuit may be thought of as the superposition of two of the previous solutions: the initial charge V_1 on the capacitor leads to a transient like Fig. 7-12*a*, and the initial current in the inductor leads to a situation similar to Fig. 7-4 where the driving current I_1 is the initial current. There are once again three possible conditions of damping, with the three types of waveform as illustrated in Fig. 7-19. The phenomenon of the "undershoot" or "backswing," i.e., the voltage going below zero (even for the overdamped case) is attributable to the current I_1 flowing in L_e. Notice that the time scale for the oscillations in Fig. 7-19 is much longer than in the case of oscillations occurring during the rise, because the

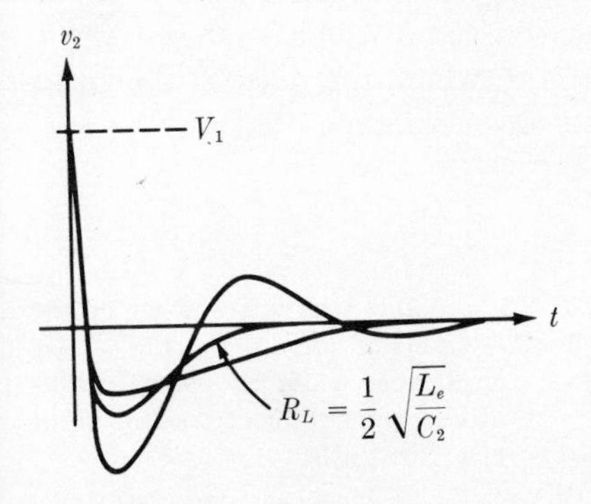

Fig. 7-19 Waveforms of response to pulse trailing edge for different degrees of damping. The middle curve is for critical damping which results when the load resistance $R_L = \frac{1}{2}\sqrt{L_e/C_2}$.

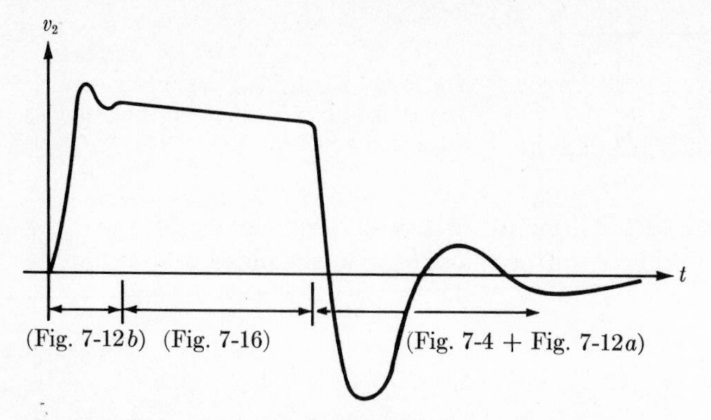

Fig. 7-20 Pulse transformer response when generator becomes an open circuit at the end of the pulse. The figure numbers give the relevant equivalent circuits for particular periods during the transient.

inductance involved in the former case is $L_e + L_2$, whereas the much smaller inductance L_2 is involved in the rise time. An approximate drawing of a complete response of the transformer to a pulse is shown in Fig. 7-20, where the longer-period oscillations after the pulse are clearly shown. These could have been damped if a smaller R_L were chosen, but the beginning of the pulse would have been overdamped also.

Another possible situation occurs when the generator resistance R_g remains connected when the driving voltage goes to zero (but R_g may have a different value than it did during the pulse). The circuit is shown in Fig. 7-21, and is somewhat more complicated than before since each of the reactive elements has a different initial condition. A simple approximate way of obtaining the output is to assume that it has a behavior immediately after t_1 just like the initial rise but inverted (or suitably modified if R_g changes). The long-term behavior after t_1 will be like the circuit without the leakage inductances and regarding V_1 as zero; therefore, the long-term behavior is like the circuit of Fig. 7-4, with the initial current in the magnetizing inductance providing the source. A typical overall pulse response for this case is shown in Fig. 7-22, where the pulse is shown to have an underdamped rise and fall, but an overdamped tail.

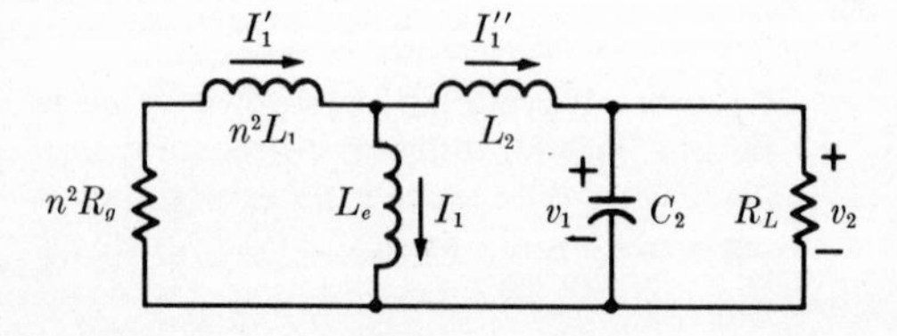

Fig. 7-21 Alternative form of the equivalent circuit of Fig. 7-18, applicable when the pulse generator remains connected at the input after the pulse.

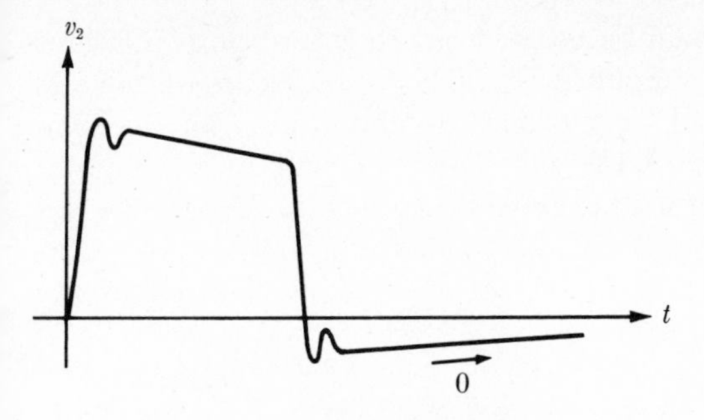

Fig. 7-22 Pulse transformer response with R_g remaining constant throughout the pulse.

7-9 THE STEP-DOWN TRANSFORMER

The case of the step-down transformer proceeds in a similar manner, only now it is assumed that the primary capacitance predominates and the secondary C_2 can be ignored. The equivalent circuit, referred once again to the secondary, is given in Fig. 7-23. We shall not go through the complete analysis but shall only state that the general nature of the waveforms is the same as for the step-up case. In particular, the conditions for the flat top of the pulse are identical for both cases.

7-10 FURTHER COMMENTS ON THE PULSE TRANSFORMER

There are a few comments to be made about pulse transformers in general. First, a rectangular pulse is representative of a whole variety of waveforms that might be used in, for instance, a video system. The abrupt rising edge is the extreme for fast transitions in level, whereas the flat top is the other extreme of essentially dc transmission. Second, a pulse transformer designed for use with rectangular pulses of a specified time duration will not perform as well for pulses of different duration. That is, if one wants the pulse output to rise in, say, $\frac{1}{10}$ of the duration and to sag only 10 percent during the pulse, this objective can be achieved by proper proportions; *but,* if the pulse is lengthened, the sag will be greater than 10 percent, and if it is shortened, the risetime will be greater than $\frac{1}{10}$ of the new duration. Hence, although we are accustomed to thinking of audio transformers as being operative over a wide frequency range, it is not correct to expect a pulse transformer to work well for a wide range of

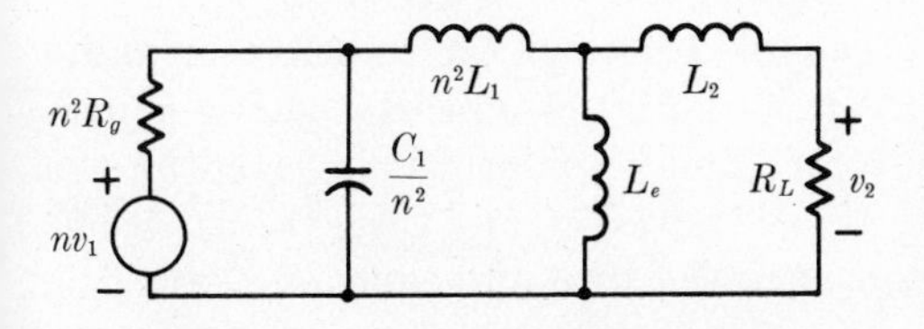

Fig. 7-23 Equivalent circuit for a step-down pulse transformer. In this case the primary capacitance C_1 dominates and C_2 is omitted.

pulse lengths. Finally, it should be pointed out that the approximations used in analysis based upon simplified equivalent circuits are quite valid for a reasonably well-designed transformer. A computation for the complete equivalent circuit (Fig. 7-13) can be carried out in a computer yielding a complete waveform approximating those in Figs. 7-20 and 7-22. The essential features will be indistinguishable from a composite waveform assembled from simple analyses of the equivalent circuits of Figs. 7-15, 7-16, and 7-18.

7-11 TRANSFORMERS IN REGENERATIVE CIRCUITS

There are numerous ways in which small pulse transformers can be used to advantage in electronic circuits. Here we shall consider two of them. The transformer is used as a coupling element between input and output circuits of a transistor in order to secure a regenerative action. Under quiescent conditions there is a complete isolation of the dc voltages in the two circuits, but under pulse-type transient conditions there is coupling between the two, following the principles discussed in Secs. 7-5 to 7-10. The transformer provides a convenient choice for the polarity of the coupled voltage, the choice being accomplished by reversing the primary or secondary terminal connections. The transformer can substitute for a coupling transistor in some circumstances, usually with fewer components required in the circuit.

7-12 REGENERATIVE PICKOFF (COMPARATOR) CIRCUITS

The first example of a transistor operating with a transformer is the regenerative pickoff circuit of Fig. 7-24. The purpose of this circuit is to produce or "pick off" an output pulse precisely at the moment an input ramp voltage (Fig. 7-26) reaches a reference or comparative level, in this case zero volts. For $t < t_o$ the transistor is in region II with $I_B \approx 75\ \mu A$, $I_C \approx 4.3$ mA, and $V_{CE} \approx 20$ V. With this set of operating conditions the transistor has considerable gain, and the output of the transistor is connected back to its input in a manner such as to give positive feedback. The negative value of v_1 keeps the diode turned off, however, so that there is presently no feedback and the circuit is stable. When v_1 reaches zero, the diode is forward biased by the amount of V_{BE}, or about 0.7 V, and the negative-going input waveform is coupled through the transformer to the base of the transistor. This begins to turn off the transistor giving a positive-going waveform at the collector. This waveform is inverted by the transformer, coupled to the base of the transistor, and turns it further off. This is a regenerative action and once the turn-off sequence starts it is

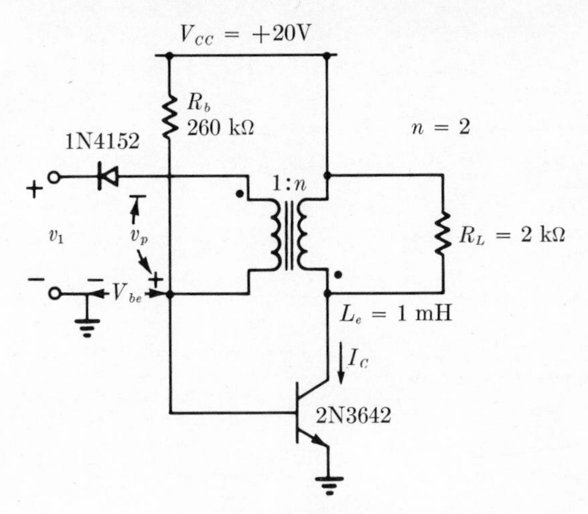

Fig. 7-24 Regenerative comparator which will deliver an output when v_1 crosses zero volts.

completed in a time depending upon the high-speed characteristics of the transistor and transformer. A typical time with a high-frequency pulse transformer and the 2N3642 would be 0.1 μs or less.

After the transistor is turned off, the behavior of the circuit may be found from the simple equivalent circuit of Fig. 7-25. In this circuit everything has been referred to the secondary of the transformer, and the transistor is assumed to be an open circuit. The current in the magnetizing inductance L_e is the current I_C for $t < t_o$. The left-hand branch representing quantities in the primary has such a high impedance that its effect is quite negligible. The collector voltage for t_1^+ is

$$V_{CE}(t_1^+) \approx V_{CC} + I_oR_L = 20 + 2\text{ k}\Omega(4.3\text{ mA}) = 28.6\text{ V} \qquad (7\text{-}20)$$

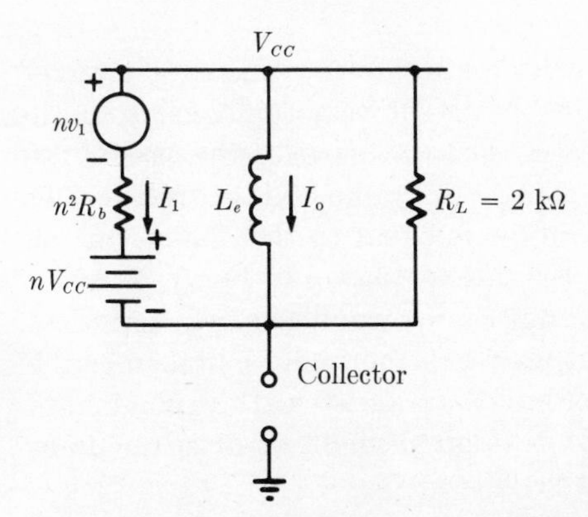

Fig. 7-25 Equivalent circuit for the transformer and circuit of Fig. 7-24 when the transistor is off.

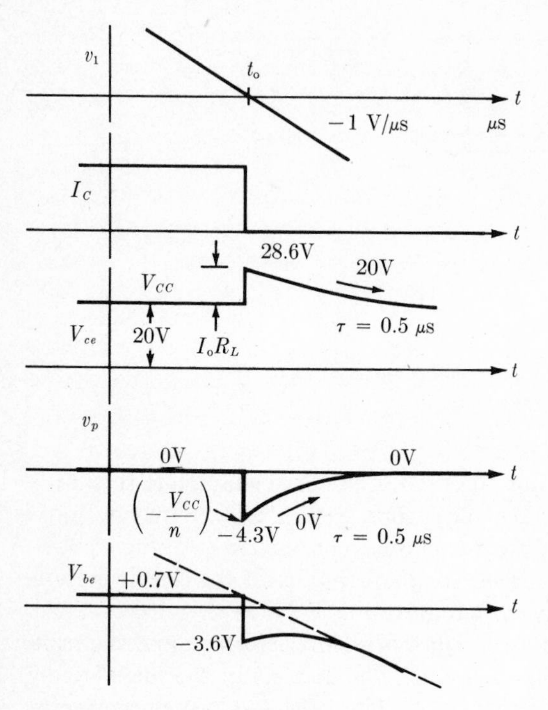

Fig. 7-26 Waveforms in the regenerative pickoff circuit of Fig. 7-24.

and the effective time constant is

$$\tau = \frac{L}{R} = \frac{10^{-3}}{2\ \text{k}\Omega} = 0.5\ \mu\text{s} \tag{7-21}$$

The waveform which occurs at the collector is shown in the circuit waveforms of Fig. 7-26. (The instantaneous jumps shown actually occur in the switching interval discussed previously.) The waveform at the base of the transistor is the sum of the negative-going input ramp and the transformer primary voltage v_p. The primary voltage is equal to the change in the secondary voltage divided by the turns ratio as shown in Fig. 7-26 by v_p. The total primary voltage plus ramp must remain sufficiently negative to keep the transistor cut off. The output would normally be taken as the positive jump occurring at the collector of the transistor at the moment of comparison, t_o. The circuit returns to its original condition after the input ramp is removed or v_1 is made positive.

7-13 THE MONOSTABLE BLOCKING OSCILLATOR

A second class of circuit which makes use of the pulse transformer and an active device is the blocking oscillator. This circuit exists in many forms including both monostable and astable versions, and versions with the transistor (or other active devices) connected in the common-emitter, common-base, or common-collector configurations. The basic ideas underlying the timing of the circuit are few and simple, but any detailed analysis of the circuit is extremely difficult because of the very nonlinear operation of both the active device and the transformer in most practical circuits.

Let us begin by analyzing a simple circuit where only one of the basic timing mechanisms is operating, as in the circuit of Fig. 7-27*a*. In this circuit a common-base transistor is used to provide power gain, and feedback is provided by the transformer, which must have $n > 1$ for the circuit to operate. The circuit is initially at rest with the transistor cut off and no currents in the transformer. If we trigger the circuit at $t = 0$ (perhaps by coupling a small pulse to the emitter to initiate current in the transistor), then collector current will begin to flow and the collector voltage will fall. The falling collector voltage is coupled through the transformer to make the emitter voltage fall also, thereby increasing the current in the transistor and further reducing the collector voltage. This regenerative action continues until the collector voltage can fall no more, i.e., the transistor is saturated. The circuit would remain in this condition if it were not for the magnetizing inductance of the transformer. To maintain the full value of V_{CC} across the secondary of the transformer requires the magnetizing

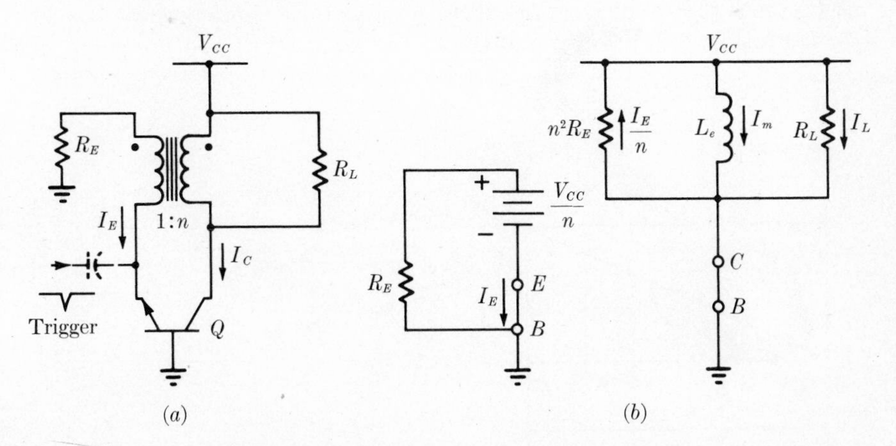

Fig. 7-27 (*a*) A basic monostable blocking oscillator using a common-base transistor. (*b*) Equivalent circuit valid after triggering. The transistor is assumed to be a short circuit when saturated.

current to rise linearly with time also. However, the value of the emitter current is limited and fixed, and a time will come when the collector current tries to exceed the value of the emitter current. Such a situation cannot occur, and the circuit will switch out of the saturated state and indeed out of conduction at this time. The circuit will generate a pulse with a magnitude equal to the entire supply voltage V_{CC} and a length determined mostly by the transformer.

To study the circuit in more detail let us draw a circuit valid for $t = 0^+$, as in Fig. 7-27*b*. The transistor is assumed to have switched to a saturated state (which would be insured in the design by the choice of n and R_E), and is represented by a short circuit. The entire supply voltage V_{CC} therefore appears across the transformer secondary, the voltage driving the emitter is V_{CC}/n, and the emitter current $I_E = -V_{CC}/nR_E$. (In actual practice the emitter-base voltage may be relatively large and a value should be included for it.) The collector current is seen to be made up of three components: the first is the emitter current as transformed through the transformer, the second is the magnetizing current I_m, and the third is the load current I_L. These three components of current are shown graphically in Fig. 7-28, where $I_E/n = -V_{CC}/n^2R_E$ and $I_L = V_{CC}/R_L$. The magnetizing current is initially zero, but increases linearly with time:

$$I_m = \frac{V_{CC}t}{L_e} \tag{7-22}$$

Since the collector current is increasing linearly with time, it is obvious that the transistor must sooner or later come out of saturation, and the equivalent circuit changes. The locus of the transistor operating point is plotted in Fig. 7-29 on a set of curves suitable for a common-base transistor (as in Fig. 3-35). For $t < 0$ the transistor is OFF, and the operating

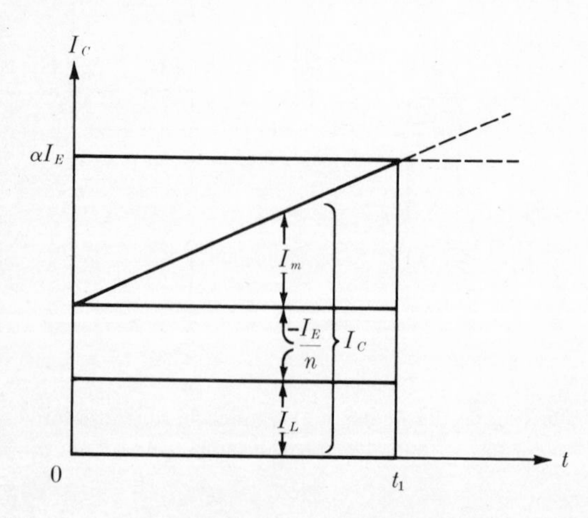

Fig. 7-28 The currents making up the total collector current I_C.

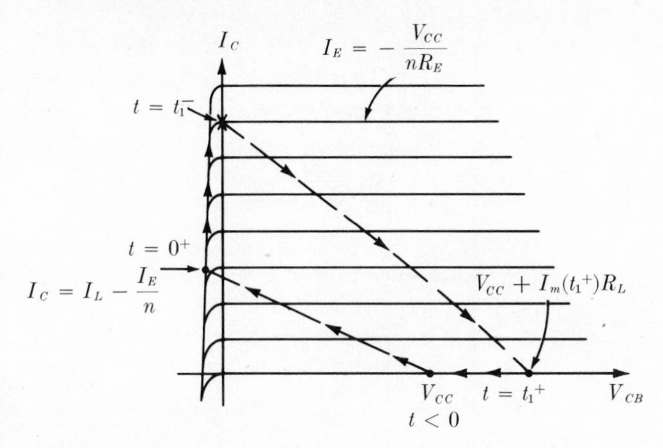

Fig. 7-29 The locus of the transistor operating point during the pulse.

point is $V_{CB} = V_{CC}$; at $t = 0$ the transistor is turned ON and the operating point moves to V_{CB} slightly less than zero and $I_C = I_L - I_E/n$. As the collector current increases due to I_m, the operating point moves up the saturation line, and the collector voltage stays near zero. However, when the collector characteristic for the I_E being maintained begins to turn a corner, and the collector voltage begins to rise, a new regenerative switching action is started. As soon as the collector voltage begins to rise appreciably, the transformer secondary voltage decreases causing a corresponding decrease in the primary voltage. The decrease in these voltages causes a decrease in the emitter current, and the device rapidly switches itself OFF. The critical point is therefore where the transistor goes from Region III (saturation) to Region II (active). In terms of currents this transition point is where

$$I_C = -\alpha I_E \approx -I_E \qquad (\alpha \approx 1) \tag{7-23}$$

To determine the switching point all we have to do is to equate the emitter current to the total collector current and solve for t.

$$-\alpha I_E = \frac{\alpha V_{CC}}{nR_E} = I_C V_{CC}\left(\frac{1}{n^2R_E} + \frac{1}{R_L} + \frac{t_1}{L_e}\right) \tag{7-24}$$

$$t_1 = L_e\left(\frac{\alpha}{nR_E} - \frac{1}{n^2R_E} - \frac{1}{R_L}\right) \tag{7-25}$$

For $t > t_1$ the transistor will have switched OFF, and the equivalent circuit becomes that of Fig. 7-30, which also includes the waveforms for the entire cycle of operation. When the energy stored in the magnetizing inductance is dissipated in R_L, the circuit is ready to begin another cycle of operation. In this case the required time is about $5\tau = 5L_e/R_L$.

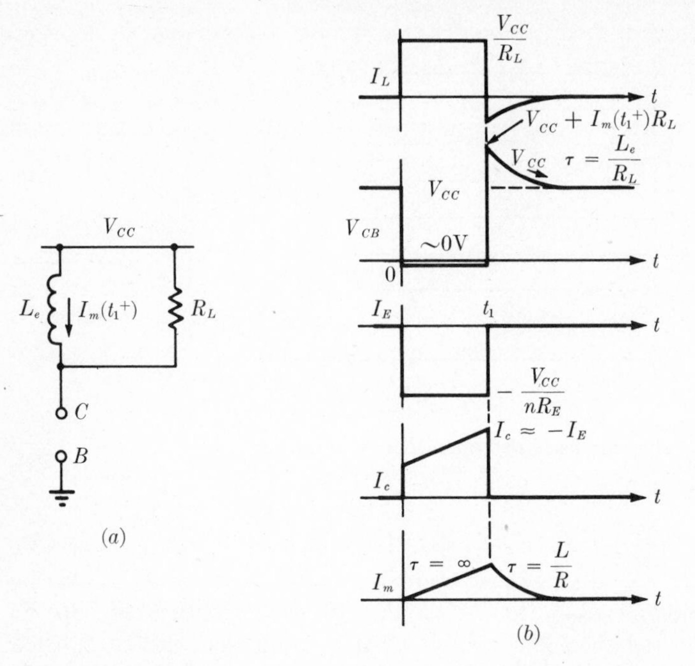

Fig. 7-30 (*a*) The equivalent circuit for the blocking oscillator (Fig. 7-27) for $t > t_1$. (*b*) The waveforms generated by the circuit.

The preceding analysis of the blocking oscillator has completely neglected any circuit or transformer capacitances and the leakage inductances. These affect both the rise and fall times of the generated pulse in a manner similar to that discussed in the sections on pulse transformers. However, any reasonably exact analysis of the situation during the rise and fall of the generated pulse is extraordinarily difficult because of the wide excursions of the operating point of the transistor, i.e., the circuit is a very nonlinear one during the transition periods.

The effect of the capacitances is also felt after the transistor turns off and the energy stored in the core of the transformer is being dissipated. The capacitance may cause ringing if the load is too small (giving rise to high Q—note Fig. 7-20), and in an extreme case the circuit may be retriggered by one of the negative-going peaks. This undesirable or false triggering may be prevented by either sufficiently loading the output or providing enough fixed bias on the emitter to prevent the retriggering.

7-14 THE ASTABLE BLOCKING OSCILLATOR

The preceding circuit produces one output pulse for each trigger pulse, and is therefore a monostable circuit. The circuit may be made astable simply by biasing the emitter so that it cannot be permanently kept off. (Notice that a bistable form of the blocking oscillator cannot be built because the transformer cannot transmit a dc level change.) A practical astable, common-base circuit is shown in Fig. 7-31. To make the circuit astable a capacitor has been added in series with the primary of the transformer, and a bias resistor R_1 is added in the emitter circuit to bias the transistor ON in steady-state conditions. As an additional refinement, which is not really necessary to make the circuit astable, the load is connected to the circuit through a separate transformer winding. This permits adjusting the load voltage during the pulse by selecting the turns ratio n_2. In addition, the load may have one end connected to any voltage independent of the oscillator proper if that is necessary.

If the circuit were turned on from rest (all voltages zero), the supply V_{EE} would cause C to charge and allow the emitter to become negative. When the emitter becomes more than a few tenths of a volt negative, the transistor begins to conduct, and the regenerative cycle starts that culminates in the transistor being full ON and saturated. A timing operation somewhat similar to the one discussed in Sec. 7-13 begins, but in this case the pulse length is determined primarily by R_E, C, n_1, and L_e. During the pulse most of the emitter current flows in C and R_E (because $R_E \ll R_1$) and charges C to the polarity shown. At the end of the pulse the emitter junction is kept turned off both by the backswing of the transformer, as before, and also by the charge on C. In the usual circuit the transformer returns to its quiescent condition long before the charge on C has time to change appreciably; therefore the time the transistor is in the OFF state is determined by the time for C to discharge through R_1 and change polarity to allow the transistor to return to an ON state. A suitable place

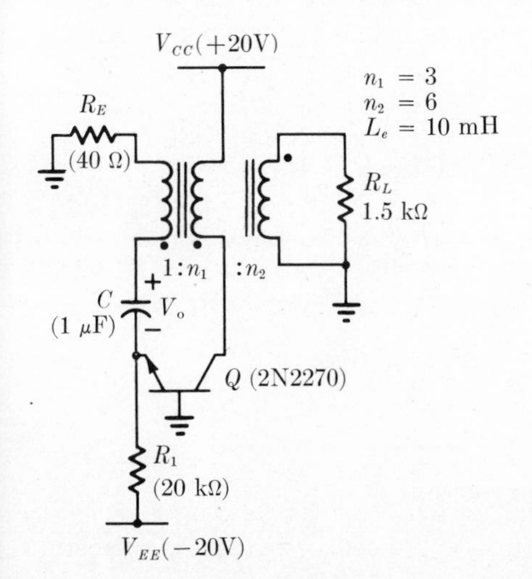

Fig. 7-31 An astable blocking oscillator with the transistor in the common-base connection.

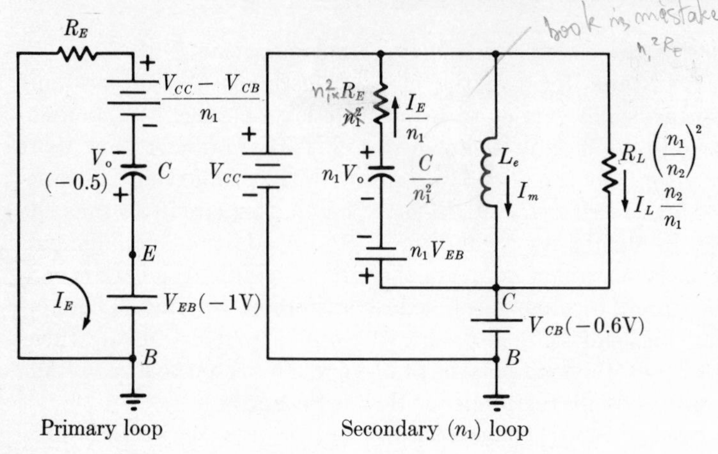

Fig. 7-32 Equivalent circuit for the circuit of Fig. 7-31 when the transistor is saturated.

to begin an analysis is just at the time that the emitter enters the ON state or $V_{EB} \approx -0.5$ V.

An equivalent circuit for both the primary and secondary can be drawn as in Fig. 7-32 for the time during which the transistor is saturated. In this case we have not assumed that the transistor is a perfect short circuit, but instead have allowed for both an emitter-to-base and collector-to-base voltage. (The value of the emitter-to-base voltage is made rather large—1 V—because of the large currents flowing in the transistor.) The elements in the primary loop determine the emitter current, which will be much larger than the current flowing in R_1. Therefore the latter current is omitted. To insure proper operation of the circuit the initial emitter current must be considerably larger than the collector current to saturate the transistor and allow for a reasonable pulse length. The equations for the initial values of the current are

$$I_E(0^+) = -\left[\frac{(V_{CC} - V_{CB})}{n_1} - V_o + V_{EB}\right]\left(\frac{1}{R_E}\right)$$
$$= -\left[\frac{(20 + 0.6)}{3} + 0.5 - 1\right]\left(\frac{1}{40}\right) = -160 \text{ mA} \tag{7-26}$$

$$I_C(0^+) = +\frac{V_{CC} - n_1 V_o + n_1 V_{EB} - V_{CB}}{n_1^2 R_E} + \frac{V_{CC} - V_{CB}}{R_L}\left(\frac{n_2}{n_1}\right)^2$$
$$= \frac{20 + 1.5 - 3 + 0.6}{9 \cdot 40} + \frac{(20 + 0.6)}{1.5 \text{ k}\Omega}(4) \tag{7-27}$$
$$= +108 \text{ mA}$$

The time constant for the RC primary circuit is

$$\tau_p = R_E C = 40 \cdot 10^{-6} = 40\ \mu\text{s} \tag{7-28}$$

The transistor is seen to be in a highly saturated state because the emitter current is much greater than the collector current. In this circuit two circumstances will cause the transistor to come out of saturation: the first is the magnetizing current, which will increase linearly from an initial value of zero; the second is the emitter current, which will not be maintained at the initial value because the capacitor C charges and reduces the current. The switching point will again occur when the transistor comes out of saturation, i.e., when the emitter becomes equal to the collector current.

$$\underbrace{I_E(0^+)\epsilon^{-t_1/\tau_p}}_{I_E(t_1)} \;=\; \underbrace{\frac{I_E(0^+)}{n_1}\epsilon^{-t_1/\tau_p} + \frac{V_{CC}-V_{CB}}{R_L}\left(\frac{n_2}{n_1}\right)^2 + \frac{V_{CC}-V_{CB}}{L_e}t_1}_{I_C(t_1)} \tag{7-29}$$

The left-hand side of Eq. (7-29) is the emitter current at time t_1, the switching time, and the right-hand side is the collector current, which is composed of three separate components: the emitter current as transformed through the transformer, the load current, and the magnetizing current—which is assumed to rise linearly with time. This transcendental equation is difficult to solve for t_1, but we may plot the four currents as in Fig. 7-33 to find the solution. We see that the emitter current is falling because of the charging capacitor, and the collector current is rising

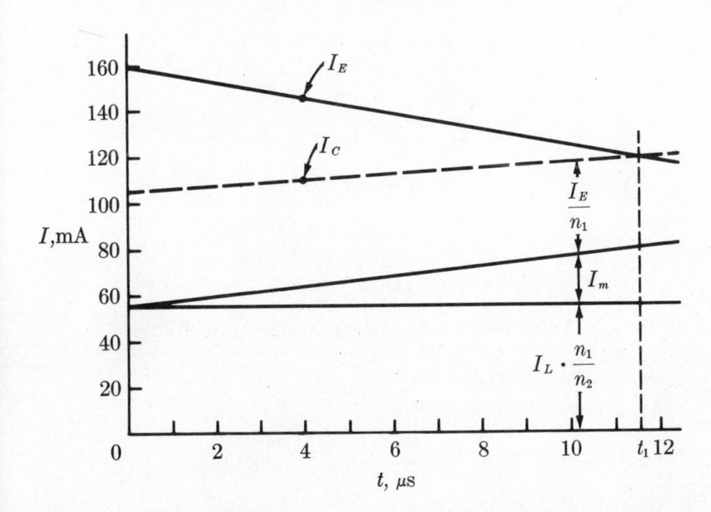

Fig. 7-33 The current components in the circuit of Fig. 7-31 during the pulse. The total collector current I_C is made up of three components: I_E/n_1, I_m, and $I_L \cdot n_1/n_2$. The emitter current I_E is shown as it is in the emitter circuit. All other currents are referred to the collector winding.

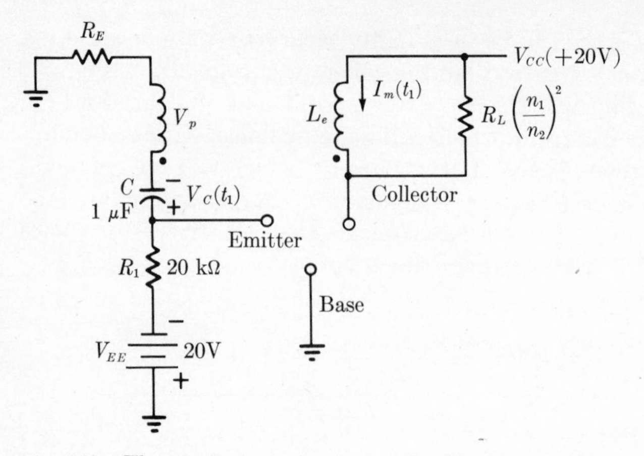

Fig. 7-34 The equivalent circuit for the blocking oscillator with the transistor off.

because of the magnetizing current; equality of the two currents occurs at 11.6 μs, and the transistor would switch OFF at this time.

After the transistor is switched OFF, the equivalent circuit is as in Fig. 7-34 where the transistor is shown as an open circuit. Two circuit elements are storing energy at time t_1, and the initial conditions of these elements must be found to solve for what happens after t_1. Reference to Fig. 7-32 gives the initial and final values of capacitor voltage $V_C(t)$ while the transistor is ON.

$$V_C(0^+) = V_o = -0.5 \text{ V} \tag{7-30}$$

$$V_C(\infty) = V_{EB} + \frac{V_{CC} - V_{CB}}{n_1} = -1 + \frac{20 + 0.6}{3} = 5.9 \text{ V} \tag{7-31}$$

The equation for the capacitor voltage during the pulse is then

$$\begin{aligned} V_C(t) &= (-0.5 - 5.9)\epsilon^{-t/\tau_p} + 5.9 \\ &= -6.4\epsilon^{-t/\tau_p} + 5.9 \\ &= +1.1 \text{ V} \qquad (\text{at } t = 11.6\ \mu\text{s}) \end{aligned} \tag{7-32}$$

During the pulse the transformer magnetizing current is rising linearly with time and is

$$\begin{aligned} I_m(t) &= \frac{t(V_{CC} - V_{CB})}{L_e} = (2.06 \times 10^3)t \text{ A} \\ &= 24 \text{ mA} \qquad (\text{at } t = 11.6\ \mu\text{s}) \end{aligned} \tag{7-33}$$

These values for capacitor voltage and inductor current are the initial values for use in the circuit of Fig. 7-34 at $t = t_1^+$. Two exponentials occur simultaneously; the first in the collector circuit is

$$\begin{aligned} \tau_1 &= \frac{L_e}{R_L(n_1/n_2)^2} \\ &= 26.6\ \mu\text{s} \end{aligned} \tag{7-34}$$

The second time constant is in the emitter circuit and is

$$\tau_2 \approx R_1C = 20 \text{ ms} \tag{7-35}$$

Since the two time constants are widely different, the discharge of the capacitor is hardly affected by what is going on in the transformer, and the waveform at the emitter of the transistor may first be calculated by ignoring the magnetizing current (also the effect of R_E may be neglected in comparison to R_1). The emitter voltage after t_1 is

$$\begin{aligned} V_{EB}(t) &= [V_C(t_1) - V_{EE}]\epsilon^{-(t-t_1)/\tau_2} + V_{EE} \\ &= (1.1 + 20)\epsilon^{-(t-t_1)/20\text{ms}} - 20 \end{aligned} \tag{7-36}$$

The currents flowing in the emitter circuit are so small at this time as to have practically no effect upon the collector circuit; therefore the voltage appearing at the collector is mainly that caused by the magnetizing current.

$$\begin{aligned} V_{CB}(t) &= V_{CC} + I_m(t_1)R_L\left(\frac{n_1}{n_2}\right)^2 \epsilon^{-(t-t_1)/\tau_1} \\ &= 20 + 9\epsilon^{-(t-t_1)/26.6\,\mu\text{s}} \end{aligned} \tag{7-37}$$

This exponential in the collector circuit appears across the secondary, and therefore also appears in the primary of the transformer reduced by the factor n_1.

The timing of the period between pulses is determined by the waveform appearing at the emitter; when the emitter returns to -0.5 V the circuit will again regenerate and form another pulse. The time for this to occur is

$$T = \tau_2 \ln \frac{21.1}{19.5} = 1.6 \text{ ms} \tag{7-38}$$

(The total period is 1.6 ms plus the length of the pulse, but this is so short as to be lost in the inaccuracy of the overall period calculation.)

The waveforms for the entire circuit are shown in Fig. 7-35 where the extra exponential which appears on the emitter waveform is shown dashed. The effect of this extra exponential is to increase the current flowing through R_1 in the period between pulses. The increased current would discharge C faster and therefore reduce the period. However, the waveform in Fig. 7-35*d*, which is drawn to a uniform time scale, shows that the time that the extra exponential affects the emitter waveform is small compared to the total time between pulses; consequently the change from the calculated period is small.

7-15 THE EFFECT OF THE CORE MATERIAL UPON THE PULSE LENGTH

Up to this point we have considered the transformer to be linear, that is, we have represented it by a constant inductance L_e. In actual practice the

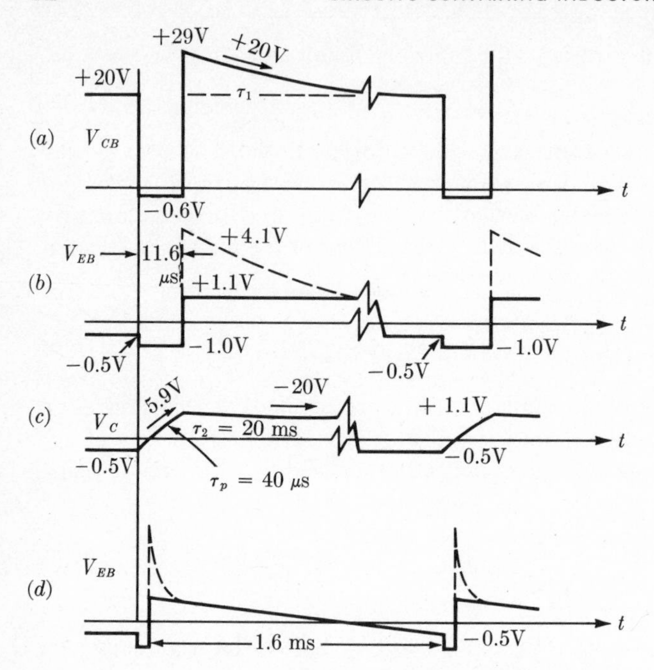

Fig. 7-35 The waveforms produced by the astable blocking oscillator of Fig. 7-31. The waveforms in (*b*) and (*d*) are the same, except for the time scale.

transformers used for blocking oscillators will not be linear in the range of voltages and currents used, and in fact may be intentionally designed to be nonlinear to control the pulse length. The effect of moderate nonlinearity in the transformer is to make the magnetizing current increase more rapidly than the linear rise shown in the preceding examples. Such an increase in the rate-of-rise of I_m is caused by partial saturation of the iron in the core, which in turn reduces the effective inductance.

To illustrate what happens in a rather extreme case, let us consider a transformer made with a special core material that has the rectangular hysteresis loop shown in Fig. 7-36. We shall consider this transformer to be in the circuit of Fig. 7-27*a* and the core to be initially saturated with a flux density $-B_m$. Before the circuit is triggered there are no currents flowing so that the flux density $-B_m$ represents the residual magnetism remaining from some previous operation of the transformer. When the circuit is triggered, the current in the magnetizing inductance rises instantaneously (as limited by the high-frequency characteristics of the transistor and transformer) to the value I_1. This is possible because the transformer has almost zero inductance until the current reaches this value. When the current reaches I_1 the inductance becomes very large because of the steepness of the *B-H* curve at this current value. (The incremental inductance would become infinite with the *B-H* curve shown.) The voltage across the transformer is approximately V_{CC} when the transistor is ON, therefore the rate of change of flux in the core is

$$V_{CC} = \frac{N_2\,d\phi}{dt} \tag{7-39}$$

where N_2 is the number of turns in the secondary winding, and ϕ is the total flux linking this winding. The current in the magnetizing inductance will stay constant at the value I_1 instead of increasing linearly as long as ϕ does not increase beyond ϕ_m. The time for ϕ to increase from $-\phi_m$ to $+\phi_m$ is

$$t_1 = \frac{2\phi_m N_2}{V_{CC}} \tag{7-40}$$

At t_1 the transformer core saturates, L_e drops to zero, the magnetizing current rises abruptly, and the coupling from primary to secondary in the transformer drops to nearly zero. The last effect removes the drive to the emitter at the same time that the collector is called upon to supply a large current to the transformer. The combined effects of the transformer saturating cause the pulse to terminate abruptly at t_1, and all the circuit currents drop to zero. In this case the saturation characteristic of the transformer and the applied voltage are the determining factors of the pulse length. We note that the pulse length would be quite accurately inversely proportional to the supply voltage, a fact which can be used.

In this case there is no recovery transient after the pulse because the core stays in the state $+\phi_m$ after the pulse is over, and the magnetizing current changes abruptly from I_1 to zero after the pulse. The circuit cannot now be retriggered because a pulse of the polarity that originally triggered the circuit cannot now be coupled through the transformer which

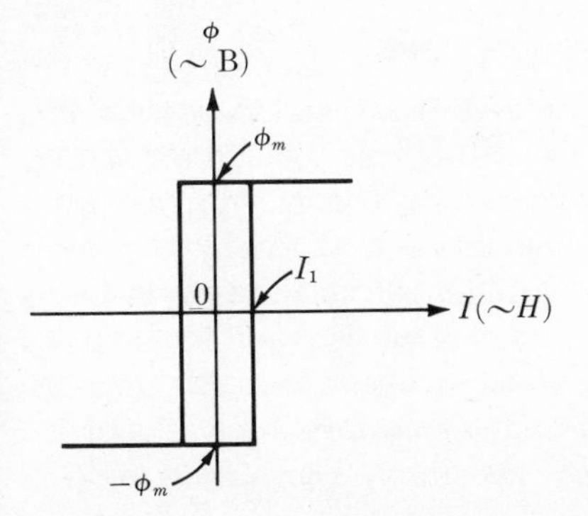

Fig. 7-36 The *B-H* loop for a hypothetical square-loop magnetic material.

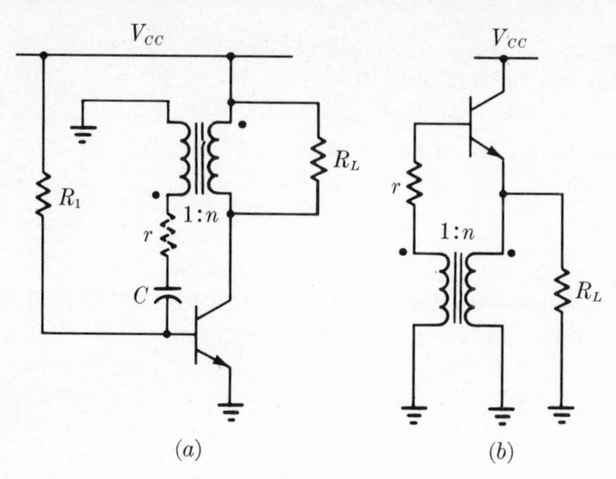

Fig. 7-37 Other forms of blocking oscillators. (*a*) Common emitter. (*b*) Common collector ($n < 1$).

is saturated in the + state. To return the circuit to its original state $-\phi_m$ some sort of reset circuitry is required. One possible way would be to add a third transformer winding connected to V_{CC} through a resistance to provide a reset current greater than $-I_1$.

In the usual blocking oscillator circuit the three factors determining the pulse length are in operation simultaneously. These are the rise in the magnetizing current due to a finite magnetizing inductance, the charging of a capacitor, and the saturation (at least partial) of the core. Therefore the accurate determination of the pulse length in a typical blocking oscillator requires detailed knowledge of both the transformer and transistor. The preceding analyses give one considerable insight into what will determine the operating characteristics of the blocking oscillator so that a reasonable design may be made and then modified in the light of experiment to obtain exactly the pulse length or repetition frequency desired.

7-16 OTHER BLOCKING OSCILLATOR CONFIGURATIONS

Many other forms of blocking oscillator may be used, and there are sometimes advantages to the other types. Two other basic configurations are shown in Fig. 7-37. In Fig. 7-37*a* the transistor is connected in the common-emitter manner, and the circuit is biased to give an astable oscillator. In this circuit only one power supply is required, and the operation is somewhat more efficient in that the collector circuit is required to supply only the base current rather than the emitter current as in the previous circuits. The period of the oscillation is again determined by R_1C, and the base waveform can be used as a sawtooth generator. The current limiting resistor r is not always necessary if the turns ratio n is sufficiently high. One major disadvantage of this circuit is that the pulse length is usually very dependent upon h_{FE} because the pulse length is again determined by when the transistor goes from the saturated to the unsaturated region. The dependence upon h_{FE} can be reduced by placing r in the emitter rather than the base circuit. Doing so, however, reduces the output from the circuit. The common-emitter circuit is probably the best when extremely fast rise and fall times are desired in the pulse.

The circuit in Fig. 7-37*b* has the transistor connected in the common-collector manner and is biased to produce a monostable circuit. There are no particular advantages to the circuit, and it has most of the problems of the common-emitter circuit since the pulse length is again sensitive to h_{FE}.

7-17 USES OF THE BLOCKING OSCILLATOR

The most obvious use of the blocking oscillator is in the production of short, high-power, or high-current pulses. It is not a very satisfactory way to produce pulses longer than a few microseconds because the transformer becomes relatively large and expensive. The circuit is very efficient in the generation of short pulses, however, since practically no power is consumed between pulses, and the device is used as a switch during the pulse. Small transistors used in this manner can produce pulse currents on the order of an ampere or more if the transistor is chosen to have reasonable current gain at the high current.

The circuit is also useful as a free-running oscillator, or as a synchronized-frequency divider; however, the latter function is often better and more economically produced with digital integrated circuits.

PROBLEMS

7-1. Calculate, sketch, and label the waveform appearing at the collector (v_2). What is the purpose of the diode in the emitter?

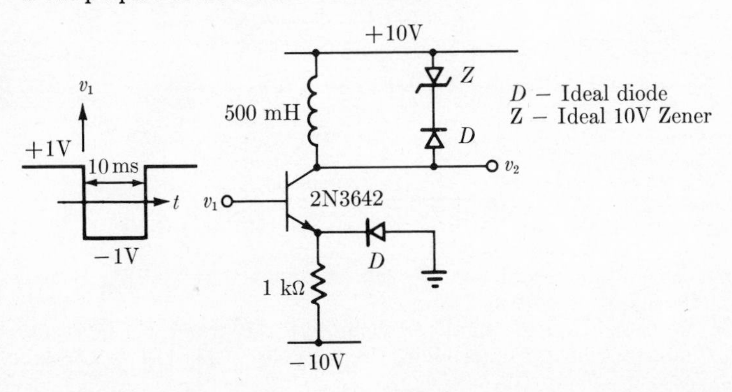

7-2. The circuit rests with $S1$ closed, and at $t = 0$ the switch opens. Find $v(t)$, sketch, and label indicating the initial value, final value, and time constant.

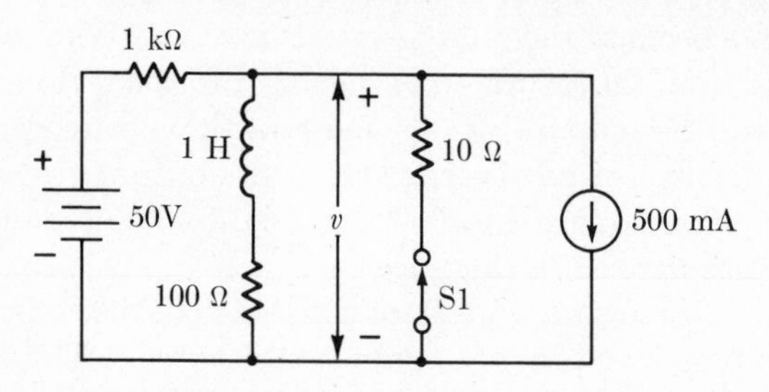

7-3. At $t = 0$ the switch is closed placing the inductor across the two-terminal box with the given V-I characteristic. Calculate, sketch, and label the resulting waveforms of $V(t)$ and $I(t)$.

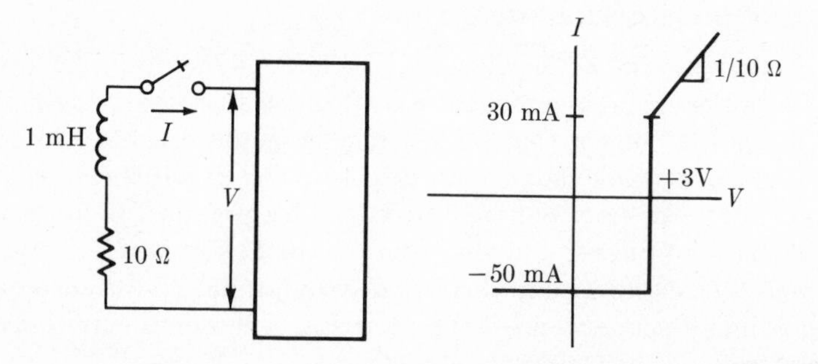

7-4. In the circuit shown the transistor conducts for $t < 0$ establishing I_L. At $t = 0$ the transistor is turned off, and the energy initially stored in the inductor is ultimately transferred to the capacitor. Calculate and sketch I_L and v_2.

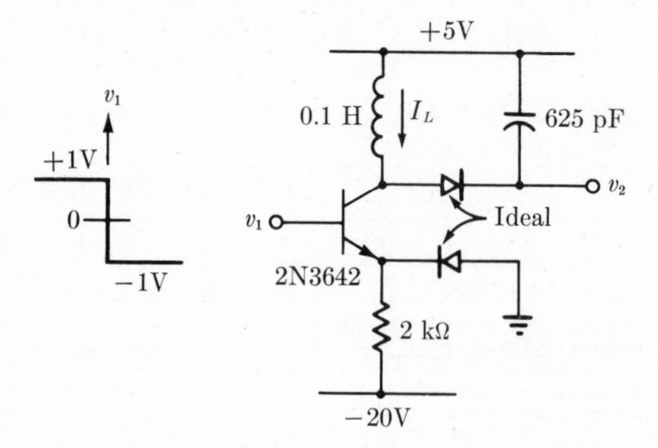

7-5. (a) Calculate plot, and carefully label the inductor voltage and current and the output voltage v_2.

(b) How would the waveforms be modified if the stray capacitance across the inductor were 500 pF (including the distributed capacitance of the inductor)?

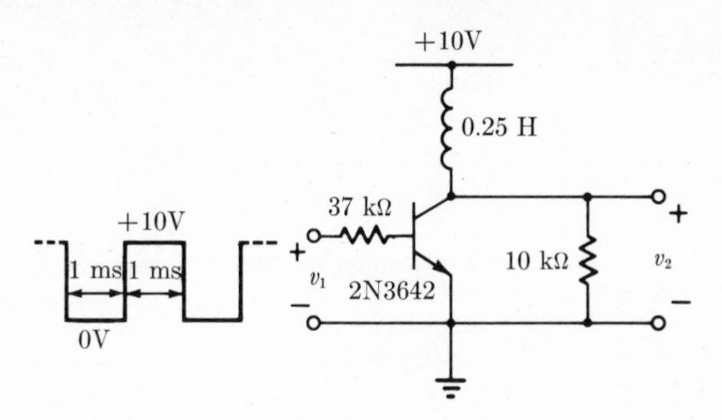

7-6. (a) The circuit is shown in a ringing circuit in which damping is added after one-half cycle of operation. Calculate and sketch the waveform of the output voltage v_2 and current I_L.

(b) At what time (t_1) is the circuit substantially in a steady-state condition and ready for another change in v_1?

(c) Assume the input v_1 returns to -1 V at $t = t_1$. Calculate and sketch the output v_2 for $t \geq t_1$.

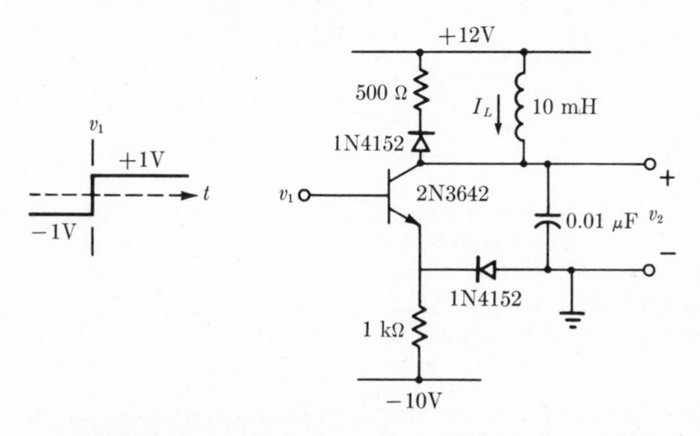

7-7. The ringing circuit of Fig. 7-11 is to be used to generate a precise 1 MHz square wave by taking the output from the ringing circuit and clipping it. Design the ringing circuit (give all the circuit values) assuming the following parameters: $V_1 = +10$ V; $-V_2 = -5$ V; $h_{FE}(Q1) = 50$; v_o is to be the maximum possible consistent with keeping the transistor always turned off; $V_{CC} = 10$ V; $R_C = 1$ kΩ. The transistor should be saturated no matter at what time the gating pulse returns to $+V_1$.

7-8. A capacitor charging circuit is frequently used to take power slowly from a source during charging and then to deliver high peak powers in a quick discharge. A simple example is in a high speed flash circuit where the capacitor supplies the energy to a gas discharge tube. If the charging is done with an RC charging circuit, the energy lost in the resistor is equal to the energy stored on the capacitor. The LC charging circuit shown overcomes this problem. The diode is present to prevent premature discharging of the capacitor through the inductor.

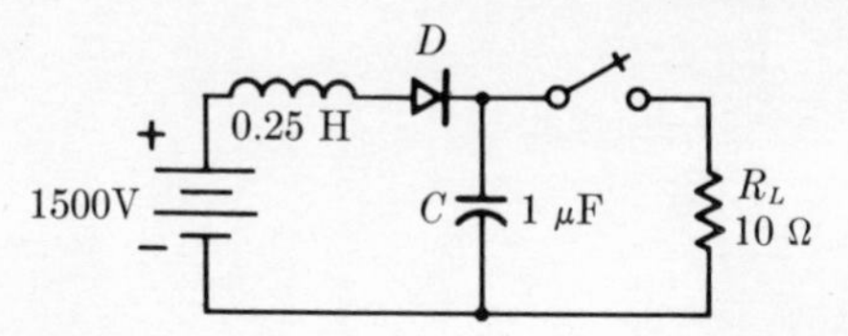

Calculate the waveform across C assuming the switch closes once per millisecond, and that it opens when $I < 0.1$ A. What is the peak power delivered to the load? What is the average load power?

Assume the diode is ideal in the forward direction. What reverse breakdown voltage rating is required in the diode?

7-9. The circuit shown is to generate a 5 V, one millisecond pulse across the 10 ohm load. Assume that the transistor saturates during the entire pulse. What is the value of N, and what magnetizing inductance, L_e, is required to keep the rise in collector current within 20 percent of the initial value during the pulse?

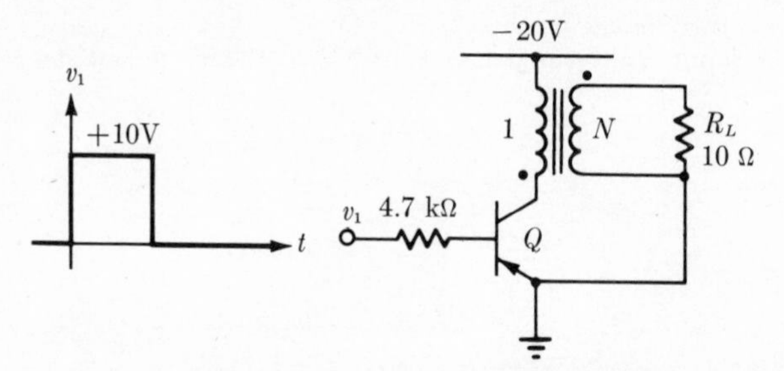

Using the previous value of L_e, plot the waveform appearing across the load, R_L. What is the peak transistor collector voltage? What is the minimum required value of h_{FE}?

7-10. Using the blocking oscillator circuit of Fig. 7-27 give the required element values to produce a 0.5 μs pulse. Assume the pulse transformer is designed for this pulse length and has $n = 3$. The load to be driven is $R_L = 100$ Ω, and the supply voltage is $V_{CC} = +12$ V. Since many combinations of R_E and L_E are possible, fix the value of L_e such that the collector current rises 50 percent above its initial value at the end of the pulse.

8
Negative-Resistance Switching Circuits

8-1 NEGATIVE-RESISTANCE AND TRIGGER DEVICES

A very useful class of trigger circuits can be made from devices which exhibit two-terminal negative resistance characteristics. These are devices which in some part of their *V-I* curves give a decreasing current for an increasing voltage. Very simple bi-, mono-, and astable circuits may be made with these devices, and in the usual circuit only one such device is required in combination with simple *RC* or *RL* circuits. Since the number of types of negative-resistance devices (NRD) and applications is very large, we shall only concern ourselves with the basic principles of operation in this chapter and utilize relatively few of the available devices.

8-2 GENERAL CHARACTERISTICS OF NEGATIVE-RESISTANCE DEVICES

The *V-I* curves of negative resistance devices fall into one of the two categories shown in Figs. 8-1*a* and 8-2*a*. In the curve of Fig. 8-1*a* the current is everywhere uniquely determined by the value of *V*, but there are

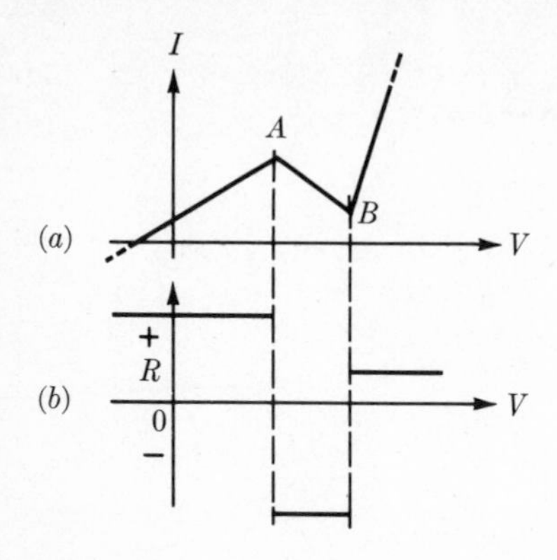

Fig. 8-1 The characteristics of a voltage-stable negative resistance device (NRD).

three possible values of V for some values of I. For this reason devices characterized by Fig. 8-1*a* are known as *voltage stable* since voltage is the determining factor for the operating point. The negative-resistance region is between A and B, as shown by Fig. 8-1*b*, where the incremental resistance R looking into the two terminals of the device is shown as a function of the applied voltage V.

Devices which are characterized by the curve of Fig. 8-2*a* have an operating voltage which is uniquely determined by the applied current; thus such devices are *current stable*. The incremental resistance is plotted versus current for this type of device and again is negative between A and B. The transition between negative and positive resistance in an actual device will be more gradual than in the idealized cases shown.

In addition to the property of negative resistance which is obtained over some range of the V-I characteristics, a voltage-stable NRD also has an internal capacitance which appears in parallel with the negative resistance (both the negative resistance and capacitance may be functions of frequency). The current-stable device has dual characteristics, that is, the

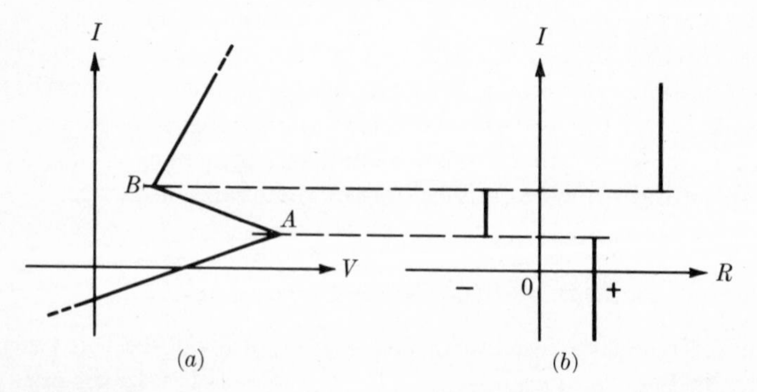

Fig. 8-2 The characteristics of a current-stable negative resistance device (NRD).

equivalent circuit looking into the device terminals is a negative resistance in series with an inductor. These energy storage elements, which approximately represent some physical process going on inside the device, determine what external conditions will be stable or unstable.

Figure 8-3 shows an equivalent circuit for the device represented by the idealized *V-I* characteristic of Fig. 8-3*b*. The device is paralleled by a conductance G in series with a voltage source V_1. The circuit can represent the situation for either the negative or positive resistance region by taking suitable values for G_i and I_i. The internal conductance G_i is a negative number when the device is in the negative-resistance region. The differential equation giving the currents and voltages in the circuit is

$$(V_1 - v)G = i + C_i \frac{dv}{dt} \tag{8-1}$$

and

$$i = I_i + vG_i \tag{8-2}$$

The general solution to these two equations is

$$v = V' + K\epsilon^{-t(G+G_i)/C} \tag{8-3}$$

where K is an arbitrary constant dependent upon the initial conditions, and V' is the solution of Eqs. (8-1) and (8-2) for $dv/dt = 0$, that is, it is the intersection of the load line and the *V-I* characteristic. If the value of the parenthesis in Eq. (8-3) is positive, the exponential is a decaying one, and the final solution as t becomes large is V'. This would be the case if both G's were positive, or if the external conductance G were greater than the internal negative conductance. If, on the contrary, the external conductance is smaller than the internal negative conductance, the exponent will be positive and any initial transient will grow with time. The ultimate solution for large time will depend upon the nonlinearities of the circuit, not Eq. (8-3). A voltage-stable negative-resistance device is therefore stable if the external source resistance is kept sufficiently low (large values

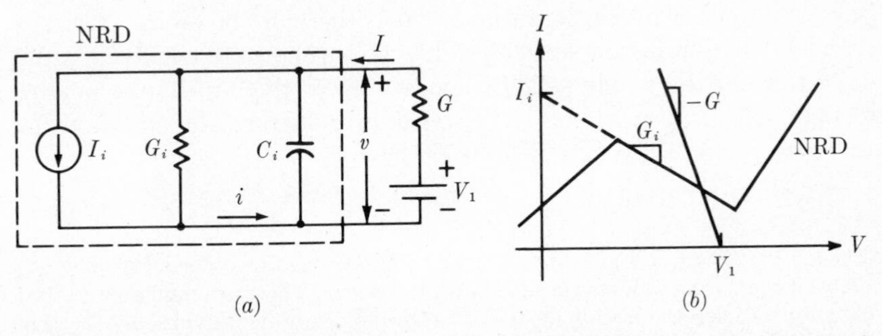

Fig. 8-3 (*a*) An equivalent circuit for a voltage-stable NRD. (*b*) A typical load line illustrating a stable intersection.

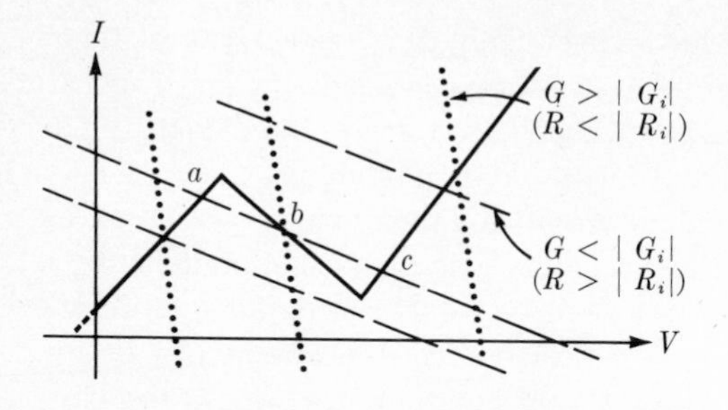

Fig. 8-4 A voltage-stable NRD with various load-line possibilities. The dotted curves represent the low-load resistance cases which are all stable.

of G) but will be unstable for large values of source resistance (low G) if biased into the negative resistance region $G_i + G < 0$. The possible situations are depicted in Fig. 8-4 where the dotted curves show the stable situation $G > |G_i|$. Such a condition is stable no matter where the device is biased[1] and has only one intersection with the V-I curve. The dashed curves represent the "high" source resistance case $G < |G_i|$ and have two possible types of intersections. The bottom and top curves intersect the V-I characteristic in only one place and in the region where the device has a positive input resistance; therefore these are stable operating points. The middle curve intersects the V-I characteristic at three points; the two outer ones are stable intersections, but the middle one represents an unstable operating point. If the device were initially forced to exactly the middle intersection b, it would remain there because such a position corresponds to $K = 0$ in Eq. (8-3). Any minute circuit disturbance which would momentarily disturb the equilibrium (for example, a noise pulse from the inherent noisiness of the resistor G) produces a finite K, and the growing exponential will begin. The operating point will move to either the right or the left of b depending upon the direction of the starting stimulus, and continue moving until the negative resistance region is left. At that time the input resistance of the device becomes positive, the exponential becomes a decaying one, and the final operating point becomes either point a or c. In practice the device operating with this load line will always be at either a or c, and the circuit is therefore bistable.

The situation for the current-stable device shown in Fig. 8-5a is the dual of that for the voltage-stable device, that is, the impedance looking into the device is a resistor in series with an inductor. The equation for this, which is similar to Eq. (8-3), is

$$i = I' + K\epsilon^{-t(R+R_i)/L_i} \tag{8-4}$$

[1] Although the circuit of Fig. 8-3 is stable under these conditions, the device may not be stable if there is an inductor in series with the source. There is a minimum critical inductance to make the circuit unstable, which may even be provided by the lead inductance of the device. See H. J. Reich, "Functional Circuits and Oscillators," sec. 45, pp. 197–204, D. Van Nostrand Company, Inc., Princeton, N.J., 1961.

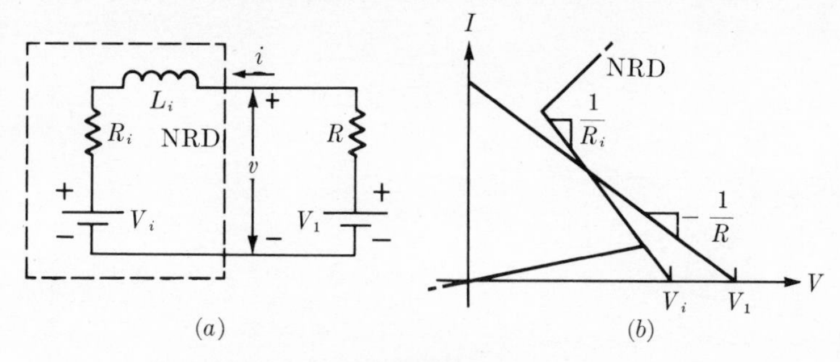

Fig. 8-5 (*a*) An equivalent circuit for a current-stable NRD. (*b*) A typical load line representing a stable intersection.

where I' is the equilibrium value of current (the intersection of the V-I curve and the load line), and K is an arbitrary constant depending upon the initial conditions ($K = \Delta i$, where Δi is the initial departure in i from I'). A stable point will be reached if $R + R_i > 0$, that is, if the external resistance is sufficiently large. The three possible situations for the two cases $R < |R_i|$ and $R > |R_i|$ are shown in Fig. 8-6. The dotted lines represent the case where the source resistance is higher than the negative resistance, and the three cases shown are all stable even though one load line has an intersection in the negative-resistance region at point b.[1] The dashed lines represent stable operating points when they intersect the V-I curve where it has a positive slope, but give an unstable operating point where the intersection is with a negative resistance slope as at b. The dashed curve with the intersection at b has three intersections with the V-I curve. The outer two at a and c are stable points, and would be the actual quiescent operating points of the device since point b is unstable. Therefore the current-stable device can be used to make a bistable circuit if the source resistance is high, and the device is suitably biased.

8-3 TYPES OF NEGATIVE-RESISTANCE DEVICES

There are many available devices which produce useful negative-resistance characteristics. We shall select a few to illustrate their use and the important operating characteristics for analysis. The rather bewildering array of devices can be conveniently divided into two categories: two-terminal devices in which the negative resistance is inherent in the device and cannot be subsequently changed, and three- (or more) terminal devices in which the negative resistance characteristics may be changed by the third terminal. In addition, various combinations of active elements may be com-

[1] The intersection at point b will not be stable if there are appreciable amounts of stray or added capacitance across the terminals of the device. *Ibid.*, p. 197.

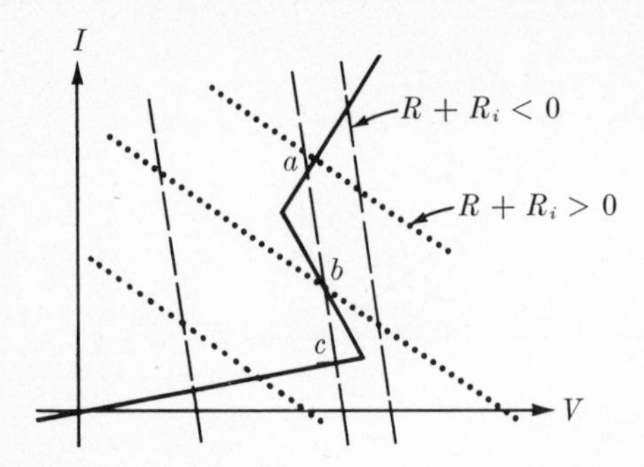

Fig. 8-6 A current-stable NRD with various load-line possibilities. The dotted lines represent the high-resistance load cases (all stable).

bined to give circuits which exhibit negative resistance between two terminals. These circuits will not be discussed except in connection with an integrated circuit which forms a useful two-terminal NRD.

A. TWO-TERMINAL NRD

1. The neon bulb and other gas-filled diodes The neon bulb and other gas-filled diodes are probably the oldest devices to be used to give a negative-resistance characteristic. The V-I curve of a typical bulb is given in Fig. 8-7. The actual breakdown voltage V_p and sustaining voltage V_v vary depending mostly upon the type of gas used and the pressure. For an actual neon bulb like an NE-2, the breakdown voltage is in the range of 70 to 90 V, and the valley voltage is in the range of 55 to 65 V. The V-I curve is reasonably symmetrical, that is, the first and third quadrants are similar but reversed in sign. The voltage drop for currents higher than I_v is reasonably constant, and the bulb may be used as a voltage regulator. (Special gas diodes are made specifically for regulator service such as the OA2.)

The main advantage of the neon bulb as a negative-resistance device is its low cost—only a few cents for an ordinary bulb. The disadvantages are that the negative-resistance characteristic tends to be unstable and varies considerably from one bulb to the next unless the bulbs are specially designed and selected. The gas tube is also relatively slow to switch, par-

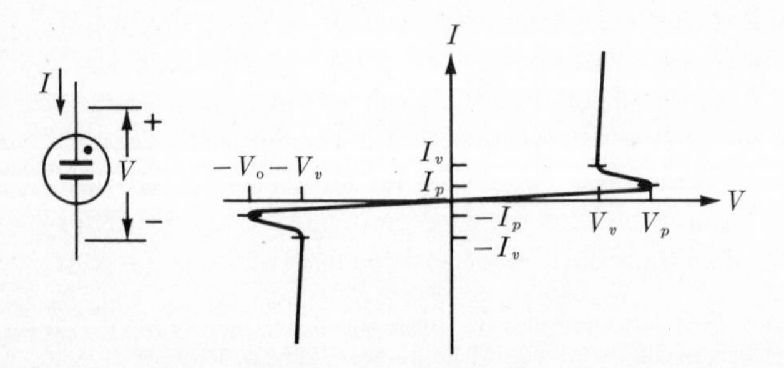

Fig. 8-7 A neon bulb and its V-I characteristic.

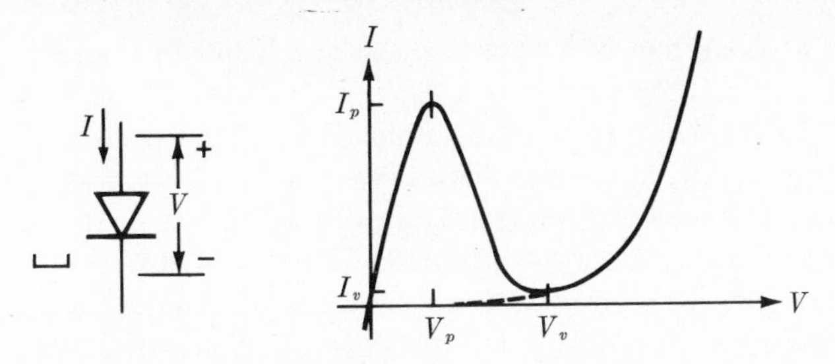

Fig. 8-8. A tunnel diode and its voltage-stable V-I characteristic.

ticularly off, since the deionization time may be on the order of milliseconds. Another way of saying the same thing is that the equivalent circuit for the device has a large inductance in series with the internal negative resistance.

2. The tunnel diode The tunnel diode is a PN junction device in which the gradient between the P and N regions is very abrupt. This leads to conduction across the junction at very low voltages by quantum tunneling rather than by an injection process as in an ordinary PN junction diode. The resulting diode has a V-I characteristic as shown in Fig. 8-8. The dashed curve in the same figure shows the current which would flow in the same type of PN junction but with reduced doping levels, thereby making an ordinary diode. As the curve shows, the tunnel diode is a *voltage*-stable device and is the only outstanding representative of the voltage-stable class.

For reverse and small forward voltages the diode is a very low resistance, but when the voltage in the forward direction is increased to V_p, the current begins to decrease and a negative resistance region begins. The current decreases with increasing voltage until the valley point V_v is reached where the current is a minimum I_v. From this point on the current increases much as in an ordinary diode. Typical values of the important parameters for diodes made of germanium (Ge), silicon (Si), and gallium arsenide (GaAs) are given in Table 8-1. The value of the peak-point current is largely controlled by the area of the junction in a given type of

Table 8-1

	Ge	GaAs	Si
I_p/I_v	8	15	3.5
V_p	0.06 V	0.15 V	0.065 V
V_v	0.35 V	0.50 V	0.42 V
V_f	0.50 V	1.1 V	0.70 V

diode, and currents in the range of 1 to 100 mA are normal in a small computer-type diode.

The most important advantage of the tunnel diode is the potential speed of operation. The tunneling phenomenon takes place at the speed of light, so that the speed of switching is controlled primarily by the capacitance of the diode, the circuit, and any series lead inductance. Switching times of less than one nanosecond are relatively easy, and switching times of tens of picoseconds (micro-microseconds!) are possible. The advantage of the extremely fast switching time is balanced by the disadvantages of low signal swing, the two-terminal nature of the device, and the extreme difficulty of incorporating the device with other types in an integrated circuit.

3. The silicon unilateral switch This is more a small integrated circuit than a device, but since it is commonly used for a two-terminal negative resistance it is proper to discuss it here. The circuit is shown in Fig. 8-9*a* and is very similar to the NPN–PNP astable oscillator circuit shown in Fig. 5-35*b*. The integrated circuit operates in the same manner as described for the latter circuit, that is, if a current is fed into terminal A some appears as base current in $Q1$ and some as collector current. A very small amount of base current is sufficient to bring the zener diode into its reverse breakdown condition so that the potential between A and C becomes approximately $7\text{ V} + V_{EB} \approx 7.7\text{ V}$. The collector current of $Q1$ forward biases $Q2$, but for small values of I the conduction in $Q2$ is negligible. Larger values of I begin to turn $Q2$ on, however, and the base current for $Q1$ is supplied by $Q2$ collector current rather than current through the zener, so that the current through the zener actually begins to decrease as the current I is increased. Further increases in I completely turn off the zener, and the voltage at G (and consequently the voltage at A) begins to decrease, so that for increasing values of I the voltage A-C decreases. This, of course, is a negative resistance characteristic. The actual V-I characteristic is shown in Fig. 8-9*c*, where the peak-point voltage and current are shown to be about 7.7 V and 150 μA, respectively. The voltage across the device when fully on is only about 1 V even at relatively large currents. The device does not break down in the reverse direction until the reverse breakdown voltage of the emitter-base junction of $Q1$ is reached at about -30 V.

The unilateral switch is reasonably fast and will discharge a 0.01-μF capacitor in less than 1 μs; however, the turnoff of the device is slower, but typically less than 25 μs.

The extra lead at G may be used to modify the characteristics of the device, but is not needed for the applications to be described; therefore the device really is used as a two-terminal device.

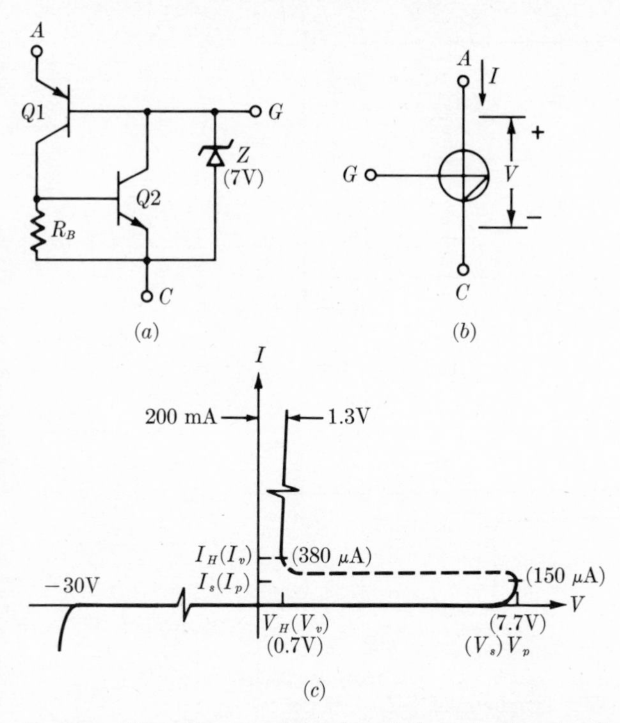

Fig. 8-9 The silicon unilateral switch. (*a*) The circuit which is integrated on the silicon chip. (*b*) A symbol for the device. (*c*) The *V-I* characteristic of the device. The values in parentheses are alternative terms: switching point voltage and current, and holding voltage and current.

4. The bilateral switch The bilateral switch is basically two of the unilateral switches just described connected back to back as shown in the diagram (Fig. 8-10*a*). The *V-I* characteristic of the device is, as the name implies, symmetrical about the origin as shown in Fig. 8-10*c*. This type of device is very useful in circuits operating directly off the ac line without rectification because the device can be made to trigger on both the positive and negative half cycles of the input.

5. The diac and the four-layer diode Both of these negative-resistance devices are multiple-junction structures. The diac is a three-layer, two-junction device, which has the characteristics shown in Fig. 8-11*a*. The negative resistance obtained is less than with the preceding devices; the ON resistance is higher, and the switching voltage is rather high, 32 V for the device shown.

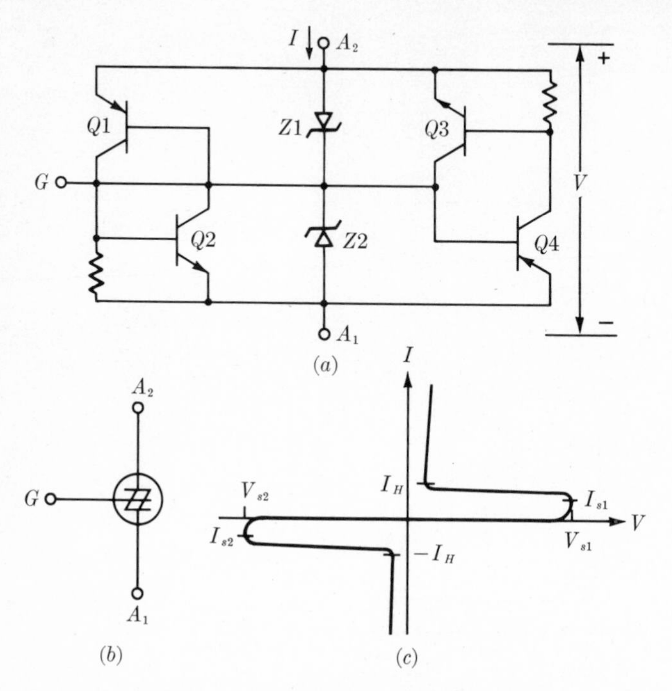

Fig. 8-10 The silicon bilateral switch. (*a*) The integrated circuit. (*b*) The symbol. (*c*) The symmetrical *V-I* characteristic.

The four-layer diode is an NPNP device, and gives somewhat better negative-resistance characteristics as shown in Fig. 8-11, and can be obtained over a wide range of switching voltages with 15 to 200 V being common. Small four-layer diodes are capable of switching amperes and conducting several hundred milliamperes in steady state. Larger four-layer diodes have been made that will conduct many amperes of current in steady-state. The same four-layer structure, but with leads connected to three or even four of the layers, gives a considerably more flexible device, and will be described in the following sections.

B. THREE-TERMINAL NEGATIVE-RESISTANCE DEVICES

1. The unijunction transistor The unijunction transistor (UJT) in its original form was a bar of *n*-type silicon with a *p*-type alloy junction alloyed to its middle as shown in Fig. 8-12. Ohmic contacts are made to each end of the bar. In normal operation the lower ohmic contact (base one, *B*1) is grounded. If the emitter junction is left open circuited, the voltage between *B*1 and *B*2 sets up a current proportional to the conductance of the bar. In the device shown in Fig. 8-12*a* the potential in the *n*-type material under the junction would be somewhat greater than half the

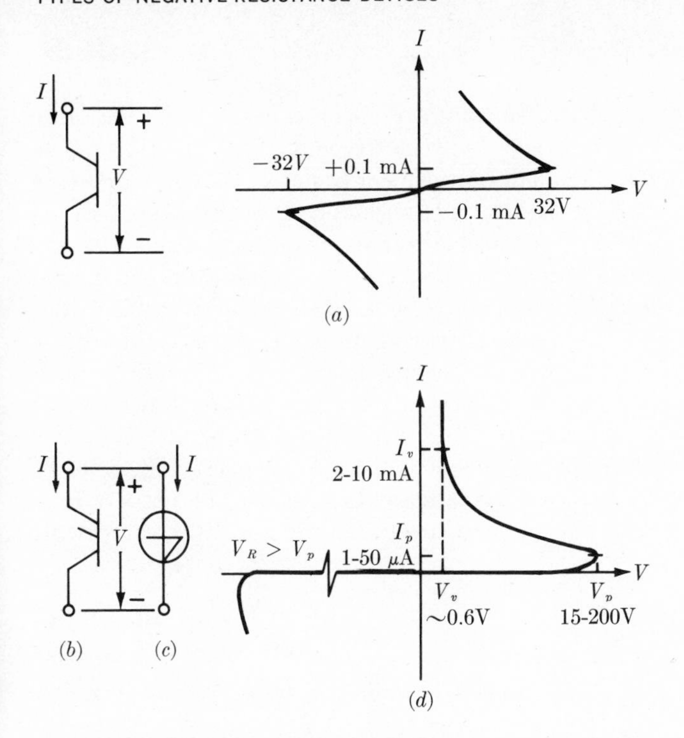

Fig. 8-11 (*a*) The diac symbol and *V-I* characteristic. (*b*), (*c*) Symbols for the four-layer diode. (*d*) The *V-I* characteristic.

voltage applied to $B2$ because the PN junction is placed above the center of the bar. Under the conditions of no current in the emitter, the UJT between $B1$ and $B2$ has the characteristics of an ordinary resistance called the *interbase* resistance, R_{BB}. At 25°C this resistance typically has values between 4 and 10 kΩ, and increases linearly with temperature up to about 140°C. The temperature coefficient at 25°C is about +0.08%/°C.

The normal biasing conditions for the UJT and its symbol are indicated in Fig. 8-12*b*. As shown in Fig. 8-12*c* very little emitter current flows until the emitter voltage V_{EB1} is raised sufficiently to forward bias the PN junction. This occurs when the lower edge of the emitter becomes more positive than the bar under the emitter. This emitter voltage is called the peak-point voltage V_p. As shown, the peak-point voltage is equal to a fraction of V_{BB} plus the PN junction drop.

$$V_p = \eta V_{BB} + V_D \qquad (8\text{-}5)$$

For the curves shown $\eta = 0.6$ and $V_D = 0.7$ V. The parameter η is called the *intrinsic standoff ratio* and is independent of bias conditions and temperature. The value of η lies between 0.47 and 0.75 according to the particular type of UJT. The voltage V_D is about 0.7 V at 25°C and decreases about 2 mV/°C.

When current begins to flow in the emitter, carriers are injected into the bar and reduce the resistance between the emitter and $B1$, thus reducing the potential drop in the bar between emitter and $B1$. This increase in bar conductance leads to the negative-resistance characteristic observed at the emitter (Fig. 8-12*c*), and also lowers the resistance between $B1$ and $B2$ so that the interbase current is increased. For large values of emitter current, the emitter-to-$B1$ resistance is quite small as shown in Fig. 8-12*c*. (Note that the V and I axes in this figure are interchanged as is the custom for drawing this curve on UJT data sheets.) When measured with currents over 30 mA, the ac (incremental) resistance between emitter and $B1$ lies in the range from 5 to 20 ohms.

For some applications we would like the peak-point voltage to be independent of temperature. This is possible if we use the positive temperature coefficient of R_{BB} to compensate for the negative temperature

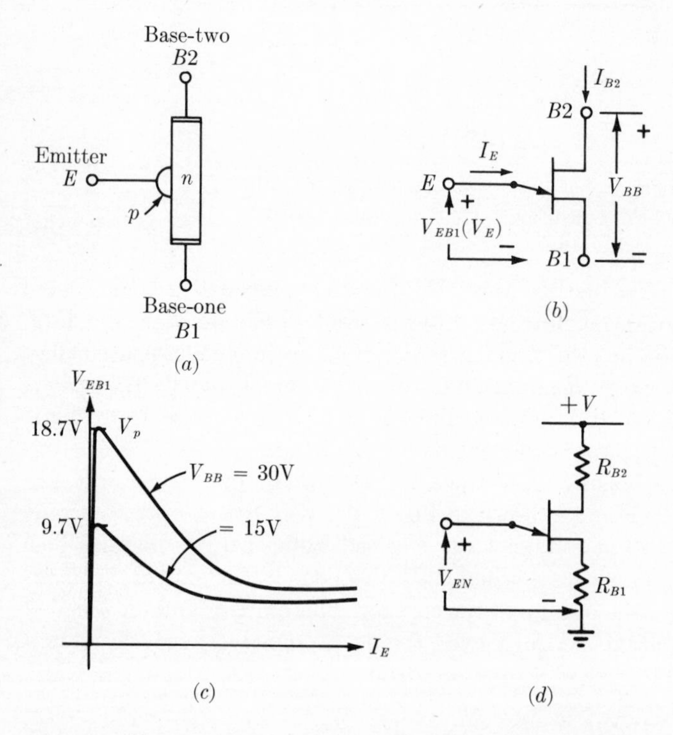

Fig. 8-12 The unijunction transistor (UJT). (*a*) The construction. (*b*) The symbol for the device. (*c*) The *V-I* characteristic at the emitter terminal. (*d*) A circuit to compensate for changes in V_{EN} due to temperature.

coefficient of V_D. The circuit (Fig. 8-12*d*) places a resistor in series with $B2$ so that the variation of interbase resistance causes V_{BB} to increase with temperature. (The resistor R_{B1} is not needed for the temperature compensation, but is usually present to provide current limiting and a useful output.) If R_{B2} is chosen correctly, the increase in V_{BB} will compensate for the decrease in V_D. The proper value of R_{B2} is approximately

$$R_{B2} \approx \frac{0.40R_{BB}}{\eta V} + \frac{(1 - \eta)R_{B1}}{\eta} \tag{8-6}$$

With this value of R_{B2} the value of V_{EN} at the peak point will be nearly independent of temperature. If the above values of R_{B1} and R_{B2} are used, the voltage at the peak point becomes

$$V_{EN}(\text{peak}) = \eta V \tag{8-7}$$

2. The silicon-controlled rectifier (SCR) The silicon-controlled rectifier shown in Fig. 8-13 is similar to the four-layer diode previously discussed, except that an additional lead is added to the P region near one end. This electrode, called the *gate electrode*, allows control of the device by an external current. When the current in this gate electrode is zero, the V-I characteristics of the device are precisely the same as those for the four-layer diode. As increasing currents are put into the gate electrode (note that the easy direction of current flow is into the gate, or the gate is positive with respect to the cathode) the forward breakover voltage is decreased. The device thus acts like a four-layer diode with a breakover voltage which can be controlled. The device may be operated with a forward voltage across it which is less than the forward breakover voltage, and can then be triggered by a pulse of current applied to the gate. Such a trigger signal causes the device to go from a region of high impedance where it is conducting only a few microamperes to a region of low imped-

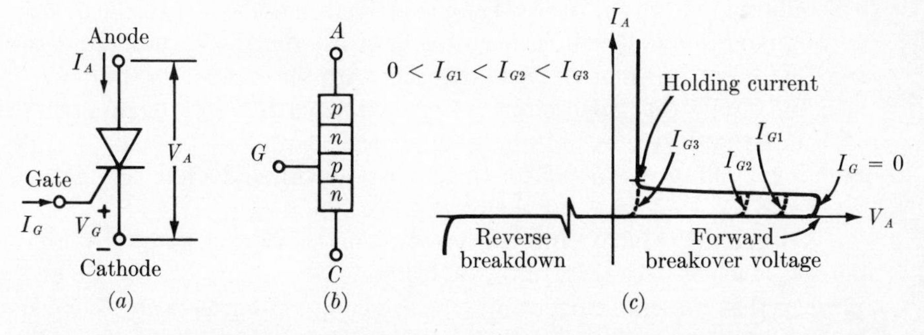

Fig. 8-13 (*a*) Symbol for the silicon-controlled rectifier (thyristor). (*b*) The construction of the SCR. (*c*) The V-I characteristic for various values of gate current.

ance where it may be conducting many amperes of current as determined by the external circuit. The SCR also can be used like a four-layer diode, that is, conduction may be initiated by applying a voltage greater than the breakover voltage with $I_G = 0$. However, this is not the normal mode of operation.

Most SCR's can be turned off only by reducing the anode current below the holding current. Some more recent SCR's can be turned off by supplying a large reverse current to the gate, but the current gain (ratio of anode current to control current) is small in this mode of operation. Small, low-current devices of this type are also known as *silicon-controlled switches* (SCS) and may have an electrode brought out from each of the four layers to give even further flexibility.

The silicon-controlled rectifier and other derivatives of the breed are available to handle currents from the low milliampere range up to hundreds of amperes on a continuous basis. The forward breakover voltage determines the maximum voltage the unit can switch and can be as high as 1,000 V. The devices can therefore control powers ranging from milliwatts to hundreds of kilowatts. Even small, relatively inexpensive devices can handle several kilowatts to control devices running directly from the 117-V power line.

3. The TRIAC The SCR previously described usually has a much higher breakdown voltage in the reverse direction than in the forward direction even with $I_G = 0$, and further, the reverse breakdown does not appreciably change when control currents are applied to the gate. Therefore the device can only conduct in one direction like a rectifier, and if an output is required for both half cycles of an input ac signal, two SCR's connected back to back must be used. One conducts on one half cycle, the other on the other half. The problem is solved in the TRIAC, which is also a four layer device, by providing a more or less symmetrical characteristic, that is, the application of a control current causes the breakover voltage to be reduced in both the forward and reverse directions, so that the device can be made to conduct in either direction. A typical V-I characteristic for a TRIAC is shown in Fig. 8-14 together with the symbol for the device. With the gate either open circuited or connected to T_1 through a small resistance the device exhibits a symmetrical characteristic with a high breakdown voltage ($>$ 500 V for the device shown), and a low-conduction voltage ($<$1.5 V at 15 A for the SC50E).

To get the TRIAC into its low-resistance state a gate current is supplied much as in the case of the SCR previously described. The TRIAC differs in that the gate current may be applied in either polarity to cause the device to trigger. Probably the most normal method would be to supply trigger pulses that are positive when T_2 is positive (all polarities are refer-

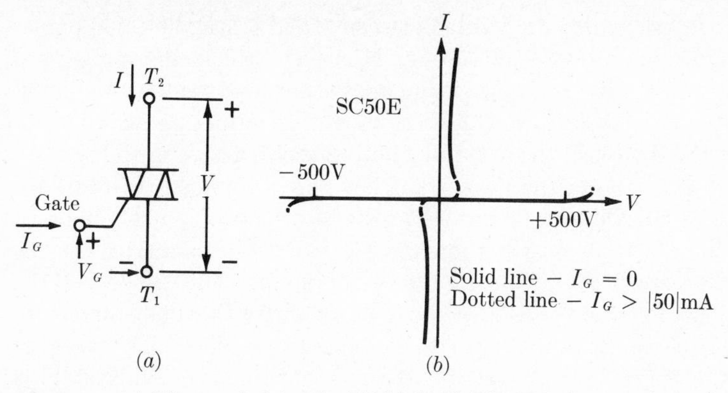

Fig. 8-14 (*a*) The symbol for the TRIAC. (*b*) The *V-I* characteristic of the TRIAC.

red to terminal T_1), and negative when T_2 is negative. However, the device will also go into the low-impedance condition when T_2 is negative when a sufficiently large positive gate current is applied. The triggering flexibility of the TRIAC allows it to be used in a wide variety of circuits.

8-4 SOME CIRCUITS USING TWO-TERMINAL NRD

The simplest circuit in which to study the behavior of a negative-resistance device is that of Fig. 8-15. Consider the operation to begin with $V_1 = 0$, and the charge on C_1 (which may be only the circuit stray capacitance) also zero. As the voltage is changed to $V_1 = A$, the only possible static operating point is seen to be a. Further increase of voltage to $V_1 = B$ gives three possible operating points: b, b' and b''. Actual operation will be at b

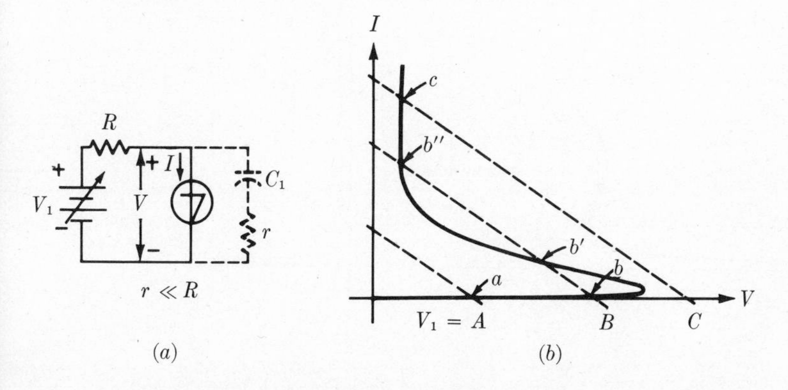

Fig. 8-15 (*a*) A circuit employing a four-layer diode. (*b*) The *V-I* characteristic of the diode and the load lines for the circuit.

because the set of operating points between a and b are all stable points and no switching actions can occur. However, when the voltage is increased to $V_1 = C$, a switching action does occur because the only possible static operating point is c where the diode is conducting heavily.

To see the details of the switching action occurring, refer to Fig. 8-16 where the region around the peak point has been expanded. In order for switching to occur the net resistance in the loop containing the diode C_1 and r must be less than zero as shown in Sec. 8-2. (The capacitor may be thought of as a short circuit during the switching interval.) Therefore the operating point must move past the peak point where the diode incremental resistance is zero to a point where this resistance is $-r$. This point is shown on the curve as d. At this point the rapid regenerative switching action is initiated, and the diode almost immediately turns full on. The dynamic switching which discharges C_1 does not occur along the static V-I curve. Indeed the static characteristics of the device are insufficient data from which to calculate the exact details of this switching action. In the cases that are discussed here the device is usually assumed to switch instantaneously, and the time to discharge the capacitor is taken as proportional to the time constant of the loop containing the capacitor, in this case rC_1. The diode does not immediately move to point d', however, because that is the static operating point obtained only after the capacitor has been discharged. Instead the operating point immediately after the switching operation is determined by V_{cap} and r (the current through R is usually quite negligible compared to the capacitor discharge current). Thus the operating point can be found as in Fig. 8-16b, where a load line for V_{cap}

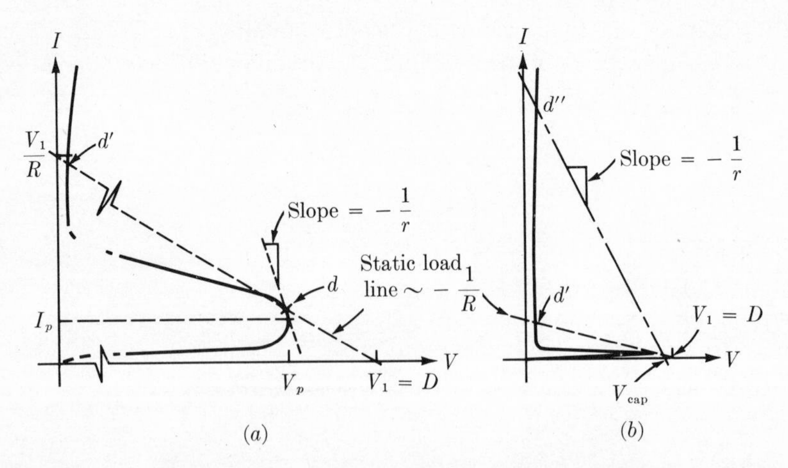

Fig. 8-16 (a) The V-I characteristic expanded near the peak point to show the details of switching from the low-current region to the high-current region. (b) The normal V-I characteristic.

and r is drawn, and the operating point immediately after the switching action is d''. Because the current in the capacitor is high, the operating point very rapidly moves from d'' to d' (the time constant is rC_1); d' is the static operating point for $V_1 = D$.

If we now decrease the input voltage from $V_1 = D$ to B, the operating point moves smoothly from d' to b''. Again with $V_1 = B$ there are three possible operating points, but nothing has yet occurred to initiate a switching action. Upon reducing the voltage further to $V_1 = A$ a switching action will have to occur because the only stable operating point is a which is in the low-current region of the device. Again the switching action occurs when the total resistance in the loop containing C_1 and the diode is zero. Therefore the device must have an internal negative resistance equal to or greater than r. The device characteristic is expanded near the origin in Fig. 8-17 to show what happens. At a current equal to the valley current I_v (also called the *holding current* in some devices) the incremental resistance of the diode is zero; for smaller currents this resistance is negative. To initiate switching the operating point must move below the valley point until the slope of the V-I curve is sufficiently negative to cancel the external resistance. This occurs at the point e' and input voltage $V_1 = E$. Rapid switching again occurs and the diode goes into its high impedance state. The operating point, however, does not instantaneously go to e because the capacitor voltage cannot change instantaneously. Therefore the diode goes off, but the voltage across it remains essentially that of the capacitor, point f. After the diode is switched off, the capacitor charges relatively slowly up to $V_1 = E$ with time constant $(R + r)C_1 \approx RC_1$. As shown, the switching action does not occur exactly at the valley point, but will be close to it if the external resistance r is small.

The intersections of the static load line and the device V-I characteristic determine whether a circuit will be monostable, bistable, or

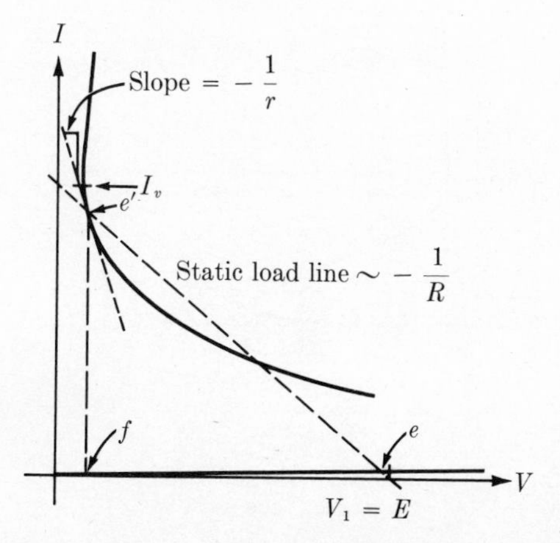

Fig. 8-17 The V-I characteristic expanded near the origin and the valley point to show the details of switching from the high-current to the low-current region.

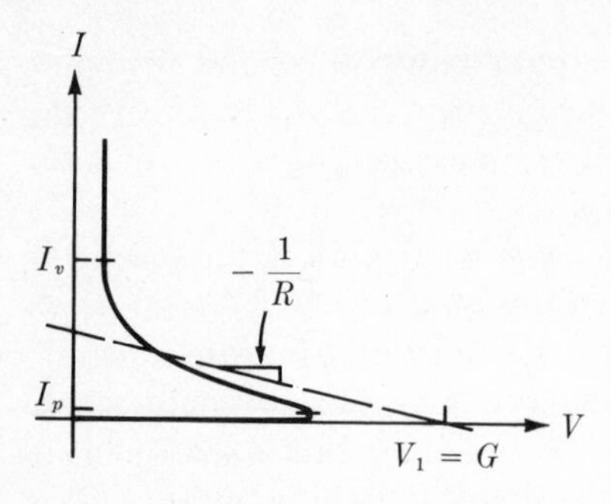

Fig. 8-18 A static load line with the intersection in the negative resistance region as required to produce an astable circuit.

astable. For example, in Fig. 8-15 a monostable circuit will result if $V_1 = A$ or C because the operating point will remain either at a or c, respectively, until some external disturbance temporarily puts the circuit into a new condition. (Such a disturbance could be a momentary change in the charge on C_1.) The circuit will eventually return to its steady-state position at a or c when the capacitor current decays to zero. Likewise, if the input voltage is changed to point b, the circuit becomes bistable with operation in steady state at either b or b'', but not at b', which is an unstable operating point.

To achieve an astable circuit or oscillator the steady-state operating point must be an unstable one, such as is shown in Fig. 8-18. With such a bias point the circuit will never rest at the steady-state point but will be operating in other regions of the V-I characteristic. An important point to notice from the preceding discussion is that it is not so much the *circuit configuration* that determines the type of operation as it is the *bias conditions* in the circuit.

To show the typical operations of a monostable and an astable circuit let us use the circuit and diode characteristic shown in Fig. 8-19. This

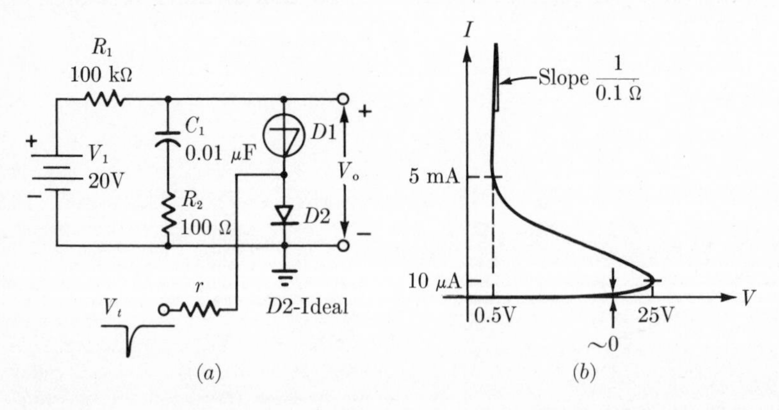

Fig. 8-19 (*a*) A monostable circuit employing a four-layer diode. (*b*) The assumed V-I characteristic of the diode.

circuit is similar to that of Fig. 8-15 with an added diode to permit triggering the circuit. In its rest state the four-layer diode cannot be ON because the maximum steady-state diode current ($^{20}/_{100}$ kΩ = 0.2 mA) is less than the valley current (5 mA). Therefore the steady-state situation is as in Fig. 8-15 with $V_1 = A$ and an operating point at a. The initial charge on C_1 is 20 V, the voltage across the diode is 20 V, and the current through it is almost zero. To make the diode $D1$ turn ON, a negative trigger pulse v_t is applied of sufficient value to raise the voltage across $D1$ to at least the peak-point voltage (25 V). Diode $D1$ then immediately breaks down and a large current flows around the loop containing C_1, R_2, D_1, and D_2. This current normally would be so large as to completely override the small current available from the trigger source so that $D2$ would effectively be a short circuit during the discharge of C_1. An equivalent circuit which is valid after $t = 0$ is shown in Fig. 8-20a. The current through the diode I_{D1} is the sum of $I_1 + I_2$.

$$I_1 = \frac{20 - V_{D1}}{100\ k\Omega} \approx \frac{19.5}{100\ k\Omega} = 0.2 \text{ mA} \tag{8-8}$$

$$I_2 \approx \frac{20 - 0.5}{100\ \Omega} \epsilon^{-t/\tau_1} = 195\epsilon^{-t/\tau_1} \text{ mA} \tag{8-9}$$

$$\tau_1 = (100\ \Omega)(10^{-8}) = 1\ \mu\text{s} \tag{8-10}$$

$$I_{D1} = 0.2 + 195\epsilon^{-t/1\mu\text{s}} \text{ mA} \tag{8-11}$$

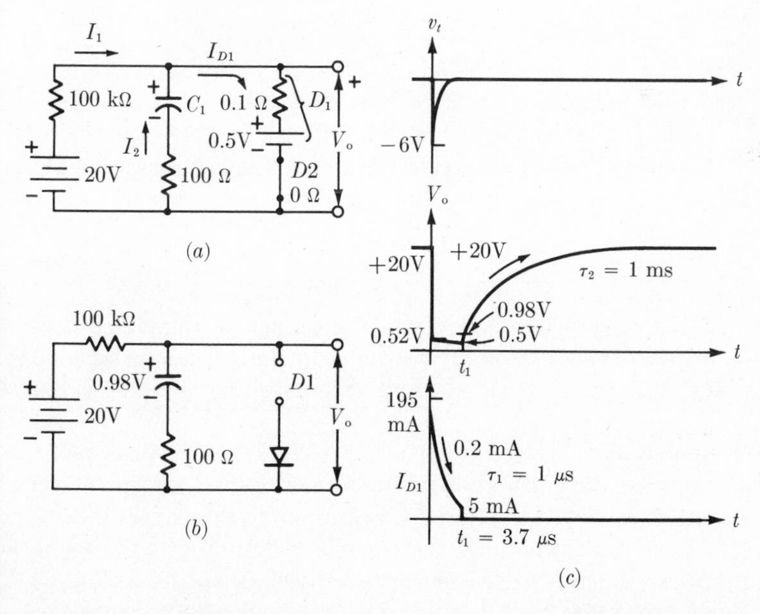

Fig. 8-20 (a), (b) Equivalent circuits for Fig. 8-19a. (c) Waveforms produced by the circuit.

With this large current the capacitor is quickly discharged; the critical point occurs when the current drops sufficiently to bring the diode to -100 ohms. At this point the loop resistance would be zero and switching would commence. For this example sufficient accuracy is obtained by assuming the switching commences when the current at the valley point is reached at $I = 5$ mA. The time to the switching point is found by equating I_{D1} to the valley current.

$$t_1 = \tau_1 \ln \frac{195}{4.8} = 3.7\tau_1 = 3.7\ \mu\text{s} \tag{8-12}$$

At this time the diode switches into its high-resistance condition, and the voltage at the output terminals is determined by the remaining charge upon the capacitor:

$$V_{C1}(t_1) = 0.5 + (4.8\ \text{mA})(0.1\ \text{k}\Omega) = 0.98\ \text{V} \tag{8-13}$$

The waveforms resulting thus far are shown in Fig. 8-20*c*. After the four-layer diode turns off at t_1, the capacitor is free to charge up again to 20 V, as the equivalent circuit of Fig. 8-20*b* shows. This is accomplished with a time constant $\tau_2 \approx (100\ \text{k}\Omega)(10^{-8}) = 1$ ms. The circuit would be ready to be triggered again after several milliseconds, and the preceding operation would repeat.

To make the same circuit astable, that is, a free-running oscillator, it is only necessary to increase the supply voltage past the peak-point voltage. In this case the minimum value of supply voltage that will make the circuit astable is that voltage that will just supply the peak-point current (i.e., the static load line intersects the *V-I* characteristic at the peak point).

$$V_{\min} = V_p + I_p(110\ \text{k}\Omega) = 26\ \text{V} \tag{8-14}$$

(This assumes the resistance in series with the capacitor is small so that the unstable region begins essentially at the peak point.) Let us assume $V_1 = 30$ V for a margin of safety, and calculate the period of the oscillation.

If we assume that the charging capacitor just reaches the peak-point voltage at $t = 0$, then the calculation of the capacitor discharge proceeds exactly as before, except that the initial charge upon the capacitor is 25 V this time. The time for discharging the capacitor is almost the same as before, and the first real difference comes when the diode switches off at t_1; the final value for the slow ($\tau_2 = 1$ ms) exponential is now 30 V instead of 20 V, as shown in Fig. 8-21. The slow exponential continues until the diode

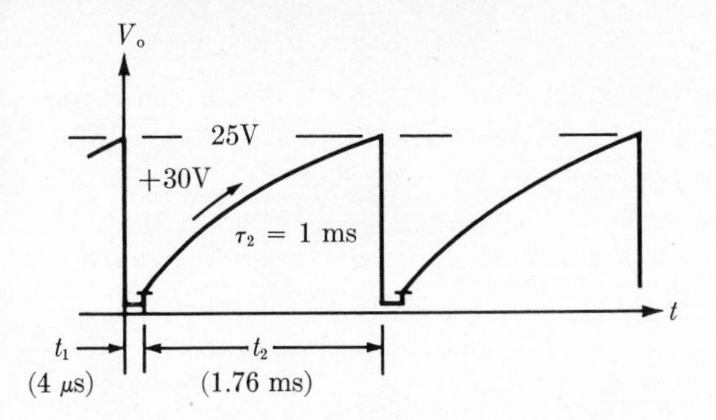

Fig. 8-21 The waveform generated by the circuit of Fig. 8-19*a* with the supply voltage raised to 30 V to produce an astable oscillator.

voltage rises to the peak point, 25 V. At that point the cycle begins again. The time for the slow rise is

$$t_2 = \tau_2 \ln \frac{29.02}{5} = 1.76\tau_2 = 1.76 \text{ ms} \tag{8-15}$$

The total period $T = t_1 + t_2 \approx t_2$, and the frequency is $1/T = 570$ Hz.

This type of negative-resistance oscillator makes an excellent sawtooth or sweep generator because it is relatively easy to obtain a large ratio of useful sweep time to recovery time. The sweep may be made more linear by using a higher supply voltage so that a smaller portion of the time constant is used. Also the linearization techniques of Chap. 6 may be used to give a large-amplitude linear sweep. Synchronization of the sweep to an external source of pulses is easily accomplished by placing a suitable synchronizing signal on the terminal v_t.

If the supply voltage is raised sufficiently, the circuit will again become monostable because the load line will move up the V-I characteristic until the intersection becomes above the valley point, as in Fig. 8-15 with $V_1 = C$. In the case of Fig. 8-19*a* the circuit becomes monostable with the quiescent condition being the ON state when the supply voltage is increased above 500 V.

8-5 A TUNNEL-DIODE ASTABLE OSCILLATOR

It is interesting to compare the operation of a tunnel-diode circuit with the previously described operation of the four-layer diode because the circuits are duals. The circuit and the tunnel-diode characteristic are shown in Fig. 8-22*a* and *b*. In this case the external energy storage element is an inductor L in series with a low-voltage V_s (about 150 mV for a germanium tunnel diode) and a low-resistance R. The resistor and V_s are chosen to give a static operating point for the tunnel diode in the negative-resistance

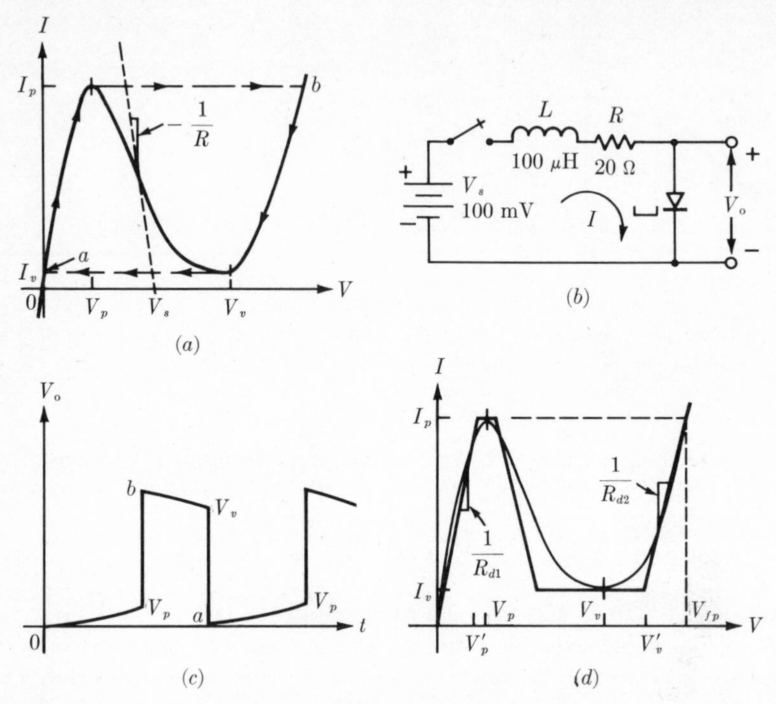

Fig. 8-22 (*a*) A tunnel-diode characteristic with the locus of operation shown for a relaxation oscillator. (*b*) A tunnel-diode astable circuit. (*c*) The output waveform from (*b*). (*d*) A linearized tunnel-diode characteristic. $V_p' = 0.75\ V_p$, $R_{d1} = 0.75\ V_p/I_p$, $V_v' = (V_{fp} + V_v)/2$, $R_{d2} = (V_{fp} - V_v')/(I_p - I_v)$.

region so that an astable circuit will result. Assume that the circuit is initially at rest and that at $t = 0$ the switch is closed. The initial inductor current is zero, the circuit current at $t = 0^+$ is also zero, and the initial operating point is the origin. As time progresses the current builds up and the operating point moves from 0 past a to the peak point V_p. At or very near this point the diode switches and the operating point very quickly moves to point b. The operating point must move horizontally because the current in the inductor cannot change during the practically instantaneous switching time (a time of under 10 ns would be typical). At point b the output voltage is greater than the supply so that the current through the inductor tries to reverse. This leads to a rapid decay of the current toward a negative value, and the operating point moves to V_v. The current cannot decrease below the I_v obtained at the valley point V_v, so a second switching action is initiated. Again the inductor forces I to remain constant so that the new operating point is a. The cycle of operation now starts over again with the operating point moving again toward the peak point.

The waveform which results is given in Fig. 8-22*c*, where the first cycle is not quite the same as those following since the circuit is assumed to start from rest. The output is a pulse of low amplitude, but with very fast transitions.

To calculate the times of the two parts of the cycle of operation a linearized model of the diode is required as shown in Fig. 8-22*d*. Equations are given in the figure for calculating the needed resistances in terms of the usually specified diode parameters I_p, V_p, and so forth. These expressions were derived with reference to a germanium tunnel diode, but are probably adequate for other types of diodes.[1] The actual calculation of the details of the figure is left for Prob. 8-8.

8-6 A UNIJUNCTION OSCILLATOR AND SCR POWER CONTROL

The unijunction transistor makes a simple and practical astable and monostable circuit which is often used in conjunction with a silicon-controlled rectifier to control large amounts of power. Let us begin with the simple oscillator circuit of Fig. 8-23*a* and proceed to a half-wave controller circuit with the SCR incorporated. Our goal is to control the time that the SCR is switched on with respect to the phase of the 60-Hz power line voltage. We would like the switching time to be late in the cycle when a control resistance is high, and early in the cycle when the control resistance is low. In this way voltage will be applied to the load through the SCR switch for most of the half cycle when the control resistance is low, and for very little or none of the cycle when the control resistance is high. The control resistance will be R_E in Fig. 8-23*a*, so let us begin by calculating the maximum and minimum values that R_E may assume and still give an astable situation. The minimum value of R_E is that which produces a load line on the UJT characteristics which intersects the V_E-I_E curve at the valley point.

$$R_{E,\min} = \frac{V_1 - V_v}{I_v} = \frac{20 - 3.5}{10 \text{ mA}} = 1.7 \text{ k}\Omega \tag{8-16}$$

The maximum value of R_E that may be used produces a load line with an intersection at the peak point.

$$R_{E,\max} = \frac{V_1 - V_p}{I_p} = \frac{20 - 11}{5 \ \mu\text{A}} = 1.8 \text{ M}\Omega \tag{8-17}$$

Any resistance within this rather large range of values will produce the desired astable oscillator.

The actual operation of the oscillator may be seen from the locus of Fig. 8-23*b* if we assume the circuit is turned on at $t = 0$. The emitter volt-

[1] *Tunnel Diode Manual*, General Electric Co., 1st ed., p. 51, 1961.

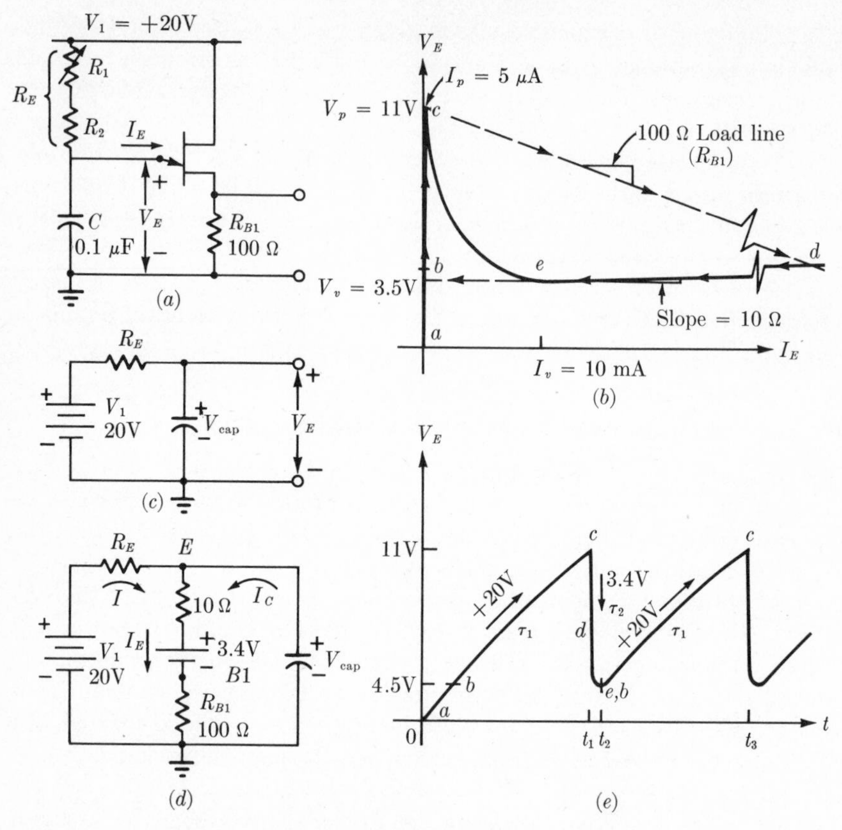

Fig. 8-23 (*a*) A unijunction astable oscillator. (*b*) An assumed *V-I* characteristic for the UJT. (*c*) An equivalent circuit during the charging interval. (*d*) An equivalent circuit during the discharging interval. (*e*) The waveform produced at the emitter terminal assuming the circuit is started at $t = 0$.

age V_E is then zero, and the operation starts at point *a*. The capacitor charges through R_E toward +20 V until point *c* is reached, at which time the UJT switches and a high-emitter current flows as limited by R_{B1} and the small emitter impedance. The operating point immediately after the switching instant is point *d* at a high current off the graph to the right. The capacitor rapidly discharges, and the current in the emitter decays quickly toward the steady-state value determined by the static load line. However, the UJT again switches when the emitter current falls sufficiently to once more place the UJT in the negative-resistance region (in this case the UJT has to present $-100\ \Omega$ for the net resistance of the loop containing the capacitor to become zero). This second switching occurs to the left of the valley point, but because of the small amount of external loop resistance added ($R_{B1} = 100\ \Omega$) the switching point will be assumed to be at the valley point: $I_E = 10$ mA. The capacitor voltage does not change during the brief switching intervals so the new operating point is at b with the UJT off. From this point the cycle begins over again.

The equivalent circuit for the charging interval is very simple (Fig. 8-23c), and the time to charge the capacitor to the peak point is

$$t_1 = \tau_1 \ln \frac{V_1}{V_1 - V_p} = R_E C \ln \frac{20}{9} = 0.8 R_E C \tag{8-18}$$

At this point the UJT becomes a low impedance, and the time for discharging may be found from the equivalent circuit, Fig. 8-23d. This time may be found most conveneintly by computing I_E and equating it to the valley-point current. The initial value of I_E is

$$I_E(t_1^+) = \frac{[V_C(t_1^+) - 3.4]}{10 + 100} = \frac{11 - 3.4}{110} = 69 \text{ mA} \tag{8-19}$$

The final value of the emitter current exponential is

$$I_E(\infty) = \frac{20 - 3.4}{R_E + 110} \approx \frac{16.6}{R_E} \tag{8-20}$$

The equation for the emitter current is therefore

$$\begin{aligned} I_E(t) &= [69 - I_E(\infty)]\epsilon^{-(t-t_1)/\tau_2} + I_E(\infty) \\ \tau_2 &\approx (110\ \Omega)C \end{aligned} \tag{8-21}$$

The actual time the UJT is conducting is found by equating I_E to the valley current.

$$t_2 - t_1 = \tau_2 \ln \frac{69 - I_E(\infty)}{I_v - I_E(\infty)} \tag{8-22}$$

A good approximation for this time would be to ignore $I_E(\infty)$, which is usually small compared with the transient component of current. In this case the discharge time becomes

$$t_2 - t_1 \approx \tau_2 \ln \frac{69}{I_v} \approx \tau_2 \tag{8-22a}$$

The capacitor voltage at $t = t_2$ is $3.4 + I_E(t_2^-)(110\ \Omega) = 4.5$ V. This is the voltage at point b in Fig. 8-23b. The charging waveform from t_2 to t_3 largely determines the period since the discharge time is relatively short.

$$t_3 - t_2 = \tau_1 \ln \frac{V_1 - V(t_2)}{V_1 - V_p} = \tau_1 \ln \frac{15.5}{9} = 0.54\ \tau_1 \tag{8-23}$$

(In many instances where R_{B1} is small a good approximate answer may be obtained by making $V(t_2) = V_v$.)

Because of the use we wish to make of this circuit, we would like the shortest charge time to be about 0.5 ms and the longest to be 8.5 ms. This will allow control over almost the full half cycle at a 60-Hz line frequency. The exact choice of R_E and C is quite arbitrary within the limits previously discovered. Therefore let us pick $C = 0.1\ \mu F$, a standard available value, and find R_E.

$$0.5 \text{ ms} = 0.54 R_{E\,(\text{min})}\, C$$

$$R_{E,\,\text{min}} = \frac{0.5 \times 10^{-3}}{(0.54)(10^{-7})} = 9.3 \text{ k}\Omega \tag{8-24}$$

$$R_{E,\,\text{max}} = \frac{8.5 \times 10^{-3}}{(0.54)(10^{-7})} = 158 \text{ k}\Omega \tag{8-25}$$

In this case the minimum value would be provided by the fixed resistor R_2, and the variation between minimum and maximum would be provided by the rheostat R_1. With these choices of R_E and C the discharge time $t_2 - t_1 \approx 1.9\tau_2 = 21\ \mu s$. Therefore the discharge time is much shorter than the period of operation.

The UJT oscillator just described is incorporated with the SCR in the circuit of Fig. 8-24*b*. In order to operate the SCR the UJT oscillator must be synchronized with the line voltage. This is accomplished by not operating the oscillator from a dc voltage, but by operating it from a clipped half-sine wave of voltage, which is produced by the half-wave rectifier $D1$. Current flows through R_3 on the positive half of the input cycle and the supply voltage for the UJT follows the input voltage until zener diode $D2$ begins to conduct at +20 V. When this point is reached the supply voltage stays constant at +20 V, and the UJT circuit operates much as before, but it begins each cycle with $V_{\text{cap}} \approx 0$. At a time determined by the setting of R_1 the UJT switches, and a large pulse of current is delivered to the gate of the SCR to bring it into the conducting state. When the SCR is "fired," the line voltage at that instant is placed across the load. Suppose that R_1 is set to fire the UJT and SCR at 6 ms; then the peak load current would be 12.7 A, and conduction would continue from about 130° to 180° of the input cycle. At 180° ($8\frac{1}{3}$ ms) the line polarity reverses and the SCR ceases to conduct because it cannot conduct in the reverse direction.

The waveforms which result from the circuit are also shown in Fig. 8-24. It is easily seen that the power delivered to the load is a function of the angle at which the SCR is fired, and is in turn controlled by the setting of R_1. The control range is such that power may be delivered for most of the cycle with $R_1 = 0$, or not at all with $R_1 \approx 150$ kΩ. The control need not be a rheostat, but could be a thermistor whose resistance is controlled by its temperature or a photoconductive diode whose resistance is con-

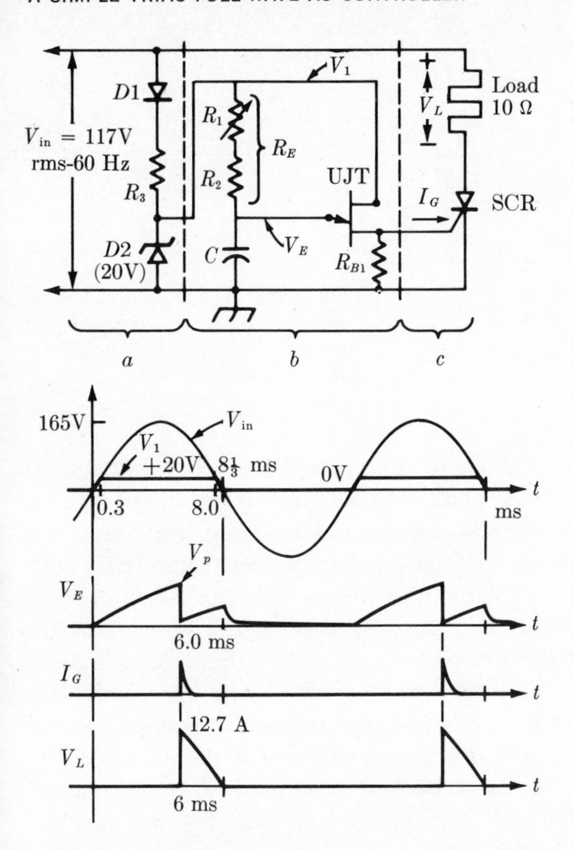

Fig. 8-24 A power controller using a silicon-controlled rectifier, which is triggered by a unijunction oscillator. The firing angle is controlled by the setting of R_1.

trolled by the light falling upon it. In other cases the current for charging C might be controlled by a transistor or other active device.

8-7 A SIMPLE TRIAC FULL-WAVE ac CONTROLLER

Very simple circuits control large amounts of ac power with the TRIAC as the power-control element and one of various negative-resistance devices to trigger the TRIAC. One such simple circuit is shown in Fig. 8-25 where the purpose is to control the light intensity by means of the small rheostat R_1. If the operation is started at $t = 0$ with both the input and capacitor voltage at zero, the TRIAC is untriggered, and the voltage across the load is approximately zero. (The load resistance is assumed to be very much smaller than the resistor R_1.) As the input voltage rises, the capacitor

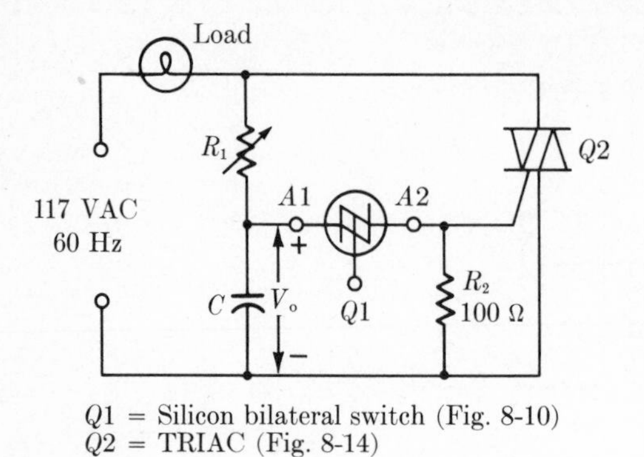

Fig. 8-25 A light dimmer employing a TRIAC. The timing waveform is derived directly from the ac power line.

charges until the switching voltage of the bilateral switch is reached. At this time the switch begins to conduct, the capacitor discharge triggers the TRIAC, and the full line voltage at that instant is placed across the load. The TRIAC continues to conduct for the remainder of the half cycle and is turned off when the input voltage drops near zero (or, more precisely, when the load current drops below the holding current of the TRIAC). The capacitor voltage stays near zero after $Q1$ conducts because the input voltage to the R_1C integrating circuit is only the small drop across the conducting TRIAC. Therefore the initial charge on C for the next half cycle is approximately zero, as it was at $t = 0$. The time conduction that begins in $Q1$ for the second half cycle will be approximately the same as during the first half cycle so that the output to the load will have a nearly zero dc component.

Since the timing circuit $R_1C \triangleq \tau$ is driven by a sine wave $V_1 \sin \beta t$, the timing waveform is not a simple exponential as before, but has the form

$$V_0(t) = \frac{V_1\beta}{\tau}\left[\frac{\epsilon^{-t/\tau}}{(1/\tau)^2 + \beta^2} + \frac{\sin(\beta t - \psi)}{\beta[(1/\tau)^2 + \beta^2]^{1/2}}\right] \qquad \psi \triangleq \tan^{-1}\beta\tau \qquad \beta \triangleq 2\pi f \tag{8-26}$$

Curves giving $V_0(t)/V_1$ can be plotted against the fraction of the sine-wave period as in Fig. 8-26. These curves and Eq. (8-26) assume that the capacitor has zero initial charge, and that the circuit begins operation at $t = 0$. The parameter for the different curves is the ratio of the RC time constant to the period of the sine wave, $\tau/T = \tau'$.

If we assume that the switching voltage of the bilateral switch is 8 V, then the normalized output voltage V_0/V_1 required to cause switching is $8/(\sqrt{2})(117) = 0.048$. Therefore if we pick $\tau' = 0.1$, or $\tau = 1.67$ ms, switching will take place at about $t/T = 0.04$ or $0.04(16.67 \text{ ms}) = 0.67$ ms after the zero crossing of the input sine wave. Increasing the time con-

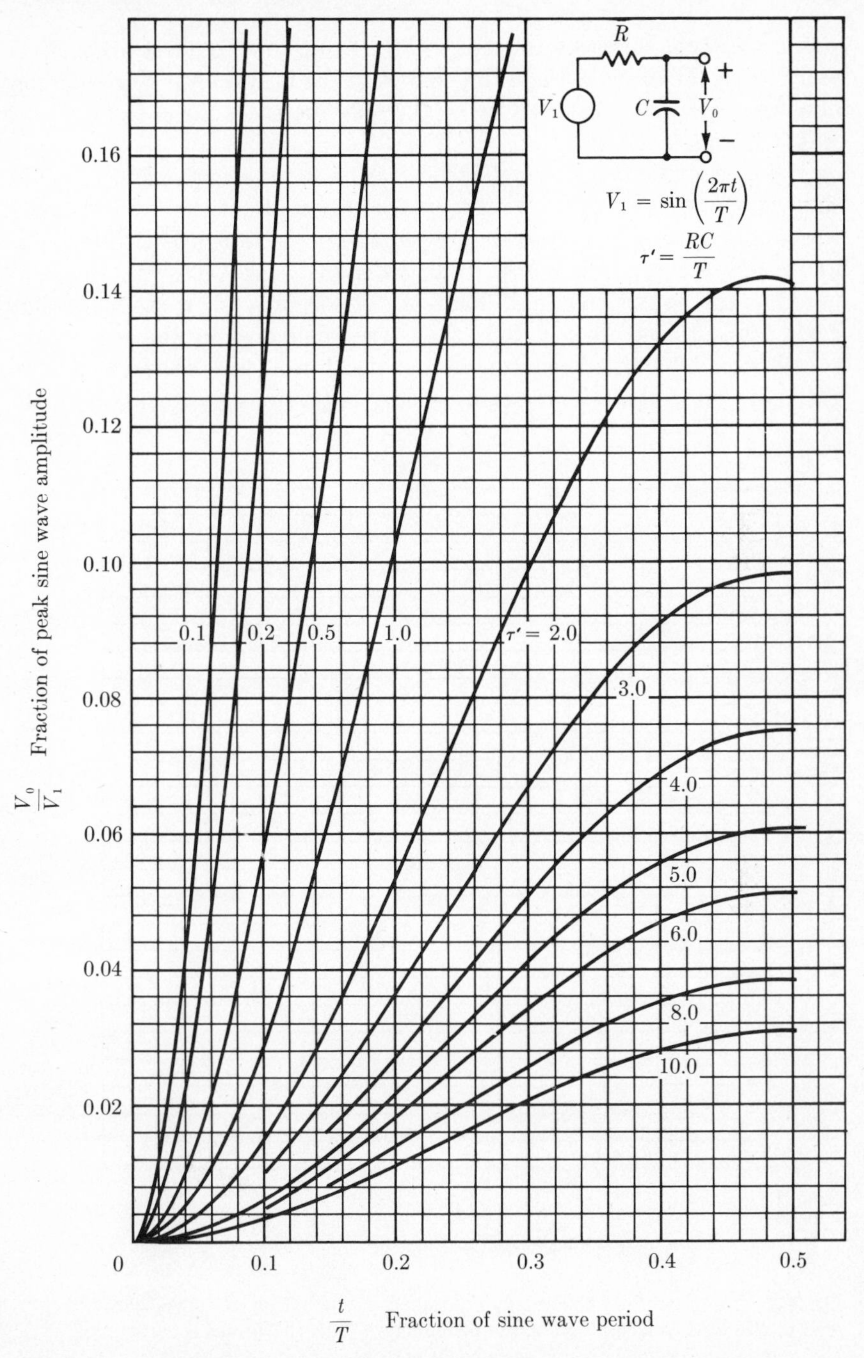

Fig. 8-26 The timing waveform given by an RC circuit driven by a sine wave. The capacitor is assumed to be initially discharged, and the sine wave is applied at $t = 0$.

stant, of course, increases the delay in the firing time, but not proportionately when the input voltage is a sine wave, as in this case. For example, doubling the time constant to $\tau' = 0.2$ only increases the switching time from 0.67 ms to 1.0 ms.

The circuit cannot be made to switch on at the exact beginning of the sine wave even though τ is reduced to zero because of the finite time for the supply voltage to rise to the switching point. Thus it is of interest to know how much power is lost from the slightly delayed switching compared to the power that could have been transmitted had the whole half cycle been utilized. In Fig. 8-27 the energy delivered by the circuit to a resistive load is plotted as a function of the time at which the TRIAC (or SCR) is switched on. The output energy is given as a percentage of the

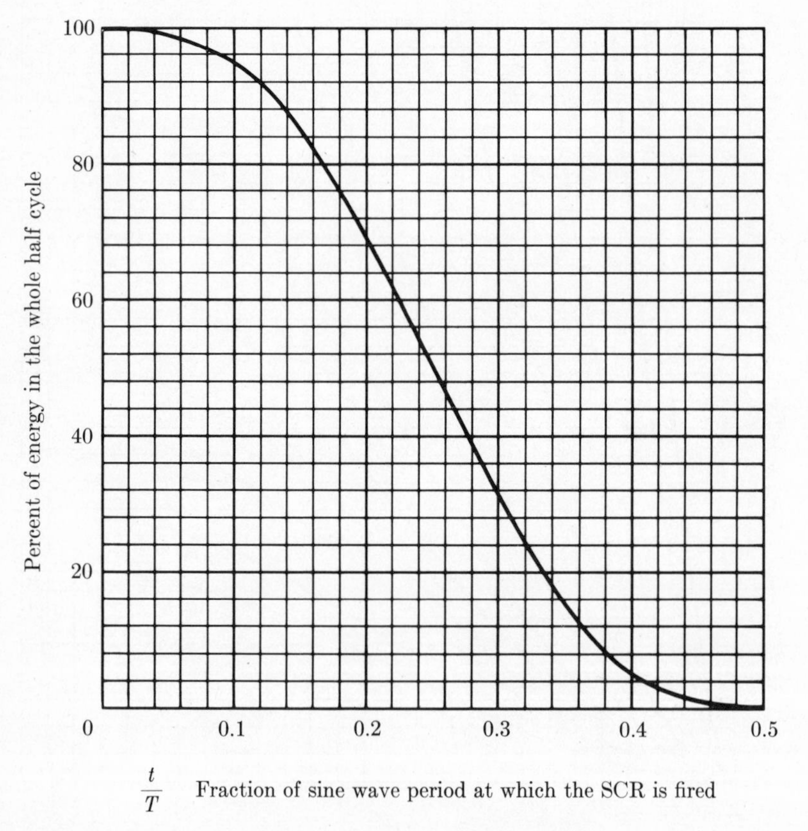

Fig. 8-27 The percent of the energy contained in one half-cycle to that delivered to the load as a function of the firing time t/T. If the circuit is half-wave the maximum power delivered is only 50 percent of the power assuming the switch is closed all the time.

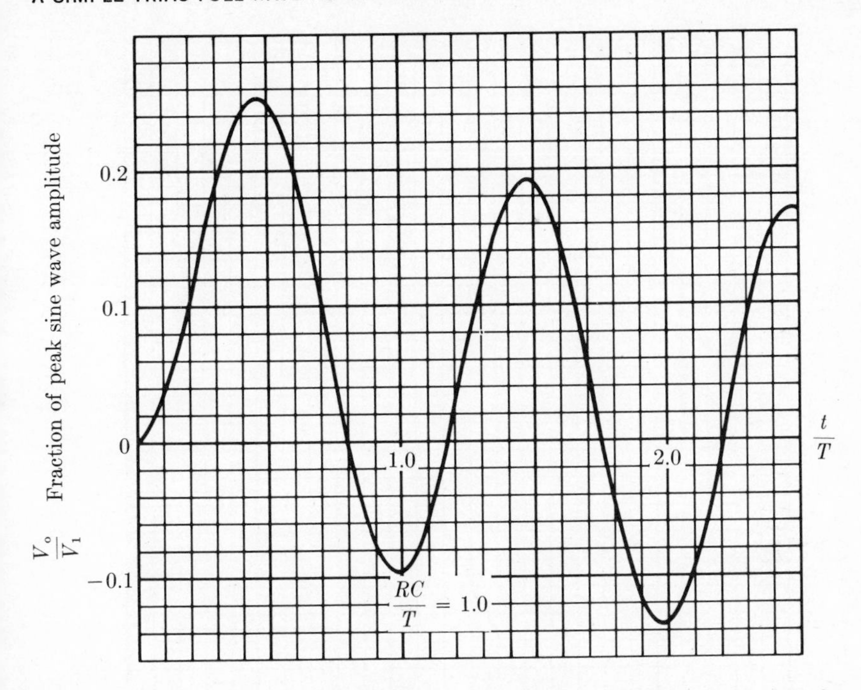

Fig. 8-28 The waveform appearing across C in Fig. 8-25 if firing does not take place. The circuit is assumed to start with C discharged at $t = 0$.

energy that would be transmitted to the load if the switching occurred at the beginning of the cycle. As may be seen from the curve, more than 99 percent of the possible energy is delivered if the switching occurs as late as 5 percent of the period. Also the output from the circuit is reduced substantially to zero if the firing point is between 45 and 50 percent of the period.

A circuit like that shown in Fig. 8-25 has a rather odd hysteresis effect: if the RC time constant is slowly increased, the time of switching increases until a point is reached where V_o does not rise sufficiently to cause switching at any time within the cycle. At this point the power delivered to the load is actually zero. Reversing the process and decreasing the time constant will not turn the circuit back on until RC is decreased a great deal from the value that caused the circuit to stop conduction. Therefore the observed phenomenon is that the bulb can be dimmed slowly to extinction, but trying to turn it back on does nothing until suddenly it turns on to considerable brightness; then the bulb may be dimmed slowly back down again.

To discover what happens to cause this hysteresis phenomenon we must look at the waveform given by Eq. (8-26) for a longer time, as in Fig. 8-28. This figure is similar to Fig. 8-26, except the scales are changed

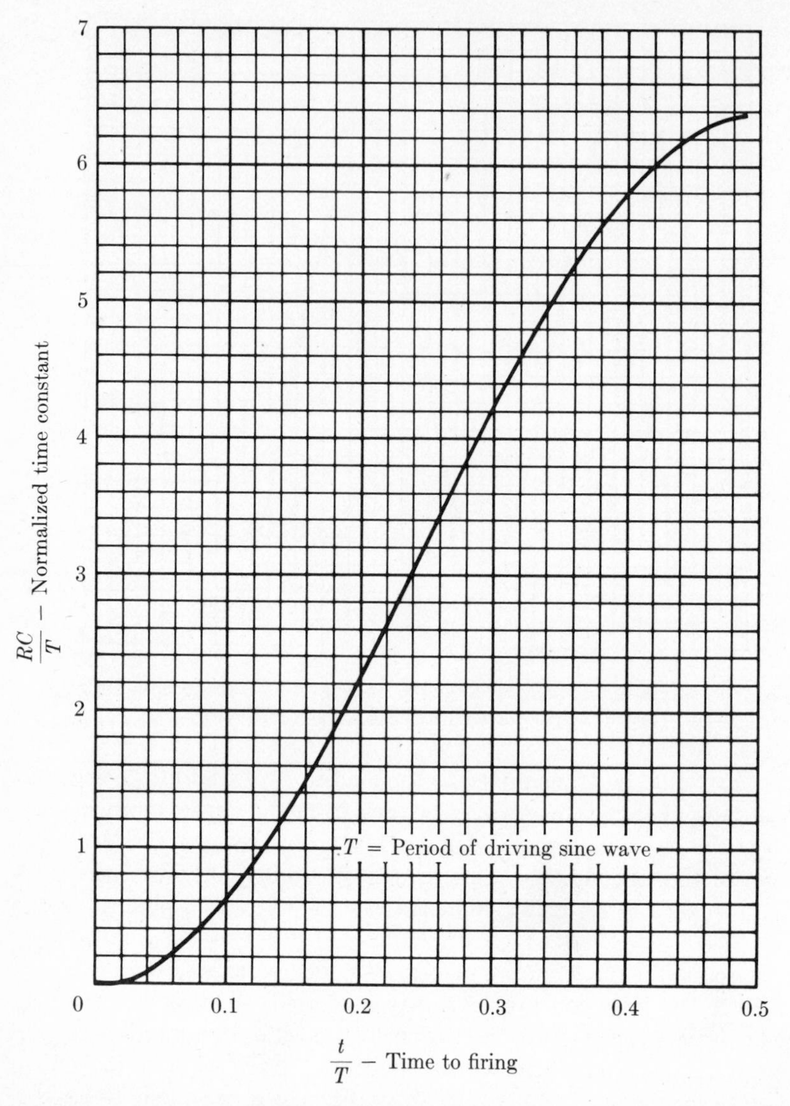

Fig. 8-29 Firing angle versus normalized time constant. The trigger point is assumed to be 0.048 of the peak value of the input sine wave. This is the case for V = 165 V peak and a switching point of 8 V.

so that several cycles of operation can be plotted. The curve is plotted for the time constant equal to the period of the sine wave, but the conclusions to be drawn are independent of the time constant chosen. The important thing to be noted from the curve is that the first peak has a considerably larger amplitude than succeeding peaks. Suppose the first peak were just large enough to fire the switch, then a small increase in the time constant would cause the circuit to stop firing. To see quantitatively what will occur we can replot the data given in Fig. 8-26 to give the time for triggering as a function of the normalized time constant. This curve (Fig. 8-29) is plotted assuming that the trigger point is 0.048 of the sine-wave peak amplitude as before. The circuit will cease to trigger when $\tau/T > 6.4$. Once the circuit ceases firing, the peak voltage across the capacitor quickly assumes the steady-state value:

$$V_o = \frac{V_1}{1 + j2\pi fRC} = \frac{V_1}{1 + j2\pi\tau'} \tag{8-27}$$

To make the circuit resume firing, the value of τ' must be reduced until the peak value of the sine wave across C exceeds the switching voltage of the bilateral switch. As before, this corresponds to a ratio V_o/V_1 equal to 0.048. Solving Eq. (8-27) for τ' gives $\tau' = 3.3$. Therefore the circuit time constant must be reduced from $\tau' = 6.4$ to $\tau' = 3.3$ to make the circuit start again. Reference to Fig. 8-29 shows that once the circuit begins to operate again the firing time will be about 0.25; therefore the hysteresis amounts to about a quarter of a cycle. There does not seem to be any really simple way to get rid of this problem, although restricting the range of τ' to less than six would prevent the circuit misfiring. A partial solution is to use two integrating circuits in series in place of the single RC shown.

A somewhat different problem arises in a half-wave circuit. Such a circuit could be similar to that shown in Fig. 8-25 with the bilateral switch replaced with a unilateral switch, and the TRIAC replaced with an SCR. In this situation the capacitor has a chance to charge during the unused cycle and does not begin the timing operation with zero charge. Under these conditions the circuit may trigger at a submultiple of the line frequency if the RC product is increased past some critical value. The explanation of this effect is left for Prob. 8-9, but an essential clue is in Fig. 8-28. This effect may be completely eliminated by placing a diode across C so that it cannot change in the negative direction during the unused half cycle. With the diode the current always starts the timing cycle with zero charge.

PROBLEMS

8-1. This is a unijunction oscillator with a rather large value of R_2 which will somewhat modify the waveforms shown in Fig. 8-23. Assume that the unijunction has the characteristics shown in figure 8-23 and that the resistance looking into the emitter equals -1 kΩ at $I_E = 8$ mA and $V_E = 4$ V.

(a) What value of R_1 is required to give a frequency of 25 Hz?

(b) Sketch and label the waveforms appearing at the emitter and across R_2.

(c) Assuming that $R_1 \gg R_2$ show that the UJT switches off at that capacitor voltage which causes the load line to become tangent to the negative resistance region. (By load line here is meant the line on the V_E, I_E characteristic beginning at the capacitor voltage and with slope $-1/1\ \text{k}\Omega$. This line mainly determines the emitter current from instant to instant.)

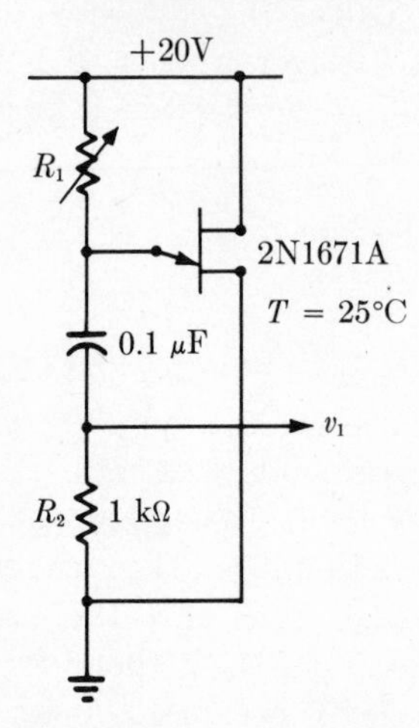

8-2. The V-I characteristic shown is an idealized tunnel diode. The circuit can be made mono-, bi-, or astable by changing the bias conditions. On a sketch of the V-I characteristic show:

(a) The quiescent load line and operating point ($I_1 = 0$).

(b) The operating point at the moment a pulse $I_1 = 4$ mA is applied.

(c) The remaining locus of operation after the pulse is removed. (Assume the pulse is so short that the initial inductor current is unchanged.)

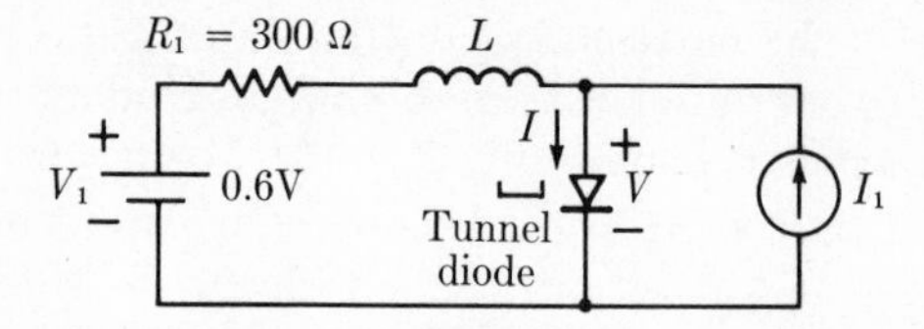

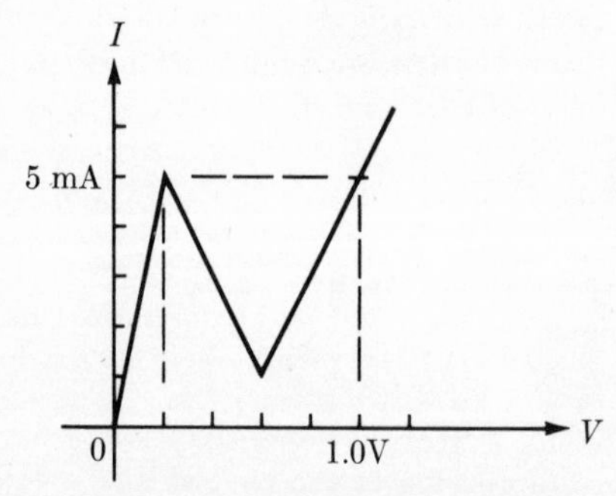

Idealized tunnel diode characteristics

(d) What is the minimum value of V_1 that will make the circuit bistable?
(e) Is the circuit mono-, bi-, or astable if $V_1 = 0.5$ V and $R_1 = 0$? Sketch the output waveform from the circuit giving all pertinent values.

8-3. The circuit for an oscillator using four-layer diodes (or any other current-stable, two-terminal negative resistance device) is shown. In operation one diode will conduct and the other will be off. The voltage across the OFF diode rises as C charges until the peak-point voltage is reached. At this voltage the OFF diode begins to conduct, the anode of the ON diode is forced negative and the diode first ON goes OFF. The cycle then begins again with the ON and OFF diodes reversed.
Calculate, sketch, and label the voltage across $D1$ and across the capacitor.

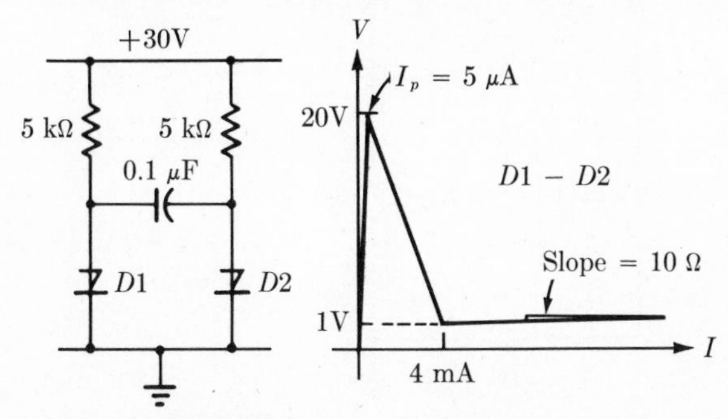

8-4. Using the TRIAC circuit of Fig. 8-25, find the value of R_1 to give firing of the TRIAC at 90° and 270°. Assume as in the text example that the bilateral switch fires at ±8 V. Sketch the waveforms appearing across the capacitor and the load. If the load is 5 ohms, what power is dissipated in it?

8-5. The circuit is a single NRD oscillator circuit which has a natural frquency near 1 kHz. The diode $D2$ allows easy introduction of a synchronizing input v_1 without interfering with the normal operation of the circuit.
(a) For $v_1 = 0$ compute and sketch the output $v_2(t)$.
(b) What is the minimum peak value of the pulse v_1 that will synchronize the circuit to run at exactly 1 kHz? What is the minimum value of pulse that will cause the circuit to run at 1.5 kHz?

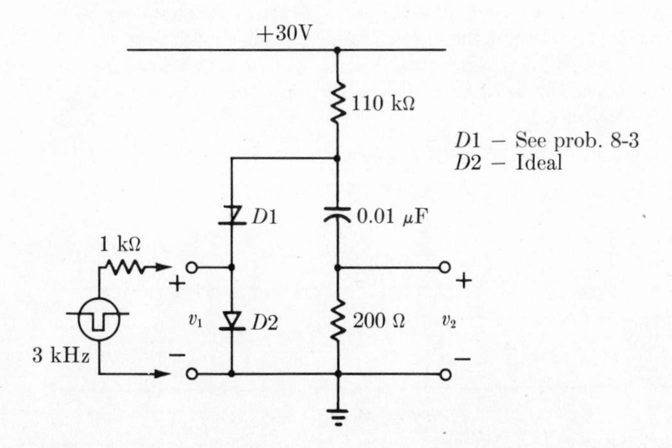

8-6. (a) The circuit with $V_1 = 20$ V is an astable unijunction oscillator. ($D2$ never conducts with $V_1 = 20$ V.) Assume the same unijunction characteristics as in Prob. 8-1. The OFF portion of the cycle is timed by $R2$ and C while the ON portion is timed by $R1$ and C. Calculate and sketch the waveforms appearing at the emitter and across $D1$.

(b) The circuit may be made monostable by reducing V_1. What is the maximum value of V_1 allowed to keep the circuit monostable? Show a way of triggering the circuit with a positive pulse.

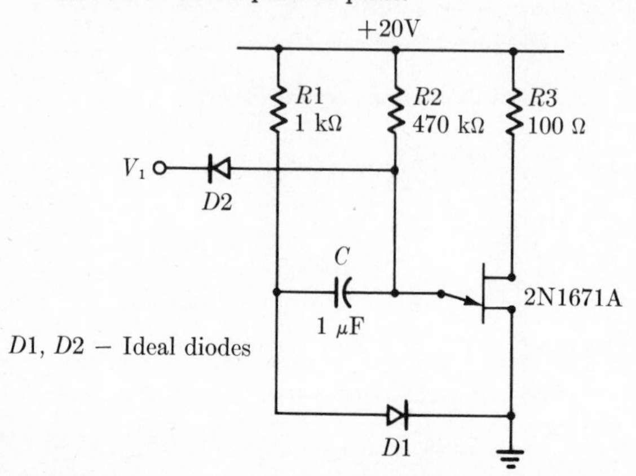

8-7. The circuit shown is really a combination of circuits whose purpose is to develop an output pulse synchronized with the line frequency but lower in frequency. The transistor operates as a clipper to produce a reasonably square wave at its collector. The square wave is the input to a step charging rectifier comprising $D2$, $D3$, and the two capacitors. The voltage across the 0.1 μF resembles a staircase. When one of the steps exceeds the peak point of the silicon unilateral switch (see Fig. 8-9) it becomes a low impedance and discharges the capacitor producing an output pulse. Thus the frequency is divided by the number of steps required to reach the peak-point voltage. A suitable time to begin analysis would be just after the capacitor was discharged.

(a) Sketch the waveforms appearing at A and B. Indicate the maximum and minimum voltages and give an order of magnitude estimate for the rise and fall times at B. Note that these times are quite short compared to the $16\frac{2}{3}$ ms period. Would the circuit operate properly without $D1$? Why?

(b) Compute the waveform appearing at C. You may assume the rise time of this voltage at the beginning of each step has the same order of magnitude as computed in (a).

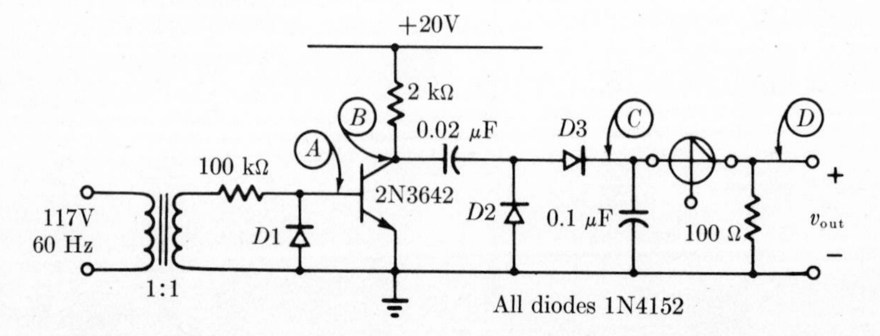

(c) Calculate the output waveform D and sketch all the waveforms to show their time relationship.
(d) Show how the addition of one variable resistor may be used to alter the number of cycles required to reach the peak and hence the division ratio.

8-8. Calculate the details of the waveform shown in Fig. 8-22. Assume $I_p = 10$ mA and the other parameters are as given in Table 8-1 for a germanium tunnel diode.

8-9. Assume that the TRIAC $Q2$ in Fig. 8-25 is replaced by an SCR with the anode end connected to the load, and that the bilateral switch $Q1$ is replaced by a unilateral switch (Fig. 8-9). Assume the SCR has conducted for the positive half cycle previous to $t = 0$ and that C begins the new cycle (which is negative) with zero charge. Explain how the circuit might not fire until the succeeding cycle, that is, why the output might occur at a 30 Hz rate rather than at 60 Hz. The waveform of Fig. 8-28 is useful. Show how an added diode removes the problem.

9

Lines and Pulse-Forming Networks

All the preceding discussion has been primarily concerned with lumped-element circuits associated with semiconductor devices, rather than with distributed systems such as transmission lines and wave guides. This chapter, therefore, constitutes somewhat of a departure, although even here we shall be interested only in the terminal effects of a length of transmission line, i.e., only in its use as a two- or four-terminal network.

In working with circuits for the generation of nonsinusoidal waves, and in particular short rectangular pulses, one finds sooner or later that some jobs can be done better by means of a length of transmission line or an equivalent network. Not only are there special pulse-generating circuits in which the line is the principal element, but also the multivibrator and the blocking oscillator can be improved by the addition of a length of line in the proper manner. In general, it might be said that line elements are most useful when (1) maximum precision of pulse duration is required or (2) pulses of large current or voltage amplitude must be generated from moderate-sized components and from a power source of limited capacity. Knowledge of the behavior of transmission lines to pulses is also useful in circuits involving digital components because the interconnections from one circuit to another are, of course, transmission lines in one sense. Therefore, reflections which occur on these transmission lines should be understood to avoid arrangements producing false responses.

9-1 TRANSIENT BEHAVIOR OF TRANSMISSION LINES

In order to understand the terminal characteristics of line elements as used in pulse circuits, one must have a basic understanding of the transient behavior of transmission lines. This behavior can be adequately summarized for our purposes by the detailed considerations of three cases, each of which leads directly to some useful applications.

Case I Figure 9-1 shows a dc voltage source V_0 in series with a resistor R_L, a switch S, and a length of lossless transmission line, *open circuited* at the far or "receiving" end. We can confine our attention to lossless lines because lines with appreciable loss have little value in the circuits to be considered. At a time designated as t_1 the switch is closed. A wave will travel down the line, be reflected, travel back, and in some cases be reflected again. Since the far end of the line is open circuited, the sum of the incident and the reflected waves at the receiving end must add to zero current and double voltage. Therefore, the reflected wave has a magnitude equal to the incident wave but an opposite sign for the current. The exact details for the case where the source resistance equals the characteristic resistance of the line (that is, $R_L = R_0$) are shown in Fig. 9-2.

The waveforms of current and voltage and the effective input resistance R_{in} are shown in Fig. 9-3 as functions of time. The duration of the pulse is equal to twice the one-way travel time of the wave on the line. The wave moves on the line at the group velocity, which in the case of a lossless line is $\nu = (lc)^{-1/2}$ where l and c are the inductance and capacitance per unit length of line. If the line dielectric is air, the velocity is the same as that of light, 3×10^{10} cm/sec. Otherwise the velocity ν is reduced by

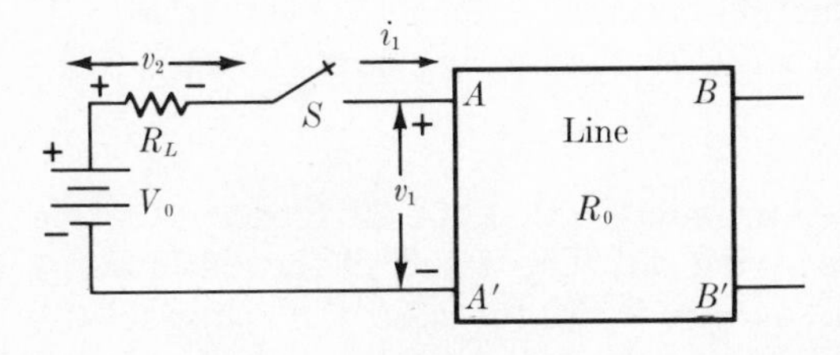

Fig. 9-1 Case I. Charging a length of transmission line in series with load resistor R_L. The line has a length D, and inductance l per unit length, a capacitance c per unit length, and hence a characteristic resistance $R_0 = \sqrt{l/c}$.

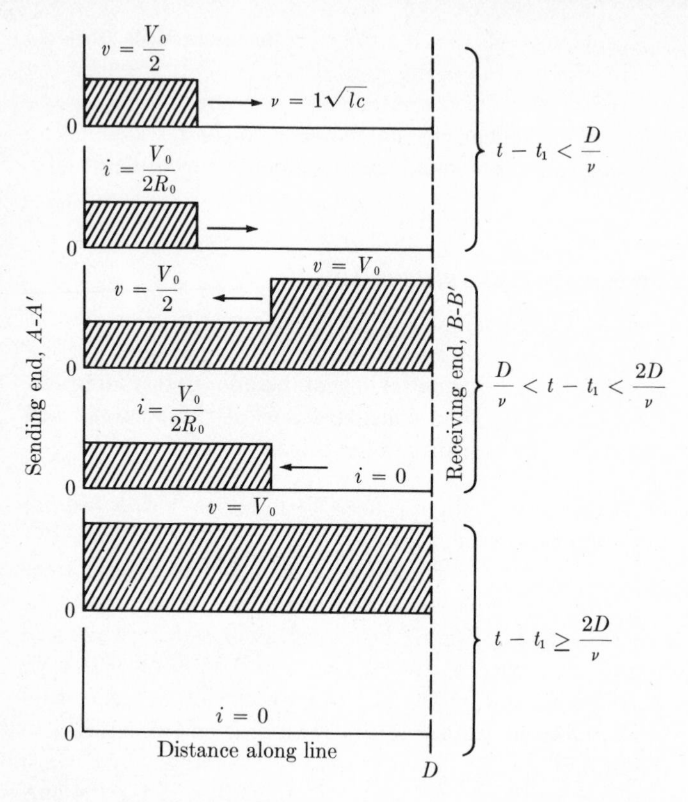

Fig. 9-2 Distribution of voltage and current along an open-circuited line where $R_L = R_0$. Arrows show direction of wavefront.

the square root of the relative dielectric constant. For instance, with polyethylene cable the velocity is about 0.6 times the velocity of light, i.e., about 650 ft/μs.

It will be further noted that after the reflected wave reaches the sending end there are no more reflections, provided that $R_L = R_0$. If this condition is not met, there will be a reflection from the sending end, and indeed many reflections will ensue before the final steady condition is achieved in which the line voltage is everywhere V_0 and the line current is zero. An example of this situation occurs in connection with Case III, during the period between pulses when the line is charging from a high-resistance source.

Case II The circuit in Fig. 9-4 is similar to that of Case I, except the line is now *short circuited* at the receiving end. The nature of the reflections is shown in Fig. 9-5 for the case of $R_L = R_0$. In this case, the voltage of the

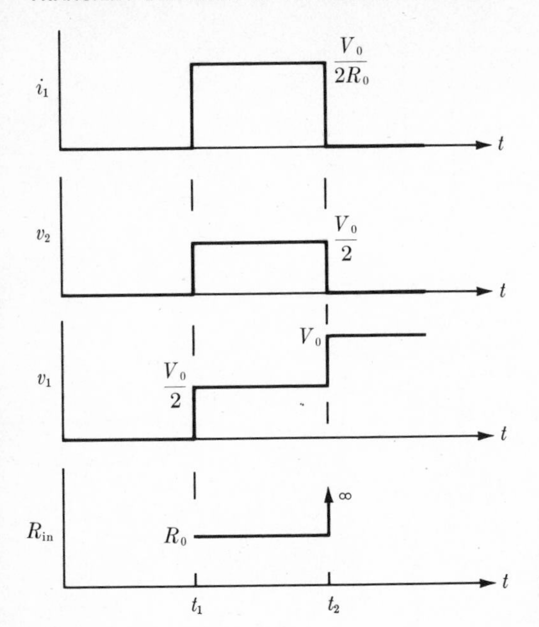

Fig. 9-3 Sending-end voltage, current, and input resistance R_{in} for an open-circuited line where $R_L = R_0$.

reflected wave is the negative of the incident wave to fulfill the end condition $V = 0$ at $d = D$; therefore, the current in the reflected wave is equal to the current in the incident wave. The corresponding waveforms of circuit voltage and current are shown in Fig. 9-6, together with the effective input resistance of the line as a function of time.

In both Cases I and II the amplitude of the voltage pulse is only half that of the voltage source. Similarly, the power divides between R_L and the line during the pulse. Thus these two cases are not used for high-power pulses. They are, however, nonetheless useful in low-power applications, and an example is given in Prob. 9-2. Case III will be seen to provide a better solution where high power is required.

The feature of special importance in both Case I and Case II is that the line itself has the unique property of being a two-terminal network element that acts like a resistance R_0 for a brief time interval $2D/v$, following which it suddenly becomes an open circuit (Case I) or a short circuit (Case II). We can use this property in applying an open- or short-

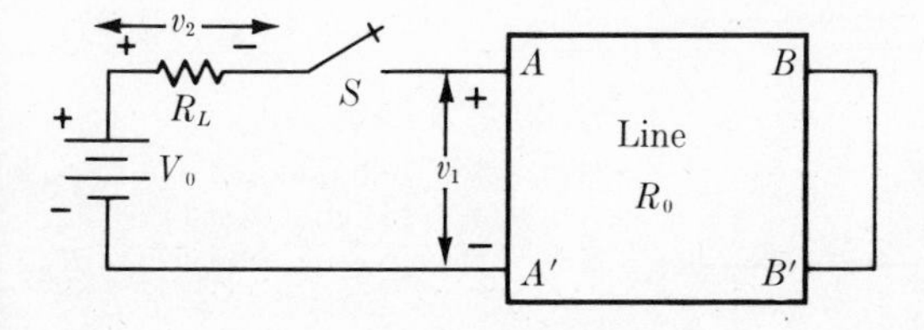

Fig. 9-4 Case II. Charging a short-circuited length of transmission line in series with a load resistor $R_L = R_0$.

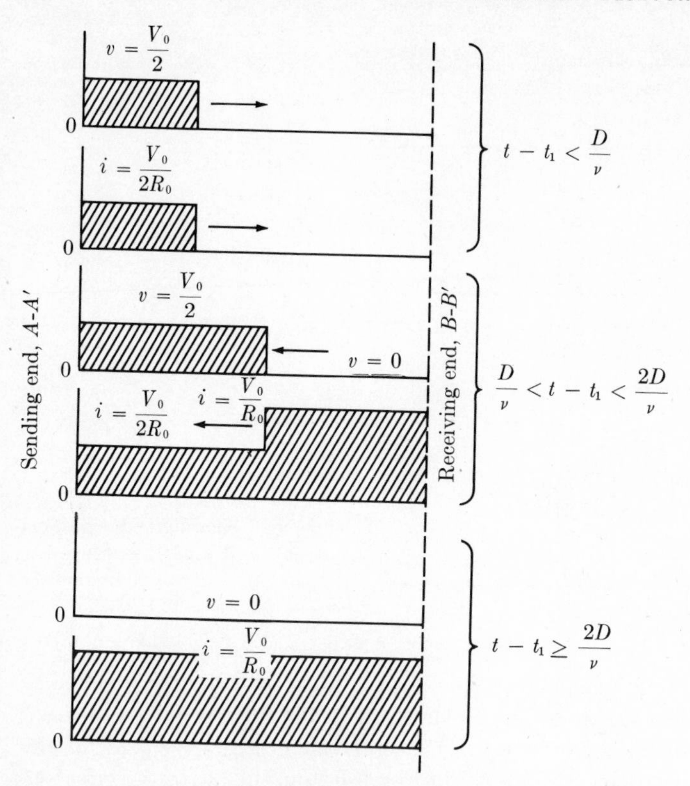

Fig. 9-5 Distribution of voltage and current along a short-circuited line where $R_L = R_0$.

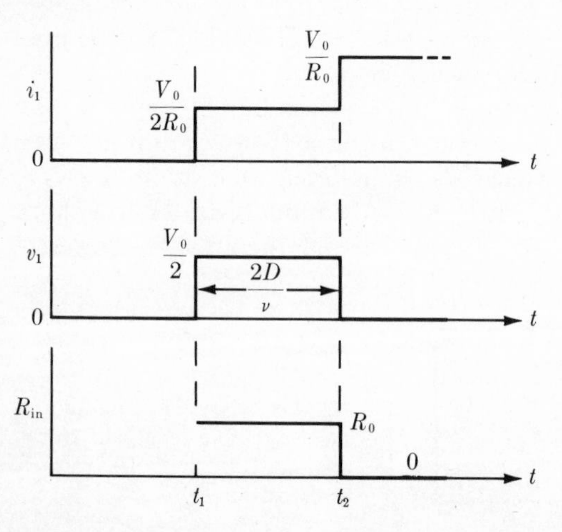

Fig. 9-6 Sending-end voltage, current, and input resistance R_{in} for short-circuited line where $R_L = R_0$.

circuited line to the multivibrator and the blocking oscillator in a manner described in Sec. 9-2.

Case III In the circuit as shown in Fig. 9-7 an open-circuited line is used as in Case I, but now the line is charged slowly through the large series resistance R_s, and the pulse is formed when the switch closes at time t_1. The resistor R_L is the load and is purposely made equal to the R_0 of the line.

Sometimes the load is placed directly in series with the line terminals; this makes no practical difference if R_s is so large that the charging current is negligible in comparison with the discharge pulse of current. An alternative arrangement has V_0 and R_s connected to the right end of the line. If R_s is much greater than R_0, the line is still essentially open circuited at the right-hand end.

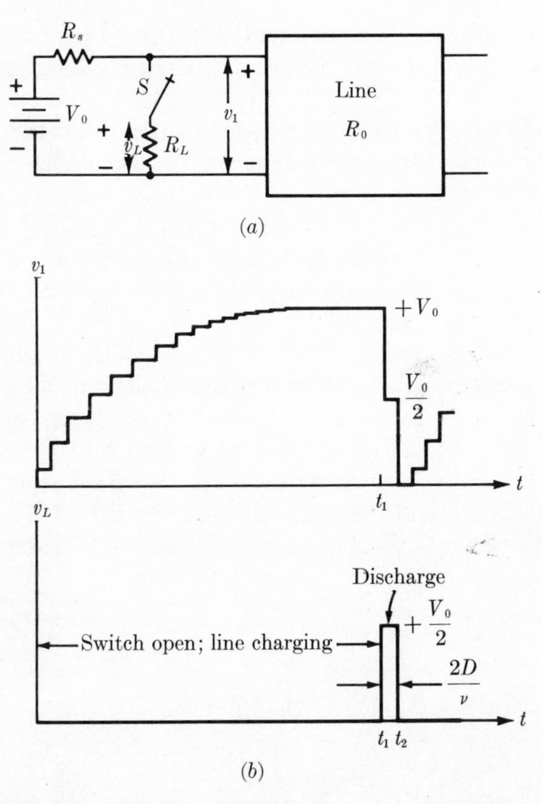

Fig. 9-7 Case III. Charging an open-circuited line through a large series resistance $R_s \gg R_0$, and discharging into a matched load $R_L = R_0$. (a) Circuit. (b) Waveforms.

Notice that during the charging period there will be repeated reflections occurring on the line. This results from the fact that the generator end and the receiving end are mismatched. The amplitude of the reflections diminishes gradually, however, and the line is eventually charged to the full battery voltage as in Case I.

During the discharge, however, the amplitude of the pulse across the line terminals is only one-half the charged level. Also, the duration of the pulse is twice the one-way travel time of the line. This can be interpreted in terms of two waves existing simultaneously on the line, one traveling to the left and one to the right, each with amplitude $V_0/2$. After the line has been fully charged, both ends of the line are essentially open circuited and these two waves are continuously reflected. There is no net energy flow to or from the line. But when the left end is connected to R_L by means of the switch, the wave traveling to the left is no longer reflected. Instead, it produces a voltage $V_0/2$ in R_L. The wave traveling to the right must first be reflected before it can reach the load, and the last element must travel a distance $2D$ before it reaches R_L. Hence, the pulse duration is $2D/\nu$.

The circuit of Case III has a practical advantage over the other two in high-power situations inasmuch as the source of power V_0 does not have to supply a large pulse of current. In addition, there is no unwanted second pulse of opposite polarity when the switch is opened as there is with the circuit of Case II.[1] On the other hand, the slow charging time for Case III may be a drawback. The smaller the charging current, the longer the time required to accomplish the charging.

In the charging of the line of Case III, the line voltage at the input terminals approaches an exponential with time constant equal to R_sC_t, where $C_t = cD$, and is the total line capacitance. Examination of a set of points on the step wave of line voltage is a convenient way to determine the time function that governs the voltage at these points. The derivation is as follows: take the points A, B, and C as suitable (see Fig. 9-8). Note that all the direct and reflected waves are shown, together with their summation, which is v_1. At $t = 0$

$$v_1(0^+) = V_0 \frac{R_0}{R_s + R_0} \tag{9-1}$$

At $t = T$ (point A, $T \triangleq$ the two-way delay of the line)

$$v_1(A) = 2V_0 \frac{R_0}{R_s + R_0} \tag{9-2}$$

[1] A similar second pulse occurs with Case I, except that here the switch and battery V_0 are usually a rectangular waveform generator, and the second pulse occurs when V_0 returns to zero.

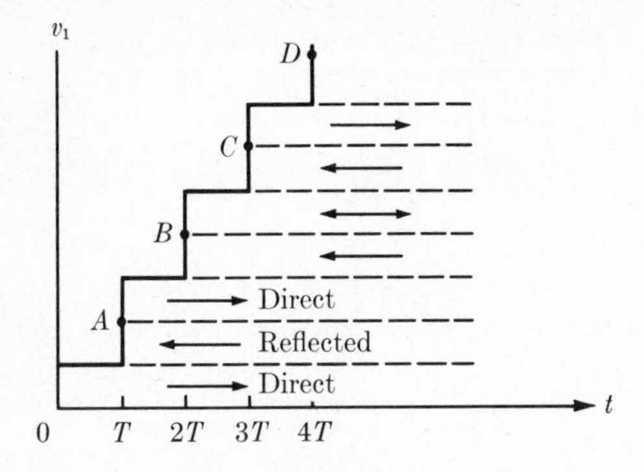

Fig. 9-8 Details of the charging waveform.

To determine the value of the next direct wave, we must determine the effective driving voltage for the line at $t = T$. The effective voltage is

$$V_0 - v_1(A) = V_0\left(1 - \frac{2R_0}{R_s + R_0}\right) = V_0\left(\frac{R_s - R_0}{R_s + R_0}\right) \triangleq V_0\Gamma \qquad (9\text{-}3)$$

The value of the next direct wave is given by Eq. (9-1), and the voltage difference $v_1(B) - v_1(A)$ is given by Eq. (9-2), both with V_0 replaced by $V_0 - v_1(A)$.

$$v_1(B) - v_1(A) = 2V_0\Gamma \frac{R_0}{R_s + R_0} \qquad (9\text{-}4)$$

Therefore the effective voltage driving the line at $t = 2T$ (point B) is

$$\begin{aligned} V_0 - v_1(B) &= V_0\left(1 - \frac{2R_0}{R_s + R_0} - 2\Gamma \frac{R_0}{R_s + R_0}\right) \\ &= V_0\Gamma\left(1 - \frac{2R_0}{R_s + R_0}\right) = V_0\Gamma^2 \end{aligned} \qquad (9\text{-}5)$$

Similarly, at later times

$$V_0 - v_1(C) = V_0\Gamma^3, \cdots, V_0 - v_1(nT) = V_0\Gamma^n \qquad (n = 1, 2, 3, \cdots) \qquad (9\text{-}6)$$

Since $n = t/T$

$$V_0 - v_1 = V_0\Gamma^{t/T} = V_0\epsilon^{[(\ln \Gamma)/T]t} = V_0\epsilon^{[-(\ln 1/\Gamma)/T]t} \qquad (9\text{-}7)$$

$$v_1 = V_0[(1 - \epsilon^{[-(\ln 1/\Gamma)/T]t}] \qquad (9\text{-}8)$$

The exponential behavior of v_1 with time is thus established, and it remains only to evaluate the coefficient of t.

$$\frac{1}{\Gamma} = \frac{R_s + R_0}{R_s - R_0} = \frac{1 + R_0/R_s}{1 - R_0/R_s} \qquad (9\text{-}9)$$

$$\ln \frac{1}{\Gamma} = \ln\left(1 + \frac{R_0}{R_s}\right) - \ln\left(1 - \frac{R_0}{R_s}\right) \qquad (9\text{-}10)$$

We can expand the expressions for the logarithms in a series as follows:

$$\ln\frac{1}{\Gamma} = \left[\frac{R_0}{R_s} - \frac{1}{2!}\left(\frac{R_0}{R_s}\right)^2 + \cdots\right] - \left[-\frac{R_0}{R_s} - \frac{1}{2!}\left(\frac{R_0}{R_s}\right)^2 - \cdots\right] \tag{9-11}$$

$$\ln\frac{1}{\Gamma} \approx 2\frac{R_0}{R_s} \qquad (\text{for } R_s \gg R_0) \tag{9-12}$$

Now $R_0 = \sqrt{l/c}$, $T = 2D/\nu = 2D\sqrt{lc}$, $C_t = cD$, so that

$$\frac{\ln(1/\Gamma)}{T} \approx \frac{2R_0}{R_sT} = \frac{2\sqrt{lc}}{R_s(2D\sqrt{lc})} = \frac{1}{R_sC_t} \tag{9-13}$$

Therefore

$$v_1 \approx V_0(1 - \epsilon^{-t/R_sC_t}) \tag{9-14}$$

9-2 APPLICATIONS

A simple oscillator using a transmission line as the timing element is shown in Fig. 9-9. In this circuit the transmission line is used to couple pulses occurring at the collector of the transistor Q to the base of the transistor with a time delay equal to the one-way travel time on the transmission line. If the circuit is turned on at $t = 0$, the base voltage of the transistor will be zero because the line is initially discharged. The collector voltage will be equal to 5 V divided by the voltage divider comprising R_L and the input resistance of the transmission line. Therefore the collector voltage initially will be 2.5 V, and a wave of this amplitude travels from the right end of the transmission line toward the base. The transistor remains off for the length of time required for this wave to reach the base or a time $T/2 = D/\nu$. When the wave reaches the transistor base, the transistor is turned on and saturated; therefore the collector voltage falls to very nearly zero volts. Again a wave starts at the right end of the transmission

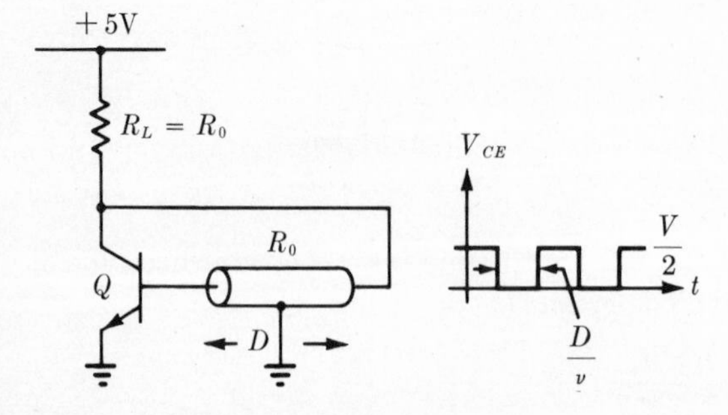

Fig. 9-9 A simple oscillator using a transmission line to determine the frequency.

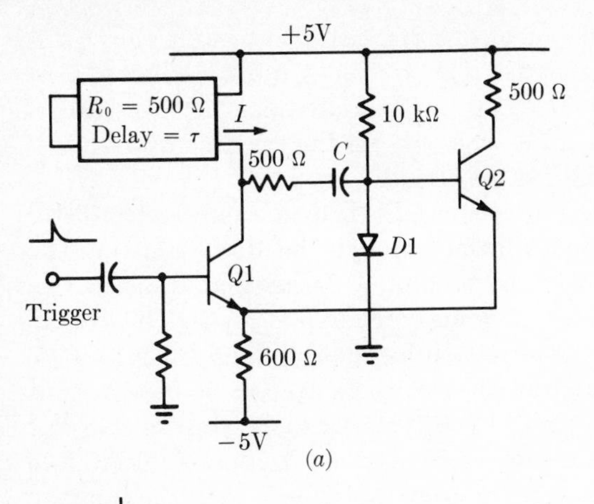

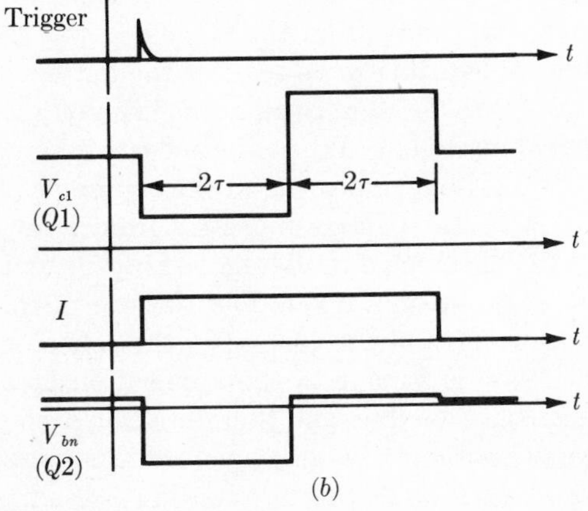

Fig. 9-10 (*a*) A monostable multivibrator using a transmission line to determine the length of the pulse produced. (*b*) The waveforms produced by the circuit.

line, this time of zero amplitude, and travels toward the transistor base. The transistor will remain in saturation with the base current being supplied by the energy stored in the transmission line until the wave starting at the right end, and of zero amplitude, has reached the base. Therefore, the on-and-off times for the transistor are equal, and both times are determined by the one-way travel time of the transmission line. The frequency generated by the oscillator is therefore determined completely by the delay time of the transmission line, and is equal to $\nu/2D$.

The circuits of Cases I and II are only occasionally used in the direct generation of pulses, but they display the valuable property of the line as a circuit element which can function as a delayed open or short circuit. This property permits the use of a line as an auxiliary element in the multivibrator or blocking oscillator to improve the timing precision. An example of such a circuit is shown in Fig. 9-10, where a short-circuited transmission line in the collector of one transistor in a multivibrator is used to determine the pulse length. This circuit is quite similar to that shown in Fig. 5-24, except that the transistors are here NPN instead of PNP. The capacitor in Fig. 9-10 is used only for the function of coupling; it does not provide the timing for the circuit.

In the quiescent state the transistor $Q1$ is held off, and the transistor $Q2$ is turned on by base current flowing through the 10 kΩ resistor. The positive trigger pulse applied to the base of $Q1$ turns that transistor on and the collector voltage drops to a value equal to $5 - R_0 I_{c1}$. The drop in the collector voltage of $Q1$ is transmitted by the coupling capacitor C to the base of $Q2$ and turns that transistor off. The decreasing base current in $Q2$ serves to increase the current in $Q1$, thereby increasing its collector current so that the circuit is regenerative once conduction is initiated in $Q1$ by the trigger signal. The voltage drop appearing across the input to the transmission line starts a wave traveling from the right end to the short-circuited left end of the line. When this wave reaches the left end, a new wave is generated, which travels to the right causing the line voltage to become zero and the line current to double. When the reflected wave reaches the right end of the line, the voltage appearing at the terminals is suddenly reduced to zero, or the collector voltage increases from a low value to approximately five volts. This change in voltage is transmitted through C to the base of $Q2$, thereby turning it on. The turn-on of $Q2$ turns $Q1$ off by the regenerative process, and the current in the collector of $Q1$ decreases to zero. This decrease is similar to the application of a negative step of current to the transmission line, and therefore causes the transmission-line voltage to reverse as shown in the waveform diagram. This new voltage causes a current to flow in the terminating 500-ohm resistance, the coupling capacitor C, and the clamping diode $D1$. This current starts a new wave traveling down the transmission line, which is in a direction to reduce the transmission-line current to zero. The reflection of this wave from the short-circuited end generates a new wave traveling from left to right, which completely discharges (dissipates the energy of the transmission line) when the wave reaches the right-end terminals. At this time, the circuit operation is completed and the reactive elements are restored to the initial energy conditions prior to triggering. The output pulse length of the multivibrator in this case has been determined by the two-way travel time in the transmission line and is quite independent of the transistor parameters and other passive elements in the circuit.

Transmission lines can be used in either the open-circuited or short-circuited condition to modify other circuits that use RC combinations to produce the timing. An open-circuited line would replace the capacitor, or a short-circuited line the resistor. In either case, the R_0 of the line should approximate the resistance of the circuit at the point of connection. If the line resistance is too high, it can be shunted by a resistor. Most coaxial lines have too low a resistance, but some of the special lines and artificial networks available have a higher impedance, say 1 kΩ.

The circuit of Case III finds its principal application in a class of devices known as *line pulsers*, i.e., circuits which generate high-power pulses of a time duration determined by a length of line, and that are used to supply power to some electron device, for example, a magnetron or klystron high-frequency oscillator. These line pulsers are particularly effective for very short pulse durations and for very high power levels. For high power use, two changes are made in the circuit of Fig. 9-7. First, a *pulse-forming network* or artificial lumped element line replaces the continuous line, especially if the pulse duration would require an excessively long line. Second, the charging resistance R_s is not satisfactory, because, as can be demonstrated, an amount of energy equal to that stored by the line is dissipated in the resistor. Hence, in its place is usually found an inductance L_s. For *dc resonance charging*, L_s is chosen so that for a period T between pulses $T = \pi \sqrt{L_s C_t}$, where C_t is the total line capacitance as measured with a direct current. The resulting waveform of v_1 is in Fig. 9-11. Not only does dc resonance charging raise the efficiency of the charging process from 50 percent for resistance charging to a value as high as 92 to 96 percent for an inductive circuit with a Q of from 10 to 20, but in addition, the pulse-forming line is charged to almost twice the voltage. That is, instead of the line being charged to a value V_0, as in Fig. 9-7, it is charged to $2V_0$ as indicated in Fig. 9-11 for the case of no loss in the inductance. If the inductance has a finite Q, as is the practical case, the voltage attained is slightly less.

$$V_{\max} = V_0(1 + \epsilon^{-\pi/2Q}) \tag{9-15}$$

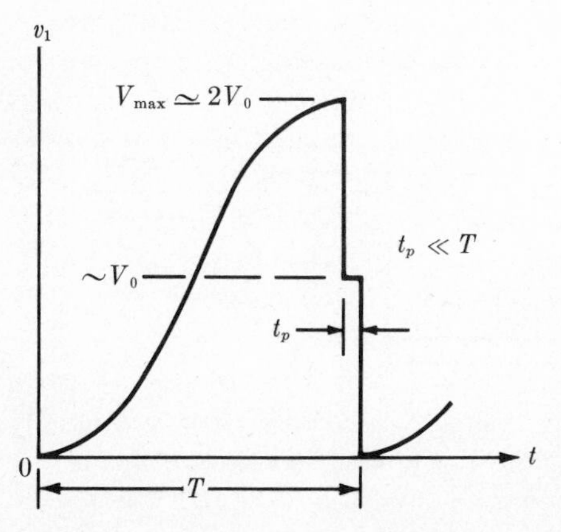

Fig. 9-11 The waveform appearing in the circuit of Fig. 9-7 when R_s is replaced by an inductor.

and the efficiency (energy stored in line/energy from source) γ is

$$\begin{aligned}\gamma &= 50\% \qquad \text{(with resistance } R_s\text{)} \\ &= (1 - \pi/4Q)(100\%) \qquad \text{(with } L_s\text{)}\end{aligned} \tag{9-16}$$

The efficiency is about 92 percent with a Q of only 10.

There are other schemes for high-efficiency charging of the line together with a wide variety of techniques pertaining to the power supply, the switch, lumped equivalents for the line, etc. Except for the last topic, we shall not explore these further. There are numerous references concerning the current techniques for high-power, line-type pulsers to aid those persons who may later need to specialize in the subject.

9-3 TRANSMISSION LINES USED TO PRODUCE PULSE DELAY

In many cases, transmission lines (usually called delay lines) are used to delay the application of one pulse with respect to another pulse. The use of delay lines for this purpose is very suitable when the delays required are small—on the order of a few nanoseconds to a few microseconds—and where the amount of delay must be quite constant over a wide range of operating conditions.

A typical application is shown in Fig. 9-12 where two pulse generators are shown driving an AND gate. The AND gate provides an output when both of its inputs are in logic state *1* or high, but produces no output when either input is low or in the *0* state. The purpose of the circuit is to produce an output pulse of width equal to the generator two but gated by generator one. Let us assume that the beginning of the output of either generator occurs simultaneously or very nearly so. Therefore, at t_1 the output of generator one is zero, the output of generator two is one; therefore the output from the total circuit should be zero. At exactly the time

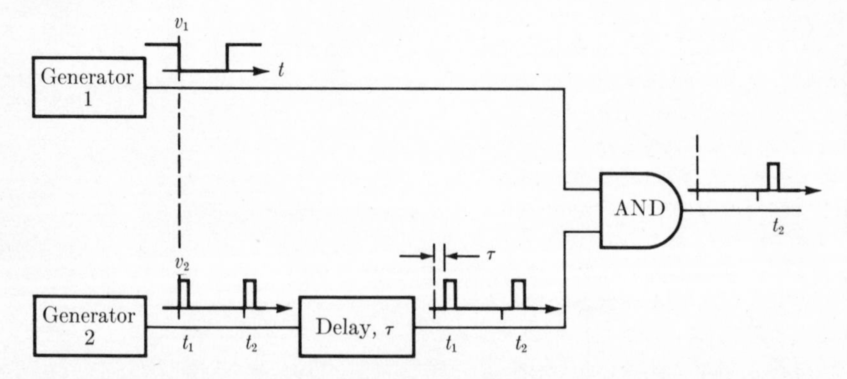

Fig. 9-12 An illustration of the use of a delay line to insure proper operation of an AND gate.

t_1, however, there is ambiguity in the output from the gate because of the finite rise and fall times from the two generators. Thus it is possible that a small sliver of an output pulse will occur at time t_1 just due to the difference in the rise and fall times of the two generators. This narrow output might be very undesirable because it could falsely trigger a circuit following the AND gate. To prevent this false output from occurring, the delay line with delay τ has been added in series with the output of generator two. The purpose of this delay line is to ensure that the pulse generated by the second generator arrives after the gating pulse has had time to occur and inhibit the AND gate. Therefore, there is no possibility that the first pulse appearing at time t_1 will appear at the output of the gate. At t_2 a second pulse occurs in generator two, but at that time generator one is also in a high state, and an output occurs from the gate at $t_2 + \tau$.

Situations such as the preceding occur in digital computers and lead to timing problems if the delay times are not properly employed. The delay line shown as τ might be a separately added group of circuits purposely employed to increase the time delay, or it might be just the natural delay of the transmission paths on the printed circuit board or in the interconnecting wiring of the computer. In either case, if the delay time is very long with respect to the rise and fall times of the pulses employed in the circuits, attention should be paid to the correct termination of the line at least at one end. Generally speaking, the input impedance of a gate such as the AND circuit employed in Fig. 9-12 is not the same in both the low and the high states of the input. Therefore, the gate cannot always terminate the transmission line correctly at the receiving end of the line. Consequently, the pulse generator should at least approximate the R_0 of the delay line so that the line is terminated at least at one end. Under these conditions, a pulse initiated by generator two will travel down the line, be reflected at the gate, travel down the line again in the opposite direction, but be absorbed in the resistance of the generator. Therefore, the reflection is not harmful to the operation of the circuit because repeated reflections will not occur. If the output resistance of generator two, however, were very small compared to R_0, then a reflection would occur from the generator end of the line, would again travel down the delay line and appear at the input to the gate. This reflected wave might well be of sufficient amplitude to change the state of the gate and thus produce a false output from the gate.

Delay lines also may be employed as storage elements. For example, if a delay line is fed a string of pulses whose duration is short compared with the delay time of the line, these pulses will be stored on the line for a brief time as they travel down it. They will appear sequentially at the output of the line delayed by τ, the delay of the line, and will be a faithful reproduction of the pulse train fed into the input of the line. In Fig. 9-13

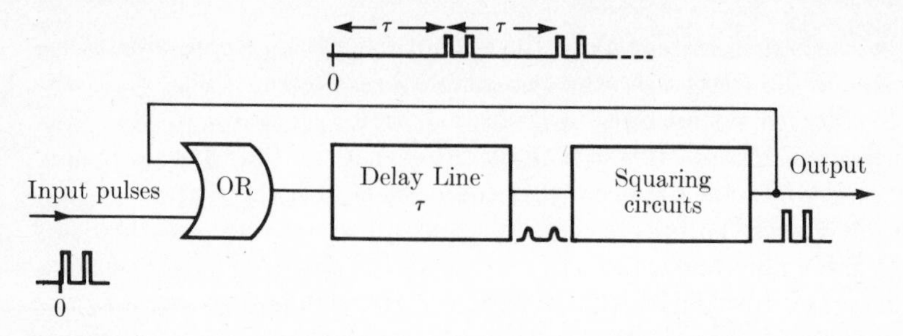

Fig. 9-13 A delay line used as a storage element. In this case the two pulses applied to the input appear delayed τ seconds at the output and every τ seconds thereafter.

such a circuit is shown. In this circuit input pulses are fed to an OR gate. The output of the gate drives the delay line, which has a time delay τ. Since the pulses which appear at the output of the delay line may be somewhat distorted with respect to the input pulses, the output of the delay line is fed into a squaring circuit which restores the original shape of the input pulse. The output of the squaring circuit is available both as an output to drive external circuits and as fed back to the second input of the OR gate. Again the output of the OR gate drives the delay line and the process is repeated, i.e., the pulses again go through the delay line, through the squaring circuits, and back to the input of the delay line through the OR gate. If, for example, we put three closely spaced pulses on the input these pulses will appear at the output delayed by τ seconds and will reappear every τ seconds thereafter. Therefore the input pulse train, in this case the three pulses, will be stored on the delay line indefinitely. In order to clear the delay line one could insert another gate in the loop so that the loop transmission could be set to zero for a time τ, thereby eliminating the pulses on the line. If the input pulse spacing is T, then up to τ/T pulses could be stored on the line.

9-4 PULSE-FORMING NETWORKS AND ARTIFICIAL LINES

In many instances it is very inconvenient, or even impossible, to use a distributed-constant transmission line for the purposes outlined in the preceding sections. For example, it is very difficult to obtain a distributed-constant line with characteristic impedance higher than a few hundred ohms; indeed in the case of a coaxial line, an R_0 of 100 ohms is about as high as may be obtained. In addition, the bulk of distributed constant lines is very great if the delay time required is large.

As an example of the incentive to replace the line in a line-type pulser by some network of lumped elements, particularly if only a few elements are required, consider the following application. To generate a

2-μs pulse, a 650-ft long coaxial cable is required (assuming the dielectric of the cable is polyethylene). For a high-power application, where a 15-kV rating might also be required, one of the high voltage cables such as RG-28/U would have to be used (650 ft of such cable would weigh about 240 lbs). By way of contrast, a lumped network with the same rating, and of the type we shall see, would weigh about 24 lbs and possess a correspondingly small volume.

The first approach to the solution of the problem of obtaining a lumped approximation to a distributed transmission line is almost an intuitive one. Since the line has the properties of series l and shunt c per unit length, let us make up a ladder network with sections of series L and shunt C, as in Fig. 9-14.

The circuit of Fig. 9-14 is a form of artificial line, or *wave filter*, and possesses some well-known properties discovered long ago in connection with telephone research. One such property is that it has a cut-off frequency f_c, given by

$$f_c = \frac{1}{\pi \sqrt{LC}} \tag{9-17}$$

Another property of the lumped approximation is that there is a delay per section equal to $\sqrt{LC}$. Therefore the line has a total one-way travel time $\delta = n \sqrt{LC}$.

For the network to simulate a line in the transient application, it must be capable of transmitting the steep wavefronts of Fig. 9-2. Such a wavefront has a Fourier spectrum extending to infinity, of course, but most of the components are at frequencies extending from zero to about $1/\tau$, where τ is the rise time of the wavefront. These frequency components should be transmitted with constant amplitude and linear-phase shift, but the network of Fig. 9-14 only does this at frequencies far below cutoff. Hence, f_c should be perhaps 10 times $1/\tau$. This means small values of L

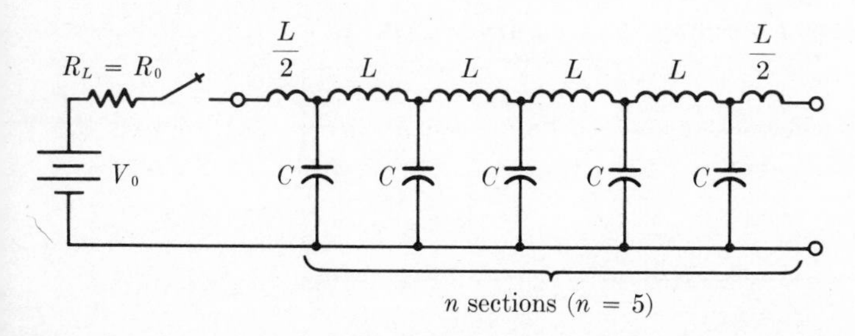

Fig. 9-14 Ladder network which is a lumped approximation to the transmission line: $Z_0 = \sqrt{L/C}$, one-way travel time $\delta = \sqrt{(nL)(nC)}$.

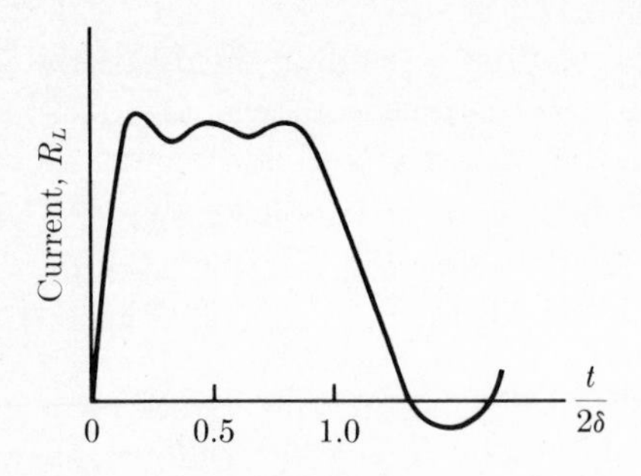

Fig. 9-15 Pulse formed with network of Fig. 9-14.

and C. Accordingly, to achieve a time delay δ of the desired magnitude, it may be necessary to use an excessive number of sections.

The appearance of a pulse generated with the five-section network like that in Fig. 9-14 is sketched in Fig. 9-15; notice the overshoot. A similar form of distortion will occur if a pulse is fed into the network at the left end and observed across a terminating impedance at the right end of the line.

The specific solution chosen for the approximation depends upon whether the line is to be used as a delay line or to generate pulses. For the former application, the main problems to be solved are the phase equalization of the line and the correct termination of the line. While the details of network analysis required for the solution of these problems are beyond the scope of this text, we can show a solution which is suitable and leaves the details to established references.[1] Fig. 9-16 shows a typical network used for a delay line. Two major modifications have occurred to the line originally proposed in Fig. 9-14. The first modification concerns the fact

[1] One of the earliest of these and a very comprehensive reference is G. N. Glasoe and J. V. Lebacqz (eds.), "Pulse Generators," M.I.T. Radiation Laboratory Series, vol. 5, McGraw-Hill Book Company, New York, 1948.

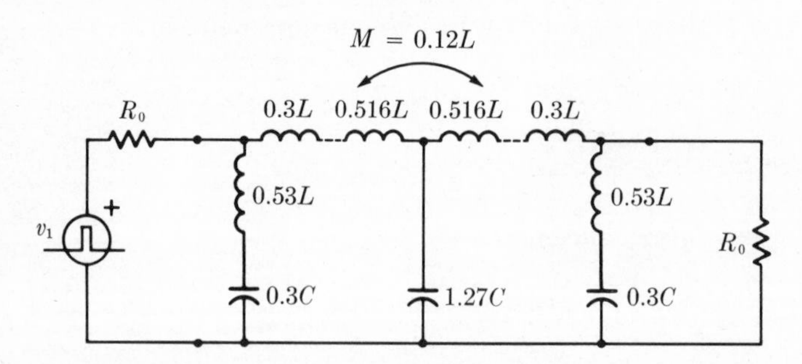

Fig. 9-16 The line of Fig. 9-14 modified to give a more constant input impedance and time delay over the pass-band frequency range. Again $R_0 = \sqrt{L/C}$, $\delta = n\sqrt{LC}$, $f_c = 1/\pi\sqrt{LC}$, and n equals the number of full sections.

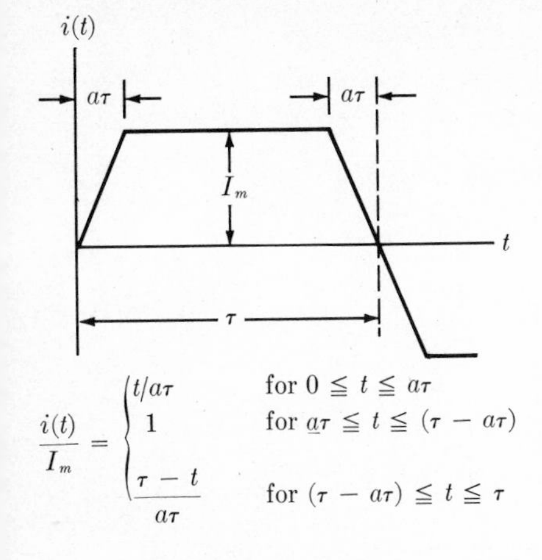

$$\frac{i(t)}{I_m} = \begin{cases} t/a\tau & \text{for } 0 \leqq t \leqq a\tau \\ 1 & \text{for } a\tau \leqq t \leqq (\tau - a\tau) \\ \dfrac{\tau - t}{a\tau} & \text{for } (\tau - a\tau) \leqq t \leqq \tau \end{cases}$$

Fig. 9-17 A trapezoidal approximation to the square wave of current that would be produced by switching a voltage source across the input to an ideal transmission line with a one-way delay time $\delta = \tau/2$.

that the input and output of the line are made by special terminating half sections, which serve to keep the input impedance of the line and the impedance seen by the load more nearly constant through the frequency range of the pass band of the remainder of the transmission line. This allows the line to be reasonably well terminated by the simple resistor R_0. The modification to the sections in the middle of the line is for the purpose of obtaining a reasonably constant time delay as a function of frequency in the line. The cutoff frequency of the line is the same as before and is given by Eq. (9-17). In practice the line is designed by picking a suitable value of L and C to first give the proper cutoff frequency and second to give the right characteristic impedance R_0. By knowing the value of L and C, we can find the delay per section $\sqrt{LC}$, and from this compute the number of sections of line required to give the desired overall delay.

When the network is to be used to generate a pulse, particularly a high-power pulse, there is no need for the artificial network to look like a transmission line. When seeking a network to generate a pulse several choices are available for the shape of pulse that one actually attempts to synthesize. The most obvious choice would be a perfectly rectangular pulse with infinitely fast rise and fall times. However, when one does obtain a network which in some sense generates such a pulse, one finds that there is considerable overshoot in the leading and trailing edges of the pulse. This is often undesirable in a high-power situation because excessive voltage may be supplied during these overshoots. A more satisfactory pulse shape for which to seek a network has finite rise and fall times or has a rather trapezoidal pulse shape as shown in Fig. 9-17. A pulse which is still easier to synthesize—in that the convergence of the waveform obtained to that desired is faster—is a parabolic approximation to the pulse as shown in Fig. 9-18. Networks can be synthesized for

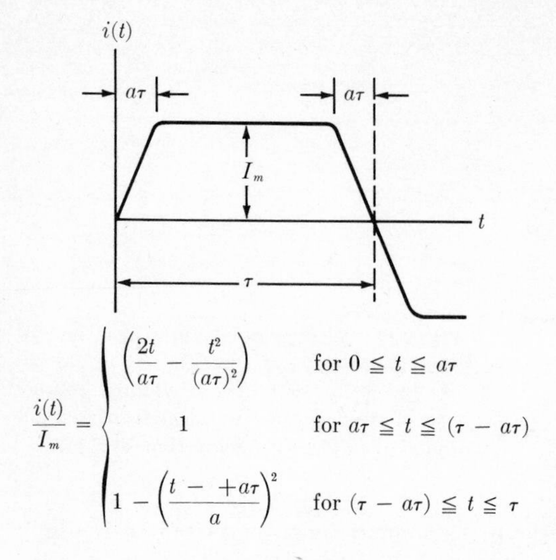

$$\frac{i(t)}{I_m} = \begin{cases} \left(\frac{2t}{a\tau} - \frac{t^2}{(a\tau)^2}\right) & \text{for } 0 \leqq t \leqq a\tau \\ 1 & \text{for } a\tau \leqq t \leqq (\tau - a\tau) \\ 1 - \left(\frac{t - \ +a\tau}{a}\right)^2 & \text{for } (\tau - a\tau) \leqq t \leqq \tau \end{cases}$$

Fig. 9-18 A parabolic approximation to the square wave of current produced as described in Fig. 9-18.

all three of these pulse shapes, and are given in the previous footnote reference.

The necessary element values L and C for the network of Fig. 9-19 may be readily determined from Table 9-1. If the applied voltage is V_0 and the desired maximum current is I_m, then a parameter Z_N is defined as

Table 9-1

Waveform	L_n	C_n
Rectangular	$\frac{\tau Z_N}{4}$	$\frac{4\tau}{n^2\pi^2 Z_N}$
Trapezoidal (Fig. 9-17)	$\frac{\tau Z_N}{4\,(\sin n\pi a)/n\pi a}$	$\frac{4\tau}{n^2\pi^2 Z_N} \cdot \frac{\sin n\pi a}{n\pi a}$
Parabolic (Fig. 9-18)	$\frac{\tau Z_N}{4\left(\frac{\sin \frac{1}{2}n\pi a}{\frac{1}{2}n\pi a}\right)^2}$	$\frac{4\tau}{n^2\pi^2 Z_N} \cdot \frac{(\sin \frac{1}{2}n\pi a)^2}{(\frac{1}{2}n\pi a)^2}$

L_1 L_3 L_5 L_n

C_1 C_3 C_5 C_n

Fig. 9-19 Network to produce trapezoidal or parabolic pulse waveforms. Element values are specified in Table 9-1.

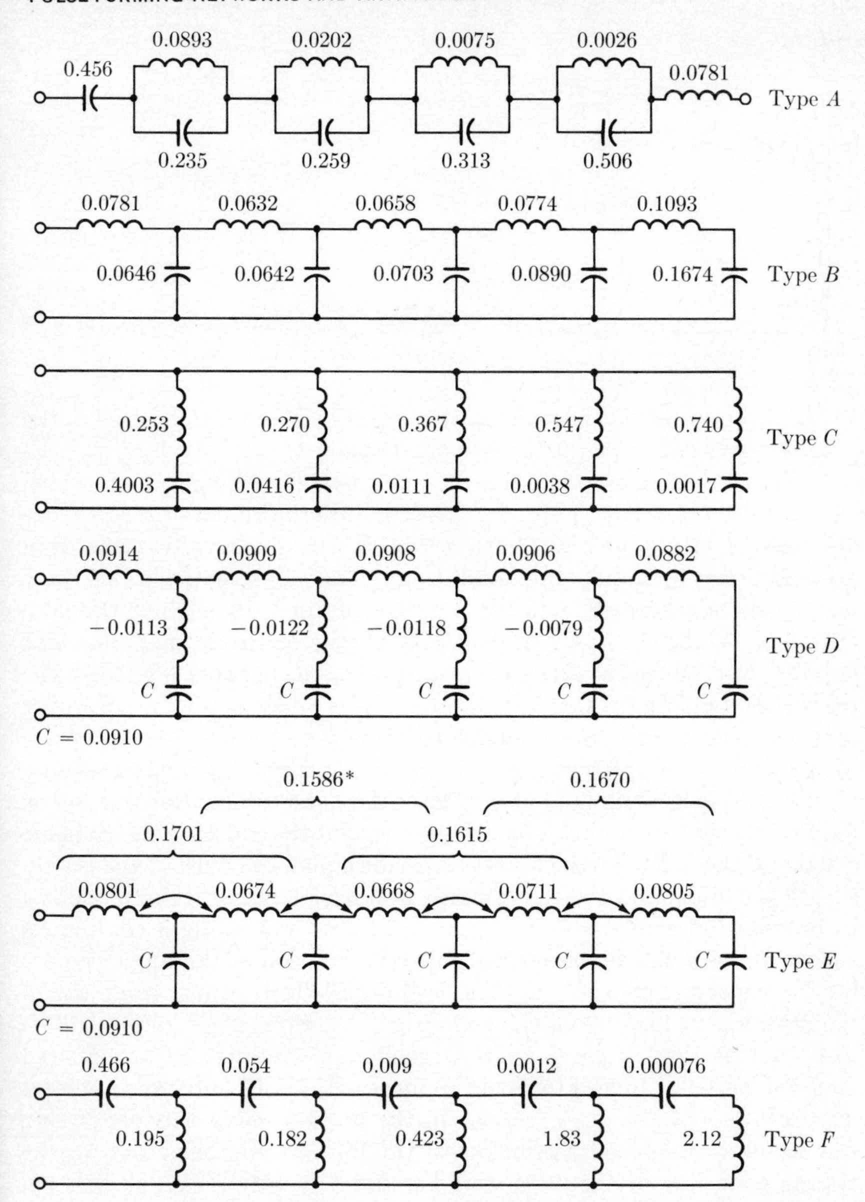

Fig. 9-20 Six alternative versions of a five-section network designed for a trapezoidal pulse (Fig. 9-17) with $a = 0.08$ [from Glasoe and Lebacqz (eds.), *loc. cit.*]. * This is $L_1 + L_2 + 2M$; $M = 0.0122 = 18\%$ of L_1, $k = 0.182$.

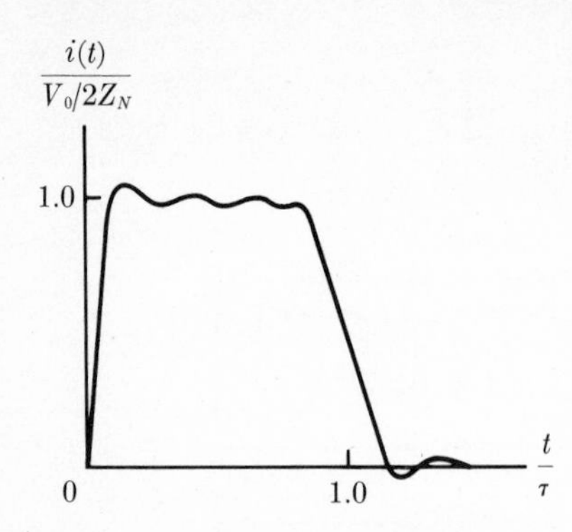

Fig. 9-21 Waveform resulting from any of the networks of Fig. 9-20.

V_0/I_m. In terms of this parameter, the values of L and C to be used in the general structure of Fig. 9-19 are as in Table 9-1.

Figure 9-19 shows only one of many possible configurations which will produce the desired pulse. In this structure the values of L and C are different for successive terms in the series. It would be an advantage from a manufacturing standpoint if the capacitors were identical, and there happens to be a network equivalent to that of Fig. 9-19 in which this condition is achieved. Again the details of the network manipulations required to produce the alternative versions of the network are left to the references, but Fig. 9-20 shows six alternative versions for the preceding network. These versions are for the particular case of a trapezoidal pulse with $a = 0.08$. In practice, the type E network is commonly used because all of the capacitors in that network have the same value. Moreover, it has been found that when the inductances of all but the end sections are made equal and the mutual inductances are made equal, the pulse shape is only slightly modified, but the construction is much simplified. It is customary to provide the complete inductance structure by winding a continuous coil on an insulated form and then tap onto the coil at the proper points for the capacitor connections. The waveform which results from any of the networks in Fig. 9-20 is shown in Fig. 9-21. From this figure it will be seen that the waveform is quite a good representation of the desired trapezoidal pulse. Indeed for some applications a sufficiently good pulse is obtained using only three sections in the pulse-forming network. Using this small a number of sections with the lumped approximation to the transmission line of Fig. 9-14 would result in a very poor pulse shape; therefore, this example shows the advantages of using the more refined synthesis procedures to obtain a pulse-forming network.

PROBLEMS

9-1. The circuit shown is an arrangement for generating precise, extremely short pulses for applying to a circuit under test—in this case merely a load resistor R_L.

For oscilloscope observation the horizontal sweep is initiated by a signal taken off at point A, whereas vertical deflection is produced by the signal at point B. Suppose line length D_1 is 3 ft and D_2 is 10 ft. The line is essentially lossless and has a velocity of 700 ft/μs. The series resistance R_s is much greater than the line characteristic resistance R_0, whereas $R_L = R_0$. The switch SW is mechanically driven, normally open but closing 100 times per second and remaining closed each time for 0.1 msec. Determine and sketch the waveforms at A and B on a common time scale. Label the voltage amplitudes.

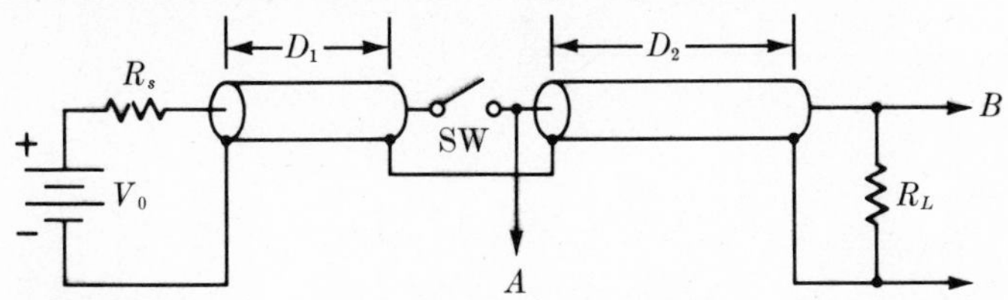

9-2. Three identical lengths of transmission line are connected as shown. All have the same characteristic impedance $R_0 = 500\ \Omega$ and a one-way delay time of 1.0 μs. The FET has an output resistance many times greater than R_0. For the input voltage v_1 shown applied to the gate of the FET, determine and plot the voltage waveforms appearing at A, B and C.

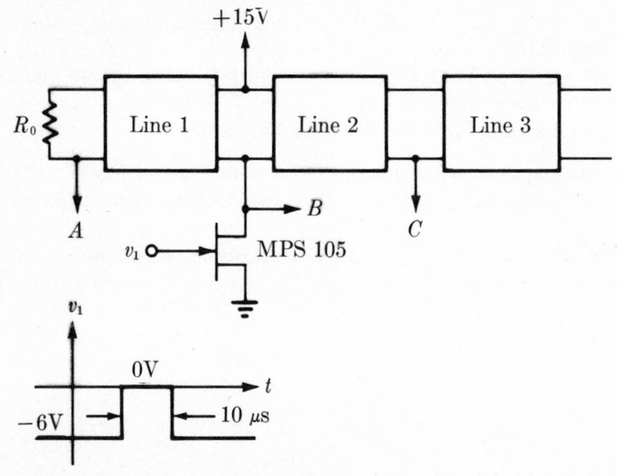

9-3. The lumped-element delay line shown may be regarded as ideal in its transmission properties with a characteristic resistance R_0 and a one-way delay time T. The switch closes at $t = 0$. Determine and plot the output voltage waveform.

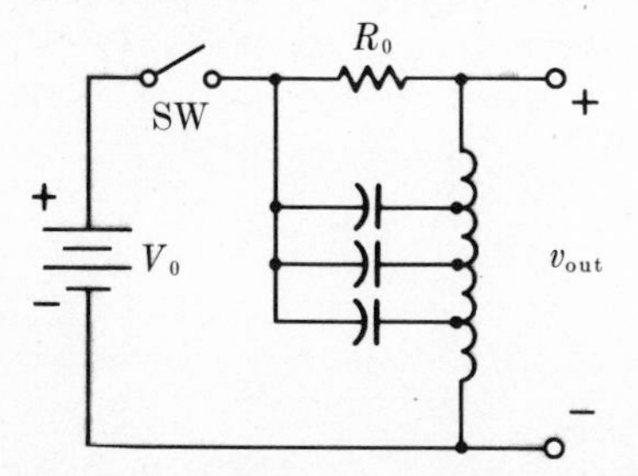

9-4. In the circuit shown the input voltage v_1 is a 1 μs pulse occurring at $t = 0$. The transmission line has a characteristic resistance R_0 and a one-way transmission time $T = 5$ μs. Determine and plot the waveforms v_2 and v_3.

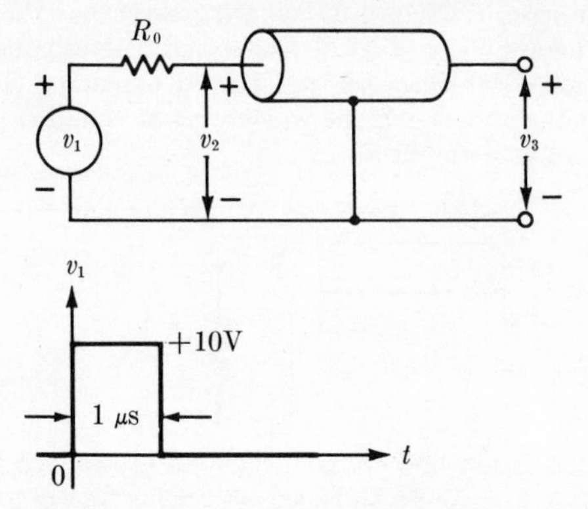

9-5. Derive the expression given in Eq. (9-15) for the maximum voltage obtained with dc resonance charging. Justify the relationship portrayed in Fig. 9-11 that the charging period is $T = \pi\sqrt{L_s C_t}$.

9-6. Derive the relationship given in Eq. (9-16) for the charging efficiency. Show that for resistance charging the efficiency must always equal 50 percent regardless of the value of R_s or R_0. Show that for inductive charging as in the dc resonance arrangement the efficiency depends only on the Q of the charging inductor (assuming the pulse-forming line or network to be lossless).

9-7. A type E pulse-forming network of five sections as in Fig. 9-20 is to provide a 2 μs pulse in a 100 ohm load resistance. The available charging time betwen pulses is 1 ms.

(a) With resistive charging, what value of series resistance R_s would be suitable?

(b) With dc resonance charging, what should the value of the series inductance L_s be?

Appendixes

A VOLTAGE REGULATOR DIODES

The data given here are for a typical group of low power diodes in the low voltage range that would be suitable for circuit work with transistors. There are many diodes available for this use with breakdown voltages ranging from about 2 to 200 volts, and power dissipating abilities up to 50 watts or more. The data in Table A-1 give the static characteristics of the diodes including the nominal voltage drop V_Z and the range of the voltage drop within a given diode type. The drop is measured at a specified dc test current, $I_{ZT} = 20$ mA, and at $T_0 = 25°C$, the ambient air temperature. The temperature coefficient α_Z gives the change in V_Z with change in temperature:

$$V_Z = V_Z(T_0)\left[1 + \frac{\alpha_Z(T - T_0)}{100}\right]$$

The coefficient α_Z is a function of the breakdown voltage as shown in Fig. A-1. Selecting a diode with V_Z in the 5-6 V region gives the lowest temperature variation of voltage.

Table A-1 Types 1N746 thru 1N759, 1N746A thru 1N759A silicon voltage regulator diodes

Absolute maximum ratings.

Average rectified forward current at (or below) 25°C free-air temperature	230 mA
Average rectified forward current at 150°C free-air temperature	85 mA
Continuous power dissipation at (or below) 50°C free-air temperature	400 mW
Continuous power dissipation at 150°C free-air temperature	100 mW
Operating free-air temperature range	−65°C to 175°C
Storage temperature range	−65°C to 175°C

Electrical characteristics at 25°C free-air temperature (unless otherwise noted).

PARAMETER	V_Z, *Zener breakdown voltage*					α_Z, *Temperature coefficient of breakdown voltage*	Z_Z, *Small-signal breakdown impedance*	I_R, *Static reverse current*	
TEST CONDITIONS	I_{ZT} = 20 mA					I_{ZT} = 20 mA	I_{ZT} = 20 mA, I_{zt} = 1 mA	V_R = 1 V	V_R = 1 V, T_A = 150°C
LIMIT→	*NOM*	1N746–1N759 *MIN*	1N746–1N759 *MAX*	1N746A–1N759A *MIN*	1N746A–1N759A *MAX*	*TYP*	*MAX*	*MAX*	*MAX*
UNIT→	*v*	*v*	*v*	*v*	*v*	%/°C	Ω	μA	μA
1N746	3.3	2.97	3.63	3.135	3.465	−0.062	28	10	30
1N747	3.6	3.24	3.96	3.420	3.780	−0.055	24	10	30
1N748	3.9	3.51	4.29	3.705	4.095	−0.049	23	10	30
1N749	4.3	3.87	4.73	4.085	4.515	−0.036	22	2	30
1N750	4.7	4.23	5.17	4.465	4.935	−0.018	19	2	30
1N751	5.1	4.59	5.61	4.845	5.355	−0.008	17	1	20
1N752	5.6	5.04	6.16	5.320	5.880	+0.006	11	1	20
1N753	6.2	5.58	6.82	5.890	6.510	+0.022	7	0.1	20
1N754	6:8	6.12	7.48	6.460	7.140	+0.035	5	0.1	20
1N755	7.5	6.75	8.25	7.125	7.875	+0.045	6	0.1	20
1N756	8.2	7.38	9.02	7.790	8.610	+0.052	8	0.1	20
1N757	9.1	8.19	10.01	8.645	9.555	+0.056	10	0.1	20
1N758	10.0	9.00	11.00	9.500	10.500	+0.060	17	0.1	20
1N759	12.0	10.80	13.20	11.400	12.000	+0.060	30	0.1	20

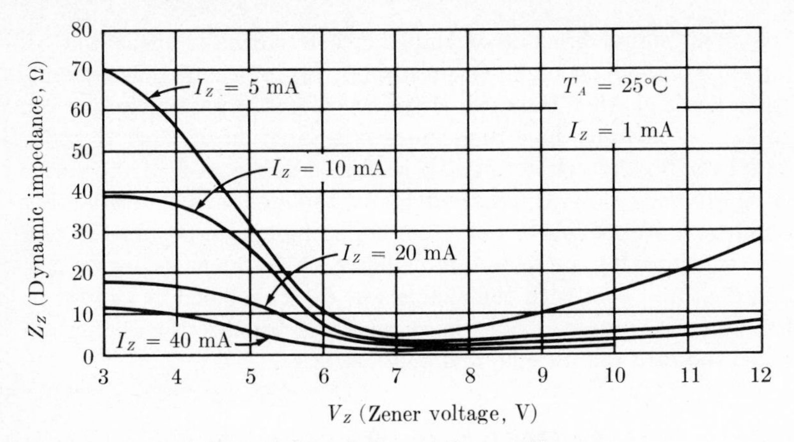

Fig. A-1 Typical dynamic impedance vs. Zener voltage.

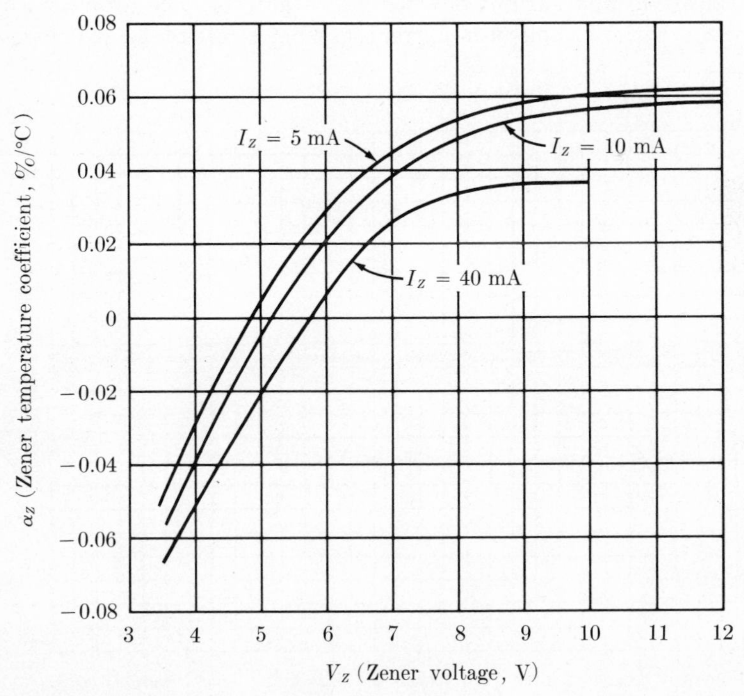

Fig. A-2 Typical Zener temperature coefficient vs. Zener voltage.

The small signal breakdown impedance is normally measured by superimposing a small ac signal (typically 60 Hz) on the dc bias current. The impedance Z_Z is then the ratio of the ac voltage appearing across the diode to the ac current. This impedance is a function of both the bias current and the breakdown voltage V_Z as shown in Fig. A-2. If the variations in I_Z are very slow, thermal effects in the diode may considerably alter the observed impedance. For example, a diode with positive α_Z will display a considerably increased impedance at frequencies where the diode can heat and cool with the changes in I_Z. This effect is important when the diode is employed as a shunt regulator because the changes in I_Z occur at the rate the unregulated supply varies.

B THE INVERTED CHARACTERISTICS OF THE TRANSISTOR

In some applications the drop across the transistor should be as near zero as possible, particularly when the current through the transistor is near zero. Such a case occurs when the transistor is used to modulate a signal as in Chap. 3. Figure B-1 shows the voltage drop across a 2N3642 as a function of the base current supplied. The collector current for each curve is constant, and curves are shown for both current flowing into the collector (positive) and current flowing out (negative). The drop across the transistor with no collector current flowing is about 13 mV with $I_B = 1$ mA.

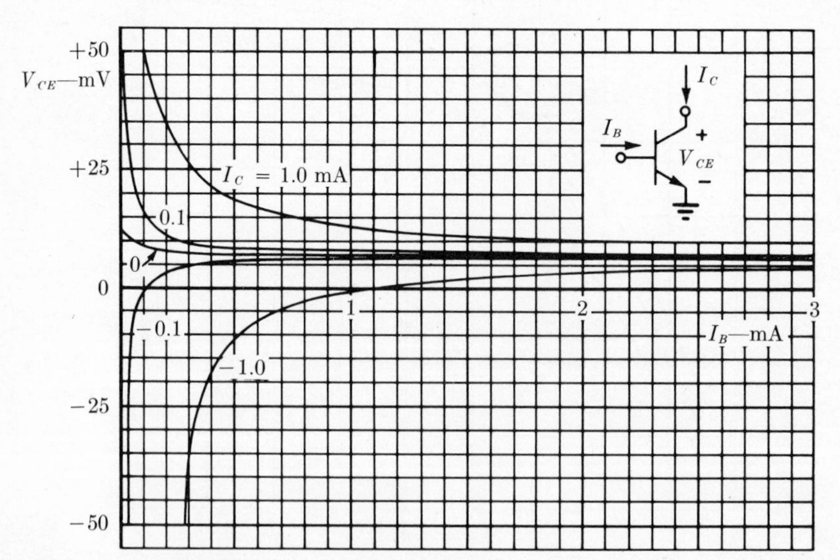

Fig. B-1 The drop across a saturated transistor (2N3642) in the *normal* connection as a function of base current I_B.

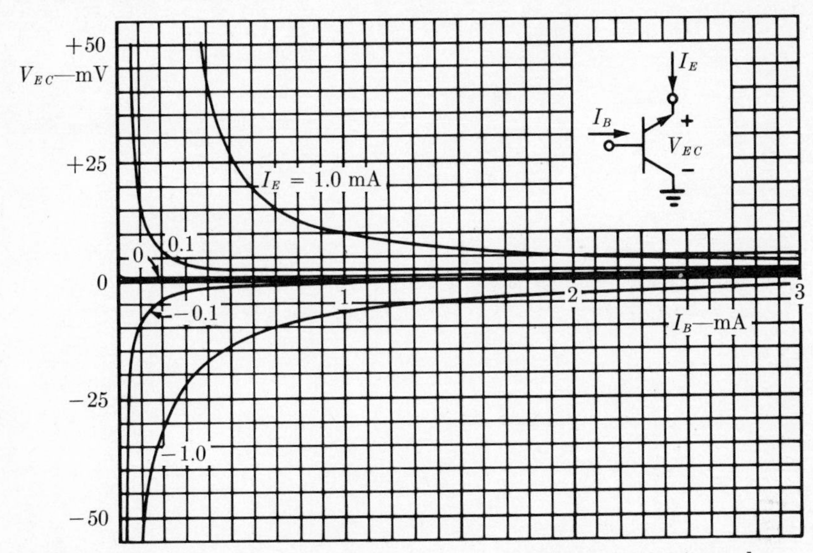

Fig. B-2 The drop across a saturated transistor (2N3642) in the *inverted* connection (i.e., with the normal transistor collector grounded) as a function of base current I_B. The voltage drop V_{EC} is at least an order of magnitude lower in this connection than in the normal connection.

The drop across the transistor may be greatly reduced by inverting it and using the nominal emitter as the collector as in Fig. B-2. Here the common terminal is the collector and the drop across the transistor is only about 1.5 mV for $I_C = 0$ and $I_B = 1$ mA. The signal handling capability of the inverted connection is not as great as the normal connection when the transistor is OFF because of the rather low V_{EB} required to cause breakdown of the emitter-base junction.

Index

Amplifier:
  operational, 163
  overdriven, 62, 88
Amplitude discriminator, 5
AND gate, 6
Astable, NRD, 235, 239, 241, 254
  (*See also* Multivibrator; Oscillator)
Astable blocking oscillator, 206
Astable multivibrator, 140, 144, 154
Avalanche diode, 39

Back diode, 40
Base-emitter breakdown, 46, 88
Bias line, 82, 165
Bilateral switch, NRD, 227
Binary (*see* Astable multivibrator)
Bistable, NRD, 235
  (*See also* Multivibrator)
Bistable multivibrator, 109, 117, 123, 147, 153, 155, 156
Blocking oscillator:
  astable, 206
  effect of core saturation, 211
  monostable, 203
Bootstrap sweep generator, 174, 179
Breakdown:
  base-emitter, 46, 88
  collector, 183
  diode, 39
  (*See also* Peak point)

Characteristic resistance R_0, 257
Clamping circuit, 55, 92, 105, 106

Clipping circuit:
diode, 34, 42
transistor, 62
Coincidence circuits, 6
(*See also* Gates)
Collector breakdown, 183
Common-base transistor, 74
Common-collector transistor (*see* Emitter follower)
Common-emitter transistor, 46
Comparator circuit, 5, 200
Complementary multivibrator, 147
Core saturation, effect on blocking oscillator, 211
Counter:
decade, 122
preset, 4
Counting circuit, 3, 116
Critically damped *RLC* circuit, 186
Current gain:
dc (h_{FE}), 51
h_{fb}, 74
small-signal (h_{fe}), 57
Current source, 91, 159
symbol, 23
Current-stable negative resistance, 220, 224, 226–228, 232
Cutoff, 52
Cutoff frequency, transmission line, 271

Dc (h_{fe}) current gain, 51
Dc restorer, 92
(*See also* Clamping circuit)
Decade counter, 122
Decoupling filter, 96
Delay line, 268
as memory, 269
Depletion mode, 86
Diac, 227
Differentiating circuit, 17, 180
Diode:
avalanche, 39
back, 40
breakdown voltage, 39
breakpoint, 32
equivalent circuit, 43
four-layer, 227

Diode (*cont'd*):
gas, 224
germanium, 36
hot-carrier, 40*n*
ideal, 31
load line, 41
point-contact, 40
regulator, 39, 279–282
reverse saturation current, 37
silicon, 36
steering, 116
storage effects, 44
switching time, 44
temperature coefficient, 279–282
tunnel, 41, 225, 252
Zener, 39, 88, 279–282
Diode clipping circuit, 34, 42
Discriminator:
amplitude, 5
frequency, 107
(*See also* Comparator circuit)
Displacement error, 158

Electron switch, 35
Emitter follower, 65
with capacitative load, 68, 72
complementary, 73, 84
Enhancement mode, 86
Equivalent circuits, linear, 7
Exponential waveform, 9

FET (field-effect transistor), 77
characteristics, 79, 80
equivalent circuit, 82
insulated gate, 85
load line, 83
Flip-flop (*see* Bistable multivibrator)
Four-layer diode, 227
Fourier series, 19
Frequency discriminator, 107
Frequency division, 119, 149, 215

Gas diode, 224
Gate electrode, SCR, 231
Gates:
AND, 6
NOR, 6, 47

Germanium diode, 36
Glascoe, G. N., 272*n*

Holding current, 227, 235
Hysteresis, 73, 127
  magnetic, 212
  power control, 249

Ideal diode, 31
Ideal switch, 31
Initial charge, 10
Integrating circuit, 15, 158
Interbase resistance, 229
International system of units, 25
Intrinsic standoff ratio, 229

Kirchhoff's rules, 18

Lebacqz, J. V., 272
Line pulser, 267
Linear equivalent circuits, 7
Linear slope, 14
Load line:
  diode, 41
  FET, 83
  NRD, 221
  transistor, 47
Locus of operation, 59, 62
Logical ZERO, ONE, definitions, 34

Magnetic saturation, 212
  (*See also* Transistor, saturation)
Memory, delay line as, 269
Miller, J. M., 162*n*
Miller integrator, 162, 178
Modulator, 54
Monostable, NRD, 235
  (*See also* Monostable multivibrator)
Monostable blocking oscillator, 203
Monostable multivibrator, 128, 134, 153, 155, 265
MOSFET, 85
Multivibrator, 108
  astable, 140, 144, 154
  bistable, 109, 117, 123, 147, 153, 155, 156
  complementary, 147

Multivibrator (*cont'd*):
  monostable, 128, 134, 153, 155, 265
  recovery time, 139
  tristable, 148
Multivibrator stability, 151
  temperature compensation, 152

Negative resistance, 147, 219
  current-stable, 220, 224, 226–228, 232
  voltage-stable, 220, 225, 239
Neon bulb, 224
Nonlinearity, sweep, 158
NOR gate, 6, 47
Norton's theorem, 21

Operational amplifier, 163
Oscillator:
  blocking (*see* Blocking oscillator)
  four-layer diode, 253
  ringing, 191, 217
  transmission line, 264
  tunnel-diode, 239, 252
  unijunction, 241, 251
Output resistance, 64
Overdamped *RLC* circuit, 185
Overdriven amplifier, 62, 88

Peak point, 225, 227, 234
Periodic solution, 99
Piecewise linear circuit, 43
Point-contact diode, 40
Power control:
  SCR, 241
  TRIAC, 245
Preset counter, 4
Propagation, velocity of, 257
Pulse-forming networks, 256, 267
Pulse transformer, 194

Q, definition, 185

Radar, 3
Ramp generator, 76, 90, 157, 174
  (*See also* Sweep circuit)
RC circuits, 7
  final value, 9

RC circuits (*cont'd*):
general solution, 99
initial value, 9
with sine-wave input, 247
time constant, 9
RC coupling circuit, 101
Recovery time, multivibrator, 139
Regeneration, 109, 124, 200
Regulator diode, 39, 279–282
Reich, H. J., 222*n*
Resetting, 112, 121
Resistance:
dynamic, 38
interbase, 229
negative (*see* Negative resistance)
output, 64
Resistor-transistor logic (RTL), 47, 88
Resonance charging, 217, 267
Ringing oscillator, 191, 217
Rise time, 64
RL circuits, 180
initial, final values, 181
time constant, 181
RLC circuit, 184
critically damped, 186
overdamped, 185
underdamped, 187
RTL (resistor-transistor logic), 47, 88

Saturation, magnetic, 212
(*See also* Transistor, saturation)
Sawtooth, 15, 174
(*See also* Sweep circuit)
Sawtooth generator, 106, 214
Schmitt trigger, 123, 154
SCR (silicon-controlled rectifier), 231
SCR power control, 241
SCS (silicon-controlled switches), 232
Set-reset multivibrator, 112
Setting, 112
Silicon-controlled rectifier (SCR), 231
Silicon-controlled switches (SCS), 232
Silicon diode, 36
Small-signal (h_{fe}) current gain, 57
Source follower, 81
equivalent circuit, 84
Speed-up capacitor, 115
Stable state, 111
Steady-state solution, 96, 99
Steering diode, 116
Step-charging rectifier, 105
Superposition theorem, 18
Sweep circuit, 58, 76, 157, 174
(*See also* Sawtooth)
Sweep nonlinearity, 158
Switches:
bilateral (NRD), 227
electron, 35
ideal, 31
separately actuated, 31
unilateral (NRD), 226
Switching point, 227
Switching speed, 114
Symbols, 22
Synchronization, 149, 253

Thevenin's theorem, 20
applied to active circuit, 66
Thyristor (*see* TRIAC)
Timing, 3
Transfer characteristic, 42, 127
FET, 80
Transistor:
base-emitter characteristics:
(2N3134), 71
(2N3642), 48
collector characteristics, 50
common base (2N2270), 75
common emitter (2N3134), 70
common emitter (2N3642), 49
common-base, 74
common-collector (*see* Emitter follower)
common-emitter, 46
description, 45
equivalent circuit, 52, 76
field-effect (*see* FET)
h_{FE}, 51
inverted connection, 281–283
linear-equivalent model circuit, 55
load line, 47

Transistor (*cont'd*):
offset voltage, 54, 282, 283
saturation, 47
saturation resistance, 53
unijunction, 228
Transistor clipping circuit, 62
Transmission lines, 256
cutoff frequency, 271
TRIAC, 232
power control, 245
Triggering, 117, 133
Tristable multivibrator, 148
Tunnel diode, 41, 225, 252
Tunnel-diode oscillator, 239, 252

Underdamped *RLC* circuit, 187
Unijunction oscillator, 241, 251
Unijunction transistor, 228
Unilateral switch (NRD), 226
Units, international system of, 25

Valley current, 224, 227, 235
Velocity of propagation, 257
V-I diagrams, 89, 91
Voltage doubler, 106
Voltage regulator, 39, 88, 224, 279–281
Voltage source, symbols, 23
Voltage-stable negative resistance, 220, 225, 239

Wave shaping, 13
Waveform, exponential, 9
Waves, incident and reflected, 257

Zener diode, 39, 88, 279–282